TISSUE CULTURE AS A PLANT PRODUCTION SYSTEM FOR HORTICULTURAL CROPS

CURRENT PLANT SCIENCE AND BIOTECHNOLOGY IN AGRICULTURE

Evans, H.J., Bottemley, P.J. and Newton, W.E. (eds): Nitrogen fixation research progress. 1985. ISBN 90-247-3255-7

Zimmerman, R.H., Griesbach, R.J., Hammerschlag, F.A. and Lawson, R.H. (eds): Tissue culture as a plant production system for horticultural crops. 1986. ISBN 90-247-3378-2

Tissue culture as a plant production system for horticultural crops

Conference on Tissue Culture as a Plant Production System for Horticultural Crops, Beltsville, MD, October 20–23, 1985

edited by

RICHARD H. ZIMMERMAN
ROBERT J. GRIESBACH
FREDDI A. HAMMERSCHLAG
ROGER H. LAWSON

US Department of Agriculture, ARS
Beltsville, MD, USA

1986 **MARTINUS NIJHOFF PUBLISHERS**
a member of the KLUWER ACADEMIC PUBLISHERS GROUP
DORDRECHT / BOSTON / LANCASTER

Distributors

for the United States and Canada: Kluwer Academic Publishers, 190 Old Derby Street, Hingham, MA 02043, USA
for the UK and Ireland: Kluwer Academic Publishers, MTP Press Limited, Falcon House, Queen Square, Lancaster LA1 1RN, UK
for all other countries: Kluwer Academic Publishers Group, Distribution Center, P.O. Box 322, 3300 AH Dordrecht, The Netherlands

Library of Congress Cataloging in Publication Data

ISBN 90-247-3378-2 (this volume)

PRINTED IN THE NETHERLANDS

TABLE OF CONTENTS

Preface ix

Introductory Lectures

Principles of Plant Tissue Culture
R. A. de Fossard 1

The Application of Tissue Culture to Plant Improvement and Propagation in the Ornamental Horticulture Industry
Irwin Y. E. Chu 15

New Technology

New Technologies in Plant Tissue Culture
W. Jan Rowe 35

Plug Systems for Micropropagules
D. D. McCown 53

Totipotent Suspensions as a Means of Multiplication
A. D. Krikorian, R. P. Kann, Stefania A. O'Connor and Mindy S. Fitter 61

Phenotypic and Genotypic Stability

Phenotypic and Genotypic Stability of Tissue Cultured Plants
D. A. Evans and J. E. Bravo 73

Summary of Panel Discussion on Phenotypic and Genotypic Stability of Tissue Cultured Plants
D. A. Evans, I. Y. E. Chu, R. D. Hartman and H. J. Swartz 95

Pathogen Detection and Elimination

Pathogen Detection and Elimination
Roger H. Lawson 97

Production of Culture Virus-Indexed Geraniums
Wendy Oglevee-O'Donovan 119

Diagnosis of Plant Viruses Using Double-Stranded RNA
Ramon L. Jordan 125

Pathogen Indexing in Large-Scale Propagation of Florist Crops
Boligala C. Raju and Jane C. Trolinger 135

Certification of Horticultural Crops - A State Perspective
L. M. Goff 139

Plant Quarantine

Plant Quarantine and International Shipment of Tissue Culture Plants
Robert P. Kahn 147

Summary of Panel Discussion on Plant Quarantines and International Shipment of Tissue Cultured Plants
R. P. Kahn, J. van Aartrijk, L. K. C. Clay, R. A. de Fossard and G. Hennen 165

Economics

Determining and Minimizing Production Costs
Alvan Donnan, Jr. 167

Determining Markets and Market Potential
Jeanne B. Jones 175

Fruit, Nut, and Vegetable Crops

Propagation of Fruit, Nut and Vegetable Crops - Overview
R. H. Zimmerman 183

Small Fruit and Grape Tissue Culture from 1980 to 1985: Commercialization of the Technique
Harry Jan Swartz and Jon T. Lindstrom 201

Temperate Fruits and Nuts
F. A. Hammerschlag 221

Tropical and Subtropical Fruits and Vegetables
Richard E. Litz, Robert L. Jarret and Madhevan P. Asokan 237

Commercial *In Vitro* Propagation and Plantation Crops
Raymond P. Oglesby and John L. Griffis, Jr. 253

Use of Tissue Culture for Micropropagation of Vegetable Crops
Timothy J. Ng 259

Workshop

Tissue Culture Techniques and Plant Introduction/Quarantine Procedures
Bruce J. Parliman 271

Ornamental Crops

Tissue Culture Propagation of Ornamental Crops: An Overview
Paul E. Read and Mary A. Hosier 283

Tissue Culture as a Plant Production System for Foliage Plants
R. D. Hartman and F. W. Zettler 293

Commercial Micropropagation of Florist Flower Crops
Dennis P. Stimart 301

In vitro Propagation of Flower-Bulb Crops
J. van Aartrijk and P. C. G. van der Linde 317

Woody Ornamentals, Shade Trees and Conifers
B. H. McCown 333

Orchid Tissue Culture
R. J. Griesbach 343

Laboratory Design and Large Scale Production

Laboratory Design
O. C. Broome 351

Summary of Panel Discussion on Laboratory Design
O. C. Broome, B. Briggs, R. Evans and G. Hennen 365

Large Scale Tissue Culture Production for Horticultural Crops
Randall A. Strode and Gerri Abner 367

PREFACE

In 1980, a conference on tissue culture of fruit crops was held at Beltsville to summarize the current status of this technology and to stimulate interest in it among research scientists, students, and commercial producers in the U. S. Interest in that conference and the proceedings from it far exceeded the expectations of the organizing committee. Since that time, micropropagation of fruit crops in the U. S. has increased significantly, but still lags far behind applications to production of ornamental plants. Within the past two years, a number of new laboratories have been established and some of the existing laboratories have expanded to a size far larger than any previously anticipated. Creation of new laboratories capable of producing more than 400,000 plants per week will test the ingenuity of laboratory managers and the skills of marketing departments.

In recent years, numerous symposia have been held on various aspects of biotechnology and genetic engineering. Although micropropagation is the key to providing large numbers of genetically engineered plants, it is a topic that has been relegated to a minor position, or ignored completely, at such meetings. Accordingly, the time seemed propitious for a conference devoted solely to all aspects of micropropagation as applicable to horticultural crops.

In planning the conference, we decided that it was essential to have both research scientists and commercial practioners of micropropagation actively involved in the program. It has been clear that research results published in the scientific literature are often of limited value to those wishing to apply the results on a commercial scale. Many factors of critical importance to large scale production are of limited interest in research. We have attempted to bring out the different needs of these two groups of tissue culturists through the selection of topics and speakers at the conference.

A panel discussion on Large Scale Production concluded the presentations at the meeting, but unfortunately the taping of this section was defective so that it was not possible to prepare a summary. We would like to thank the following members of this panel for their participation, which the registrants at the meeting found most stimulating: Dr. D. I. Dunstan, Plant Biotechnology Institute, National Research Council of Canada, Saskatoon, Sask.; S. E. Hyndman, Weyerhaeuser Tissue Culture Center, Apopka, FL, U.S.A.; and A. Martinelli, Zanzivivai, Ferrara, Italy.

We would like to thank Dr. W. Klassen, the Director of the Beltsville Area of the Agricultural Research Service, for his support and encouragement. Mr. Gordon Carpenter of the Visitors Center at Beltsville provided invaluable information and assistance. The other members of the Organizing Committee, Olivia C. Broome and Ingrid Fordham, worked diligently on many tasks to insure the success of the conference. Finally, appreciation is expressed to the numerous other staff members at Beltsville who carried out many necessary tasks before and during the meeting.

Richard H. Zimmerman
Robert J. Griesbach
Freddi A. Hammerschlag
Roger H. Lawson

PRINCIPLES OF PLANT TISSUE CULTURE

R.A. DE FOSSARD

Tamborine Mountain Plants, Eagle Heights, Queensland 4621, Australia

1. INTRODUCTION

Plant tissue culture is used to achieve many different objectives which have in common the growth of microbe-free plant material in an aseptic environment. The objective of this conference is the use of plant tissue culture as a plant production system for horticultural crops, and my role is to discuss the principles of plant tissue culture with respect to this objective. I have interpreted this to mean, to describe which plant tissue culture techniques we use and why we use them, that is, what are our reasons for proceeding in the way we do. Willy-nilly, some of the reasons are scientific and others are economic and marketplace-wise. I will not be dwelling on the latter even though it is of considerable and underlying importance. By marketplace-wise, I mean, we should have sound, realistic marketplace reasons for contemplating the use of tissue culture for a particular species. We should ask questions, such as: How is the species normally propagated in Nature and by Man? What are the advantages and disadvantages of using these methods? What are the likely advantages and disadvantages of using, instead, tissue culture methods for its propagation? Are the advantages, more rapid multiplication? All-the-year-round propagation? A bushy habit? Elimination of viruses? How much will the research and development cost to achieve these advantages and are these costs likely to be justified by attaining the perceived advantages? Part of our research is to achieve stated objectives, and another part could be classified as production-line research aimed at cost-cutting and maximizing performance.

2. THE MAIN INTERACTING FACTORS IN TISSUE CULTURE

Whatever the objective of the tissue culturist, there will be some common considerations, as depicted in Figure 1; it might also be necessary to overcome severe microbial contamination problems.

Zimmerman, R.H. et al. (eds), Tissue Culture as a Plant Production System for Horticultural Crops.

ISBN 90-247-3378-2.

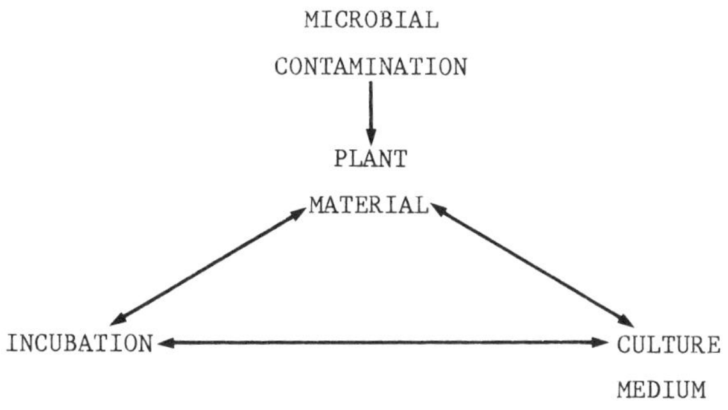

FIGURE 1. The triangle of main interacting factors that summarizes the complexity of the tissue culture situation; see text for further explanation.

Each of the main interacting factors at the corners of the triangle, in turn, consists of a number of further interacting factors. For PLANT MATERIAL, these include the type and size of explant used, the ontogenetical stage of the plant, the plant's physiological condition and the time of the year and of the day to excise explants. With repeated subculture, we also recognize that the cultured plant's requirements change, often necessitating modifications in the formulation of its culture medium. For INCUBATION, the interacting factors include temperature (constant, diurnally variable, degrees) and light (intensity, wave length, diurnally variable). For CULTURE MEDIUM, it has to be considered whether to use liquid or semi-solid medium, and to experiment not only with many different chemicals, but also with several concentrations of these chemicals. There is a considerable number of permutations of these factors, and many more than can possibly be tested. The addition of undefined chemical factors from the PLANT MATERIAL make for an undefined cultural system.

The question is: How do we proceed with such a complex situation? Bearing in mind that most of us, outside of research institutes and universities, have very few options with respect to incubation (e.g., growth rooms), we are limited to experimentation with plant material and culture medium, still a vast amount of variables. Incubation conditions are set rather arbitrarily to: 30°C/25°C (light/dark) for tropical species, 25°C/20°C for subtropical species and 20°C/15°C for temperate species, and either short days (8hr light/16hr dark) or long days (16hr light/8hr dark) depending on our knowledge of the species' requirements for vegetative growth. We are aware that our choice might inflict a selection pressure on the species and might be a limiting factor leading to a failure to realize the full potential of the species in culture. Needless to say, if we have or had the facilities for testing different types of incubation conditions, we would use them.

3. MICROBIAL CONTAMINATION OF CULTURES

When I started tissue culturing 25 years ago, initiating cultures was a major problem due to microbial contamination, but I quite soon discovered the advantage of growing mother plants in a clean and controlled environment (1). Now-a-days not only is this a routine procedure, but also overhead watering is avoided, and running water treatment of explants is done, all of which reduce the size of the microbial population. Many chemicals and antibiotics were tested in these early days, but none proved as effective and/or safe as solutions of hypochlorite. The practice then was to follow chemical treatments with several sterile water rinses. Now, we always treat the explant with hypochlorite prior to placing it on the sterile culture medium, using a concentration that is not harmful to the plant material. The following is an example of an initiation procedure:

1) Make mother plants responsive (e.g., by exposing them to either short or long days; by pruning to induce new shoot growth; by chemical treatments), and rear in a clean environment in a clean potting-mix, and avoid overhead watering;
2) Defoliate to 1/2-petiole and brush in detergent;
3) Running water for 1 hour;
4) 1% (w/v) available chlorine in 0.01% 7X-detergent (20 min);
5) Decant chlorine and hold in sterile water;
6) Dissect to explant size;
7) Dip in a concentration of chlorine tolerated by the explant;
8) Place on a simple minerals-sucrose-Gelrite medium.

Gross contamination is not entirely a thing of the past but, for many species, initiation of cultures is now much easier. Later, we were to learn that many of our apparently aseptic cultures carried microbes that developed very slowly but never the less impaired the growth of cultures. At least part of this delayed contamination problem was due to our use of agar and to the pale, filmy nature of the contaminating microbes. Use of clear gelling agents, such as Gelrite, makes detection of this type of contamination much easier. Microbiologists, it appears, are extremely skeptical of our ability to get cultures free of microbes and to maintain them in this state for any length of time. Their argument is along the lines: You might get rid of dynamic microbes, buy why bother with static microbes, i.e., microbes that are not metabolizing? I think we could be persuaded to this view if we were assured that static microbes stayed static during repeated subculturing. I think a satisfactory compromise is to initiate cultures on Gelrite (or similar) medium, followed by tests of apparently aseptic cultures on various microbial growth media. Cultures that pass these tests can be used for limited production runs, while some are repeatedly monitored and re-treated for later limited production runs. Currently, we encourage our inoculators to assume that cultures are always contaminated and our laminar flow procedures are designed with this pessimistic attitude in mind.

4. STAGES IN MICROPROPAGATION

Stage I: Select responsive plant material, disinfest it, and find suitable culture medium and incubation conditions to induce the plant material to grow _in vitro_; research culture media and incubation conditions to find a combination that will induce rapid multiplication. This stage is commonly extended over a 6-month period, but can be much longer with "difficult" species.

Stage II: The multiplication stage involves growth in culture for, commonly, 2 months followed by division of the cultures, and subculture of each piece to fresh culture medium, followed by 2 months incubation, and so on; this sequence of grow-divide-grow continues until marketing target figures have been reached. For various reasons (e.g., lack of effective research; recalcitrant variety), rootless, multi-shooted cultures might not be obtained in Stage II; instead Stage II-AD, i.e., rootless, apically dominant cultures (e.g., _Grevillea_ 'Robyn Gordon') are obtained, in contrast to the faster multiplying Stage II-MS cultures, i.e., rootless, multi-shooted cultures (Fig. 2).

Stage III: Changes are made to the constitution of the culture medium and to incubation conditions to prepare cultures for planting-out. Stage II-AD cultures are induced to form roots, and these cultures, i.e.,

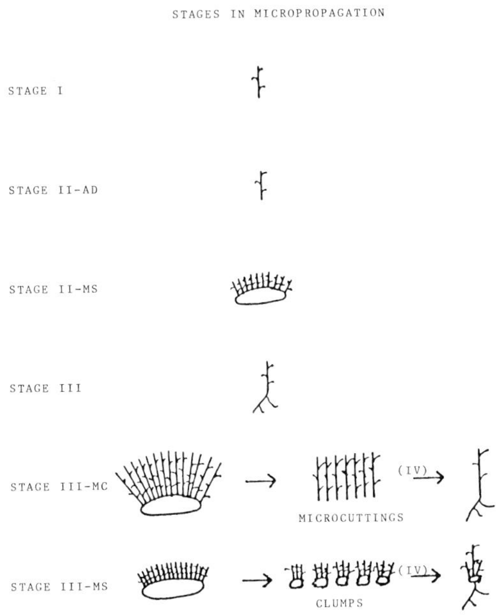

FIGURE 2. Diagramatic representation of stages in micropropagation. Stage I: Initiation of cultures from buds or nodal explants to produce an apically dominant shoot, with or without roots. Stage II-AD: subcultures growing to form apically dominant, rootless shoots. Stage II-MS: Subcultures growing to form rootless, multi-shooted cultures. Stage III: Subcultures forming rooted, apically dominant shoots. Stage III-MC: Subcultures inducing shoot elongation on rootless, multi-shooted cultures; the elongates shoots are harvested and rooted in the planting-out facility. Stage III-MS: Stage II-MS cultures are cut-up into clumps and these clumps are induced to form roots either in culture or in the plant-out facility.

apically-dominant and rooted, are described as Stage III cultures. Stage II-MS cultures in contrast might be handled in one or more of three ways:

(i) Stage III-MC: Cultures are subcultured to media that, while still multi-shoot-inducing, lead to the development of longer shoots; these shoots are then harvested as micro-cuttings and are sent to the planting-out facility for rooting; the bases of such cultures are subcultured to fresh medium. NOTE: Stage II-AD cultures might also respond to this micro-cutting procedure.

(ii) Stage III: Stage II-MS cultures are divided into individual shoots that are then induced to form strong, apically-dominant shoots with roots in culture.

(iii) Stage III-MS: Small clumps of cultures from Stage II-MS are transferred to a root-inducing, apical-dominance-inducing medium, so as to get a strong plant that also has a bushy habit; this habit is highly desirable for many foliage plants of importance to ornamental horticulture.

Stage IV: Removal of the aseptic plants from their sterile "test-tube" environment followed by planting in a relatively harsh environment requires a great deal of care and research. Planting-out procedures can be divided into several phases, but the most critical part is the initial step of inducing the formation of fully-functional roots in a "potting-mix" while ensuring that the delicate shoot system is protected against desiccation. Often Stage IV lasts for 4 months and, when completed, the culture-derived plants are ready for growing-on, i.e. handling in a manner similar to other methods of propagation.

Production costs: The cost of producing a tissue-cultured plant is influenced by whether it is an "easy" or a "difficult" species, by the multiplication rate achieved in every period of incubation, by whether it is necessary to use a Stage III for successful planting-out (in contrast to planting-out of either a Stage III-MC, Stage III-MS or even a Stage II-AD or -MS), by the ease and success rate of planting-out, and by the size of the marketing target. Effective research at all 4 stages of micropropagation greatly reduces the final cost of establishing a plant in the field. The production cost is often far less by micropropagation than by other methods of propagation but if, for whatever reason (e.g., "difficult" species; lack of effective research), the cost is the same or more, advantages such as having disease-free multiplication, bushy habit and all-the-year-round propagation, often make micropropagation the more attractive option.

Species profile chart: We use a Species profile chart (Fig. 3) for each variety in our laboratory to visualize its known responses and to highlight our areas of ignorance. Also noted on each chart is our perception of the habit desired by the marketplace, e.g., a bushy habit or a single-shooted plant. In turn, we also make decisions about whether further research and development can be justified in terms of expected commercial returns (i.e., sales).

SPECIES PROFILE CHART

NAME OF VARIETY
HABIT DESIRED:

	STAGES					
	I	II-AD	II-MS	III	III-MS	III-MC
STAGE ACHIEVED (X)						
PLANTING-OUT: WHICH STAGE TO IV? (X) WEEKS TO SALE?						

FIGURE 3. Species Profile Chart. A standard chart for each variety that indicates habit (e.g., bushy or apically-dominant) desired, stages in micropropagation achieved, which stage(s) can be used for planting-out, and the time taken for the planted-out material to attain saleable size.

5. MANIPULATIONS TO ACHIEVE DIFFERENT GROWTH FORMS IN CULTURE

The main avenues used to manipulate cultures are hormones in the auxin and cytokinin categories. Cytokinins predominate in Stage II and are used to induce multi-shooted cultures and/or adventitious bud formation, and to inhibit root formation. In contrast, auxins predominate in Stage III (along with the absence of cytokinins) to restore apically-dominant growth and to induce root formation. Stage III-MC and III-MS, on the other hand, are likely to be achieved with careful use of both auxin and cytokinin. I would like to make two observations about these manipulations: First, it should not be assumed that best responses to auxin and cytokinin manipulation will be obtained in the absence of some experimentation with other constituents of the medium, e.g., minerals. Second, it should not be assumed that using a "favorite" auxin and a "pet" cytokinin is necessarily going to be the best to induce optimal responses in a culture. With the woody species, Eucalyptus and Eriostemon, we found that of 5 cytokinins tested, only benzyladenine (BA) and 6-(benzylamino)-9-(2-tetrahydroxy-pyranyl)-9H-purine (PBA) induced multi-shooting; kinetin, N^6-isopentenyladenine (2iP), and zeatin were ineffective (Fig. 4). In other work with Eucalyptus, we found that responses to the oxygenated auxins, para-chlorophenoxyacetic acid (pCPA), 2-naphthoxyacetic acid (NOA) and 2,4-dichlorophenoxyacetic acid (2,4-D), were distinct from the responses to the auxins, indoleacetic acid (IAA), indolebutyric acid (IBA) and naphthaleneacetic acid (NAA) (Fig. 5). One species, Chenopodium rubrum, that I worked with extensively, was extremely sensitive to iron and sucrose concentrations, requiring as little as 10 μM Fe and 6 mM

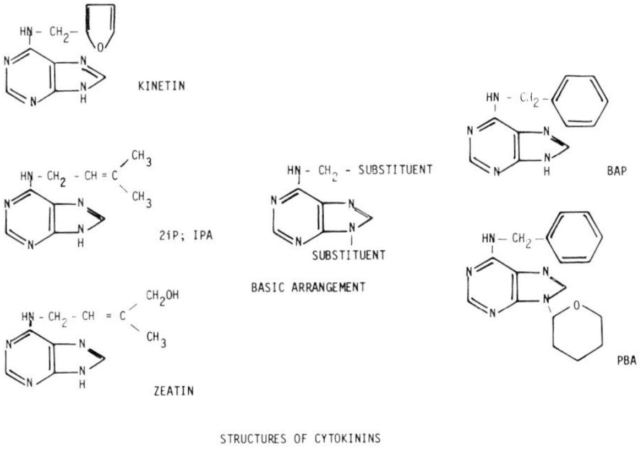

FIGURE 4. Structures of cytokinins. In the middle, the basic arrangement of cytokinins, showing two substituent groups.

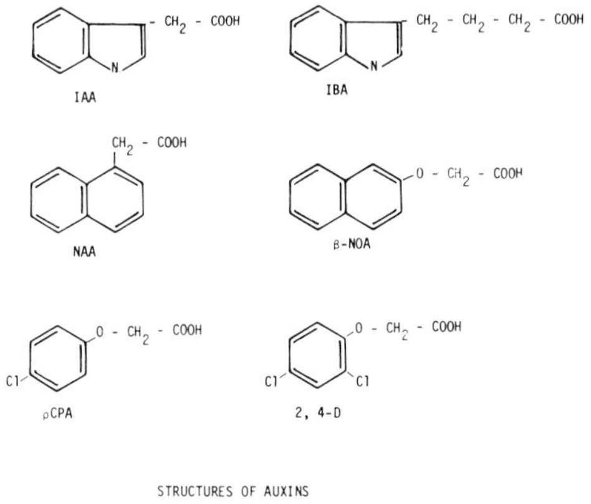

FIGURE 5. Structures of auxins. If the top righthand substituent groups are examined, it can be seen that NOA, pCPA and 2,4-D have, in common, an oxygen connecting their side chain with their rings; in contrast, in IAA, IBA and NAA, the side chains are connected directly to their rings without oxygen.

sucrose. No amount of experimentation with auxins and cytokinins would elicit responses until the basal medium had been corrected for these two limiting factors. I am not, for one moment, suggesting that any of these results with Eucalyptus, Eriostemon or Chenopodium can be extrapolated to other species. I am stressing that, for many species, we will not approach their peak performances unless systematic work on many variables in the medium and in incubation conditions is done. And, when this has been done with a lot of species, I believe we might be able to make some testable generalizations that might eventually give us far better control over our culture systems.

6. THE BROAD SPECTRUM (BS) EXPERIMENT

In the early 1970's, we tested our first theoretical model for selecting culture media for a species, using tobacco callus and, to our surprise, it was very successful (1). Basically, it was a 3^4 factorial experiment with the four factors (i) minerals, (ii) auxins, (iii) cytokinins, and (iv) growth factors, amino acids and sucrose, using three concentrations of each factor. We were looking to this model experiment to indicate interacting combinations of factors that could then be used as basal media for further experimentation. There were 81 treatments (media) in the original BS-experiment and later, with nodal explants from *Eucalyptus ficifolia* seedlings, I was amazed at the different responses of cultures that were induced under the influence of different combinations of factors (2).

Such large experiments yield a wealth of information and we can often select basal media for each of callus, apically-dominant growth, multi-shooted growth, adventitious bud formation, and rooting. Each basal medium, at least in theory, is then modified following omission-type experiments, i.e., omission of single ingredients in the medium to determine their essentiality, and, later, dose-response experiments, in which the essential ingredients are tested to find the concentrations optimizing the growth response. In theory also, we would do a certain amount of backtracking during the course of this long series of experimentation. In practice, we have approached this level and depth of experimentation only with *Eucalyptus ficifolia*. This led to an awareness of the powerful effect of riboflavin in the original BS-model, and also led to other modifications in experimental approach.

7. PATHWAY I-PATHWAY II

The sheer size of the BS-experiment, viz. 81 treatments x 5 replicates (=405 explants), was a practical problem. If we were to start with field-grown explants, we would have to initiate perhaps twice as many explants to offset contamination and injury losses. And if this were done, would the results be reproducible and useful for a production-line multiplication procedure? Initial explants, compared with later subcultured explants, add a different load of undefined chemicals to the culture medium, and we postulated that their tissues might also respond differently to the chemicals in the medium formation. The answer seemed to lie in doing all preliminary screening of variables with seedling-derived explants. Seeds are easier to get into apparently-aseptic culture, and the resulting seedlings are then a source of explants without further disinfesting treatments. But, would the information obtained from experiments with seedling-derived explants be relevant to explants from mature plants? With E. ficifolia, the answer was: "Yes-well, mostly yes". We were able to multiply adult Eucalyptus material on media selected from "seedling" experiments, but the former "remembered" they were adult when it came to rooting, and failed to root on media suitable for "seedling" rooting. We overcame this problem by wounding adult trees at their base to induce the formation of adventitious shoots with juvenile characteristics.

In summary, we have Pathway I of a project in which we screen experimental variables with either seedling explants or explants from a readily available source. Pathway II involves preparing selected plants to be responsive and to have less contamination problems. In the case of trees, we would, ideally, wound the selected tree, strike the cuttings of adventitious shoots, pot these and grow them in a controlled environment, and keep them coppiced as an ongoing source of selected, responsive material. Following standard procedures, and using a simple culture medium, apparently-aseptic cultures from these potted plants would then be used as a source of explants for testing with the formulations derived from Pathway I. Theoretically, we should then be able to build up a population of apparently-aseptic cultures clonal with the selected tree, and use these for experiments to refine their cultural requirements.

8. MINI-BROAD SPECTRUM EXPERIMENTS

It is years since I last did a full BS-experiment. Instead, I use different (and ever changing) Mini-BS experiments based on current knowledge and experience. These experiments usually have 18 treatments (3 x 3 x 2 factorials) and vary depending on the objective and type of plant. Woody species are treated differently than herbaceous species; a different array of treatments would be used to select media for callus, for multi-shoots, for roots, and so on. Follow-up experiments are used to test variables dropped from the full BS-experiments. And, finally, omission and dose-response experiments would be done--at least, in theory. In commercial practice, it is often a matter of doing enough R.&D. to get an acceptable production-line performance, rather than inflexibly pursuing an ideal systematic approach to optimize every aspect of micropropagation. For many species a short-cut approach achieves an acceptable production-line performance and, in its extreme form, we initiate cultures on a simple minerals-sucrose medium, follow this with a test of three or four media (selected on past experience) and then with an 18-treatment Mini-BS experiment. Often, one of the three or four selected media is rated as satisfactory for production-line purposes, though some fine-tuning might take place later on.

9. "EASY" IN CONTRAST TO "DIFFICULT" SPECIES

An "easy" species is one that readily responds to standard procedures. A "difficult" species is one that does not readily respond to standard procedures. *Eucalyptus* was very much a "difficult" species--now, we wonder why we had so much trouble with it. As our knowledge increases, we are having less trouble with a wide range of species, but still there are groups which are classified as "difficult," e.g., *Protea*, *Paphiopedilum*. With these, we again become aware of how ignorant we are of the basic underlying processes that we manipulate, and we yearn for more basic knowledge of developmental biology.

10. STATISTICS AND PUBLICATIONS

There is an adage "Research not published is research not done." And, in my view, "Research not designed for rigorous and appropriate statistical analysis should not only not be done, it should not be published." Before you all have apoplexy, let me be the first to admit that I have often not lived up to this maxim. My experiments are always designed for appropriate statistical analysis, but I rarely analyze "sighting" experiments of the Broad Spectrum type. These are used to give an indication of suitable basal media for follow-up omission and dose-response experiments. It is these later experiments that should be analyzed before being presented for publication. The majority of published papers on tissue culture have no statistical analyses even of the inefficient mean ± standard error type, and many are not even appropriately designed for analyses. The papers must therefore be read as unsupported statements. We all know of difficulties in attempts to reproduce the findings of published papers. One reason for these difficulties is, I believe, the lack of statistical analysis. Another reason is their lack of specific details, including media formulations. I know that HortScience now insists on authors specifying the brand of agar used. I think that this insistence should be extended to include other known variables in common ingredients. In fact, I am suggesting that we should upgrade our game to attain the level of acceptability reached in most areas of science.

11. INCREASING OUR FUNDAMENTAL KNOWLEDGE

I believe referees should insist upon these scientific standards (of statistical treatment, definition of procedural details and use of mole rather that mass units) in papers with applied objectives, as well as for papers designed to give insights into developmental processes. The former papers are statements on what was done by the author, the latter papers should take a far broader view and examine many aspects of the complex interaction between plant material, incubation conditions and constituents of culture medium. It is this type of work that will permit us to convert remaining "difficult" species into "easy" species.

12. CONCLUSIONS

Of necessity, I have had to be selective in the topics covered in this paper, but matters omitted should not be taken as an indication of less importance and concern to practicing micropropagators. Such matters as variability in organ cultures, the use of callus and suspension culture, somaclonal variation, laboratory design and the automated systems developed in Australia and Israel, virus eradication and testing, planting-out (acclimatization) of cultures, and so on. I would certainly like to see much more systematic research into ways to reduce the time from laboratory door to salable product.

The overall thrust of my paper is a plea for more systematic research into all aspects of micropropagation using the methods that are our inheritance as scientists.

REFERENCES

1. de Fossard RA: 1976. Tissue culture for plant propagators. Department of Continuing Education, University of New England, Armidale, N.S.W., Australia 2351.

2. de Fossard RA: 1981. Plant tissue culture propagation. Filmfiche Corporation, Box H222, Australia Square, Sydney, Australia 2000.

THE APPLICATION OF TISSUE CULTURE TO PLANT IMPROVEMENT AND PROPAGATION IN THE ORNAMENTAL HORTICULTURE INDUSTRY

IRWIN Y.E. CHU

Geo. J. Ball, Inc., P.O. Box 335, West Chicago, IL 60185, U.S.A.

1. INTRODUCTION

The technology of plant tissue culture for mass propagation is well developed for industrial applications. Indeed, world-wide there are more than 200 commercial micropropagation laboratories (6). Increasing numbers of universities, governmental institutions, research venture companies and botanical gardens are also actively involved in the application of tissue culture technologies.

Most commercial micropropagation labs were established by nurseries to satisfy in-house needs and many propagate only a few crops. Although difficult to assess accurately, approximately half of all commercial labs propagate orchids (Table 1). Many of these tend to be small labs with few employees. Larger labs tend to propagate other crops with the largest ones having over 100 employees and presently concentrating on the production of tropical foliage and flowering herbaceous ornamentals. Several labs, and particularly those not associated with a specific nursery, propagate plants on contract to other growers, and may provide other services including disease indexing and germplasm storage.

An increasing number of labs are exploring the potential of tissue culture for product development, including the genetic improvement of various crop species. A variety of in vitro methods are being attempted including mutant induction and selection, ploidy modification, protoplast manipulations and other methods of gene transfer. Although such sophisticated techniques as protoplast fusion and gene transfer have yet to yield commercially viable plant products, many large corporations, public institutions and research venture companies are committing major investments in anticipation of tremendous financial returns. Much progress towards reaching their goals has been made in the past 10 years and there is little doubt that tissue culture technologies will play a major role in

Zimmerman, R.H. et al. (eds), Tissue Culture as a Plant Production System for Horticultural Crops.
 ISBN 90-247-3378-2.

TABLE 1. The percentage of world and USA commercial T.C. labs producing specific crops.[1]

Crops	90 USA labs	115 World labs excluding USA[2]	205 World labs including USA
Orchids	60%	56%	58%
Ferns	15%	21%	19%
Foliage & Flowers	28%	37%	33%
Woody crops	15%	17%	16%
Tropical plantation crops	3%	9%	6%
Vegetable crops	5%	13%	10%
Fruit crops	4%	23%	15%

[1] Information source: Directory of Commercial Tissue Culture Laboratories in (6), Pp. 549-593.

[2] Australia, Belgium, Canada, Denmark, France, Germany, Greece, Israel, Italy, Ivory Coast, Malaysia, Netherlands, New Zealand, Norway, Philippines, South Africa, Singapore, Sweden, Taiwan, Thailand, U.K.

future product development, especially in horticultural crops and later in agronomic crop plants.

Geo. J. Ball, Inc. is committed to tissue culture and its application to plant improvement and propagation. Our main line of business involves flower and vegetable seed, and young ornamental plants. Plant improvement and propagation technologies are key to our success and therefore, we are keenly assessing the potential of tissue culture to satisfy real in-house needs. This report reflects our industrially oriented viewpoint that generally demands rapid benefit from the emerging technologies and a high proportion of project goals that are short-term. Techniques are not emphasized in this report. Rather, examples of applications are included. Methods used are ranked according to our belief and experience in their value for developing new value-added products quickly. This is not to say that the latter methods have less value, only a more long-term benefit.

2. COMMERCIAL MICROPROPAGATION

Orchids remain as the classical example of the successful application of tissue culture technologies for clonal propagation of elite selections. Prior to tissue culture, growers were predominantly only able to propagate orchids from seed; asexual propagation by division or rooting of cuttings being an inefficient process. The highly heterozygous state of seed-propagated Cattleya orchids, for example, resulted in a high percentage of undesirable flowers in a crop which flowered sporadically over a long season and could not be controlled. Now, award-winning Cattleya hybrids can be mass cloned to any quantity desired. The result enabled commercial orchid growers to produce a uniform high-quality cut flower crop on schedule to match particular holiday seasons.

2.1. Limitations

From its conception, tissue culture promised to provide the commercial grower with several unique propagation advantages. Whether actual or perceived, the promised advantages included: 1) Tissue culture provides a highly efficient way of clonally propagating large numbers of high value crops quickly. 2) The aseptic process results in disease-free plants. Early in the 1970's, tissue culture was heralded as a grower's panacea. Many investors entered the scene with notions of high financial rewards from a technology whose limitations at the time were poorly understood. Many problems resulted and for several years in the late 1970's tissue culture propagation had a less than favorable reputation in the horticulture industry.

People establishing commercial labs often underestimated the state of the science, which in many respects is still very much of an art. Often the technology developed at universities could not be translated adequately into commercial situations. Commercial labs often experienced crop scheduling and seasonality problems. Many have had difficulty filling orders due to poor management or lack of knowledge about the production of the particular crop.

Grower's receiving plants and microcuttings directly out of culture often had problems with the material. Many lost entire crops soon after receiving the cultured material and liabilities have been debated in the courts. Growers are becoming more aware of the often exacting environmental conditions needed to successfully acclimatize and establish tissue cultured plants in soil and greenhouse conditions.

Tissue culture has caused some industry wide problems. Despite the aseptic conditions, which generally rids bacterial and fungal organisms, tissue culture is blamed as the leading cause of a major disease problem in the orchid industry (12). Viruses were manageable prior to orchid tissue culture. Now, the widespread use of clonally propagated plants has spread viruses to all sectors of the orchid industry. Tissue culturists are urged to have their stock plants virus indexed prior to the culturing process.

Genetic instabilities have been another problem associated with the micropropagation industry. In certain crops, the culture process encourages an increased frequency of sporting (11), which can be particularly disadvantageous when the change is a reversion to a less desirable form. This has caused trouble for companies involved with the micropropagation of Dieffenbachia, for example, whose customers expect receiving true-to-type plants. The popular cultivars of Dieffenbachia, 'Compacta' and 'Camille', in culture have demonstrated a tendency to revert to the original less attractive 'Perfecta'. Reversions have been as much as 80% of a grower's crop, and detection and subsequent roguing is presently only possible later in the plant's development as it reaches maturity in 4-inch pots. Despite its disadvantage in a commercial micropropagation situation, researchers are discovering that the process that encourages genetic changes is useful for product improvement. Indeed, several plant biotechnology companies have been established to specifically exploit this phenomenon. This use of tissue culture in mutation breeding programs is discussed further in this report.

2.2. Benefits

The past decade has provided tissue culture companies with the needed settling-in period and an opportunity to discover the limitations and realistic potential of tissue culture for propagation. Their customers have also gradually discovered how the commercially available tissue cultured products best fit into their businesses. Demand for tissue cultured products is presently increasing and many established tissue culture companies are now expanding to satisfy such demand. Commercial tissue culture production is expected to more than double within the next year.

The commercial tissue culture production of Syngonium (Nephthytis) 'White Butterfly', an attractive tropical foliage plant, further demonstrates the range of benefits that micropropagation has provided commercial growers. Syngonium microcuttings are readily produced in culture and the resulting plants have several desirable qualities compared to conventionally propagated Syngonium plants. Conventionally, Syngonium are propagated by rooting stem-tip cuttings, they do not branch well and are plagued by several systemic pathogens including Erwinia chrysanthemi, the causal agent of the common 'Bacterial leaf spot' problem. Tissue culture provided the grower with microcuttings free of the bacterial disease. Unexpectedly, the culture process provided sufficient "hormone carryover effect," typical of cytokinins, which caused early multiple branching. The process also introduced an apparent invigorating juvenility-like character. The result was a vigorous, mostly disease-free, well branched plant; a far superior product for the commercial grower and ultimate consumer. Many tissue culture labs are currently mass micropropagating Syngonium and many new labs are utilizing this crop to assist lab establishment. The potential of overproduction is of concern and the predictable 'price-cutting wars' have begun, which should make tissue cultured Syngonium even more attractive to the commercial grower.

A consideration gaining further interest is the value of tissue culture in shipping plants internationally. Increasingly larger numbers of plants, as germplasm or production materials, are being shipped in vitro. The aseptic conditions, guaranteeing freedom from insects and bacterial and fungal pathogens, virtually eliminate the quarantining process at most entry points. Plant species, however, known to be specific virus carriers continue to be banned whether in vitro or not.

2.3. Selection of plants that may benefit from micropropagation

Companies consider the application of micropropagation from two directions: 1) as a propagation method, and 2) as a potential tool for "improving" the physiological state of the plant.

2.3.1. Micropropagation, when no conventional clonal propagation system exists. For certain crops, conventional methods of asexual propagation are not possible or commercially practical. Orchids were previously discussed but other crops may include runnerless strawberries, super-male asparagus and high promise male sterile inbred parental lines of certain flowers and vegetables when maintainer lines are not available.

2.3.2. Micropropagation, when it is the most efficient clonal propagation system.

2.3.2.a. When high volume is essential. Following a decision to release a new plant cultivar, companies are always interested in rapid scale up and production of commercial qualities. Micropropagation is often the fastest method and, if not sufficiently economical to produce the final product, may still be practical for the rapid production of sufficient stock plants for subsequent conventional propagation. Micropropagation methods are currently being used to mass produce new cut flower, pot plant and landscaping *Gerbera* varieties.

2.3.2.b. When micropropagation is the most economical method of propagation. Micropropagation is often an expensive process. However, as technologies are refined, efficiencies increased, and mechanization and materials handling systems introduced, micropropagation will become an increasingly more economical procedure. The cost of conventional propagation methods should continually be compared with micropropagation. Hybrid tea roses are an example where tissue cultured plants are less expensive to produce than conventionally budded plants.

2.3.3. Micropropagation for plant enhancement.

2.3.3.a. When a more disease-free condition is important. The aseptic requirement of the tissue culture process provides plants that are generally free of bacterial and fungal diseases. Certain crops have such major systemic disease problems that a tissue cultured plant is preferentially sought. *Gypsophila* cut flower producers have a serious problem with crown gall bacterium that is difficult to eradicate and is readily transmitted through conventional asexual plant propagation. The availability of clean tissue cultured *Gypsophila* stock has done much to reduce field problems. As previously mentioned, *Syngonium* is another crop that greatly benefits from a more disease-free condition that the tissue culture process provides.

2.3.3.b. When multiple branching is desirable. A well-branched plant with a "full" appearance is frequently preferred. As mentioned previously, the tissue culture process enhances the branching of *Syngonium*. Other foliage plants are physiologically "improved" by the tissue culture process including Spathiphyllum, Dieffenbachia and many ferns. A desire has been expressed for better branching in Fuchsia and Hibiscus, and research is under way to develop suitable systems for these crops.

3. DISEASE DETECTION AND ELIMINATION

As commercial micropropagation labs get closer to supplying demand, and competition increases, more will be searching for ways to distinguish their products. Applications of disease indexing and elimination procedures along with sufficient advertisement of the fact will become routine. Already, many labs selling small-fruit crops such as strawberries and raspberries advertise their products as being "virus-free." Labs in other sectors of the micropropagation industry, including those specializing in ornamental crops, will soon follow. Micropropagation companies without trained pathologists are able to send plants to any of the several specialist companies who will apply increasingly sophisticated disease indexing procedures, and if need be, undertake a disease elimination program.

Meristem-tip culture techniques are frequently being used to eliminate viruses from plants. The uppermost fraction of a millimeter of an infected plant's meristem-tip, which is often free of viruses, is used as the explant source. Heat treatment and treatments with antiviral agents such as virazole (16) prior to meristem-tip culture have been used successfully to slow or halt the spread of viruses into the meristematic tissue, thus further assuring pathogen elimination. Certification may be done through use of very sensitive serodiagnosis or DNA probes. Disease indexing and elimination techniques are being used extensively to assure the commercial availability of pathogen-free chrysanthemums and geraniums.

4. GERMPLASM PRESERVATION

Tissue culture technologies are becoming increasingly used for the storage of valuable germplasm. Many companies store their plant materials aseptically in vitro and modify nitrogen and sucrose levels to slow growth rate. Growth retardants and regulators are also being assessed, and show promise. When possible, lighted refrigerated growth chambers are used to maintain the cultures. Improved culture techniques are continually being developed to slow the culture's growth rate and reduce the need for culture maintenance.

Companies are utilizing tissue culture to maintain asexually propagated hybrids that cannot be stored through seed. Chrysanthemum cultivars are commonly stored in culture and they are often disease indexed to insure that they are virus free. Thus production stock beds can be replenished with "clean" material if they become virus infected or lost due to unexpected foul weather. Geraniums and double flowering petunia are similarly handled.

Plants that cannot be seed propagated, such as male sterile parental marigold lines, are stored *in vitro* until progeny tests reveal which ones have good recombining ability. Other composites such as *Rudbeckia* are also stored until field trials are completed.

Companies becoming actively involved in biotechnological applications are using tissue culture to store unique experimental germplasm such as albinos and other chlorophyll deficient mutants, and antibiotic resistant lines, which have markers that assist selections.

Increasingly more horticulture-based companies are involved internationally. Indeed, much of the world's F_1 hybrid seed production takes place in third-world countries. The threat of natural disasters such as hurricanes, typhoons, earthquakes, etc., and uncertain political stabilities increase the need for germplasm storage techniques as insurance.

Freeze preservation by liquid nitrogen may eventually become the ultimate method of storing elite disease-free germplasm. At liquid nitrogen temperature tissue maintenance in principle is unnecessary and germplasm could be stored mostly unchanged for an unlimited period. A suitable method when developed would also allow tissue culture companies to spread ordinarily cyclic production evenly over the entire year and store microplants, such as strawberries, until seasonal demand dictates when to revive them. Considerable research has successfully demonstrated liquid nitrogen freeze preservation of seed with subsequent reviving, germination and growth into healthy plants (9). Probably, commercial tissue culture applications will center around the freeze preservation of somatic embryos with subsequent restoration based on customer orders.

5. MUTANT INDUCTION AND SELECTION

In vitro mutant induction and selection can provide a rapid method for developing new products. This is especially true when slight variations of outstanding cultivars are desired. Several types of mutants can be produced in vitro. These include somaclonal variants and induced mutants.

5.1. Somaclonal variation

Tissue culture is a very unnatural way of inducing plant growth. Indeed, for a particular species, the callus form of plant growth along with adventitious regeneration may be scarce or nonexistent in nature. Probably a combination of factors is responsible for inducing the often permanent genetic changes that are occasionally seen in tissue culture derived plants. Certain species are prone to sport in culture and companies are utilizing this tendency to their advantage.

The mum industry utilizes this phenomenon to develop new cultivars of chrysanthemums. From an outstanding pink flowering seedling, all of the other colors such as yellow, white and bronze are derived while keeping the outstanding plant qualities of the original pink flowering variety. The tendency for the production of somaclonal variants in chrysanthemum is very genotype-dependent with some cultivars such as 'Citadel' being very stable in culture while others such as 'Parasol' frequently mutating in culture.

5.2. Induced mutations

When chrysanthemum cultivars are very stable in culture, mutation frequency is 'helped along' with ionizing radiation, gamma and x-ray (3). Stem tip cuttings are irradiated and the chimeric plants are grown to flowering. Changes affecting flower color are then expressed and subsequent petal tissue culture results in mostly non-chimeric plants. Resulting plants are then rigorously trialed and commercially released if sufficiently improved.

Chemical mutagens such as ethyl methane sulfonate (EMS) have been valuable in inducing mutations in culture (5). This is especially true when cytoplasmic mutations such as foliage variegation or male sterility are desired. Nuclear genetic changes are also induced.

6. OVERCOMING BREEDING INCOMPATIBILITIES

When transfer of genetic materials between plants is desired, conventional sexual hybridization should be attempted first. If failures occur, every attempt should be made to determine the reasons for the experienced incompatibilities. Too often, grandiose sophisticated biotechnological programs are proposed to solve what often turns out to be a simple avoidable problem. Relatively simple techniques are available to determine whether the experienced incompatibilities are pre- or post-zygotic. The extent of pollen tube growth can be determined and embryo viability simply assessed.

6.1. Overcoming pre-zygotic incompatibilities. Pollen can fail to germinate, fail to grow sufficiently through the style or fail to enter and fertilize the ovule. Such incompatibilities have on occasion been overcome by a heat treatment that is simply accomplished by dipping the attached pistil for a short time in hot water prior to pollination. This system was used successfully to obtain hybrids from sexually incompatible crosses (7).

Tissue culture systems are available to assist the obtaining of incompatible crosses. Pistils have been grafted successfully onto ovary tissue to aid hybridizations. Research programs have demonstrated successful seed production from *in vitro* pollination of ovules attached to their placenta (17). Few commercially beneficial applications have been demonstrated. The closest is perhaps the use of *in vitro* placenta pollination to obtain inbreds from self-incompatible plants of *Petunia hybrida* (14). Self-incompatibility in *Petunia hybrida* commercial varieties is rare; mostly this problem has been eliminated by breeders. Occasionally, petunia self-incompatibilities arise but this incompatibility tends to be incomplete. Nevertheless, *in vitro* techniques are available to assist overcoming pre-zygotic sexual incompatibilities when needed.

6.2. Overcoming post-zygotic incompatibilities. Failure of seed development after fertilization can be due to many reasons. Often, however, it relates to problems in endosperm development. Poor or lack of endosperm development typically results in embryo abortion. Utilizing tissue culture techniques, it may be possible to rescue the embryo before abortion takes place and to germinate the seedling in culture.

In the breeding and development of *Impatiens*, many colors exist with yellow lacking. Companies currently are building on the research of Dr. T. Arisumi, U.S.D.A., Beltsville, Maryland, who succeeded in rescuing embryos from incompatible hybridizations between the commercial *Impatiens* and a yellow flowering species, *I. auricoma*. Research has demonstrated successful interspecific hybrid plant production and currently white flowering commercial *Impatiens* inbred lines are being used to develop a more commercial product.

Embryo rescue attempts are being heavily concentrated on the introduction of virus and powdery mildew resistance into cultivated winter and summer squash varieties. Resistance was found in certain species such as Cucurbita ecuadoriensis and C. martinezii that unfortunately do not easily cross with cultivated varieties. Interspecific sexual hybridizations combined with embryo rescue techniques have been successful and development of virus and powdery mildew resistant cultivars are well along. Research continues and has been expanded to other cucurbits such as cucumbers and melons.

7. HAPLOIDIZATION AND POLYPLOIDIZATION

The possibility of a technique that produces large numbers of "instant inbreds" intrigues any breeder who has often spent years performing the seven or more generations of self hybridizations that are usually necessary to produce commercially usable inbreds. Research has demonstrated that with certain crops it is possible to tissue culture immature anthers, pollen grains or unfertilized ovules and regenerate plants with half the normal complement of chromosomes. Subsequent colchicine application can induce chromosome doubling, which results in a completely homozygous inbred line. Such doubled haploids are then ready for trials to assess recombining ability in F_1 hybrid crosses.

Today, limitations to application are mainly due to technology. Generally, large numbers of genetically distinct haploids are required and few species perform adequately in culture. Techniques, however, are being improved and additional species are continually being reported successful through haploid induction programs. Tobacco (Nicotiana tabacum) anther culture is considered the model system for the production of haploids. The ornamental industry is attempting to perfect anther and ovule culture systems for tobacco's relative, flowering tobacco (Nicotiana alata) and Petunia, to assist breeding programs through the production of instant inbreds.

In certain crops there is a distinct advantage of higher ploidy levels; larger flowers, greater flower substance leading to longer cut life, stronger stems and thicker leaves. Such plants are often in greater demand and sell at a premium compared to their conventional diploid counterparts. These may include tetraploid geraniums, snapdragons, day lilies, tuberous begonias, to name a few. In crops such as watermelons, French marigolds and fibrous-rooted begonias, higher ploidy level in the triploid form offers the advantage of seedlessness. Conventional diploids are crossed with tetraploids and the resulting triploid product's seedlessness generally leads to extension of the plant's flowering season, as well as extending flower life. The induction of ploidy increases are used as a rapid way of developing new value-added plant products.

Ploidy increases are generally easily accomplished in vivo through colchicine applications by treating vegetative cuttings or seedlings. However, combining the process with regenerating tissue cultures may be an advantage if large numbers of polyploids are desired. Previously mentioned tissue culture induced somaclonal variation may produce a higher rate of spontaneous ploidy changes making colchicine treatments unnecessary.

8. PROTOCLONAL VARIATION

An ability to isolate and culture plant protoplasts opens many opportunities for a wide range of product development possibilities. The technology requires turning a cell into a protoplast by enzymatically stripping-off its cell wall, and an adequate culture system that will induce formation of a new cell wall, stimulate divisions and induce regeneration into a complete plant. Such a process is quite traumatic to the plant cell and with certain species, frequent mutations occur resulting in "protoclonal variations."

The protoplast isolation and culture process appears to induce greater variability than what is induced by a more conventional tissue culture somaclonal variation process. Variations in protoplast-derived plants have been reported in a wide range of crops including potatoes, tomatoes, alfalfa, lettuce and petunias. Some potentially valuable genetically stable variations have resulted including disease resistance, increased yield, dwarfness, increased seed vigor, and improved nutritional and food processing qualities (15). Geo. J. Ball, Inc. is fortunate that the typical protoplast "model system" involves petunia, which for us is a commercially important crop. Our petunia protoplast cultures have resulted in a wide range of variability including flower colors, flower shape, plant form, leaf color and shape. One petunia variant has particularly attractive stable foliage variegation and we are currently assessing its potential as a commercial product.

9. SOMATIC EMBRYOGENESIS

Labor is typically the greatest single cost in U.S. commercial micropropagation laboratories. Such cost typically exceeds 60% of the entire production cost. A concentrated effort should be made by tissue culture companies to automate, and several large aggressive labs are considering various mechanization systems including robotics.

Maximum efficiency involving automation will probably require an alternative to the typical tissue culture system of organogenesis that most labs currently utilize. Several companies are considering the phenomenon of somatic embryogenesis as the tissue culture system that will best lend itself to automation technologies. As previously mentioned, somatic embryos appear to be a potentially valuable form of storing valuable germplasm.

In vitro embryo formation has been observed in a wide range of plant genera and extremely large numbers have been produced in liquid cultures. However, the phenomenon of embryogenesis including production quantity is mostly genotype specific. In most cases, each somatic embryo originates from single somatic cells and it has been possible to produce large numbers of genetically uniform plants from such tissue culture induced embryos. Under proper conditions, somatic embryos germinate readily, similar to their seedling counterpart.

Several laboratories are attempting to develop bioreactor systems to produce large quantities of somatic embryos. Efforts are being made to solve problems such as embryo synchronization, proliferation, maturation and singulation at the production level. Some laboratories are paying special attention to "delivery system" development. An adequate delivery system would allow successful transfer of somatic embryos from culture into soil at a high germination rate. Fluid drilling and capsulation systems are currently being considered. As technologies develop, increasing numbers of commercially micropropagated plants will be derived via automated somatic embryogenesis systems.

10. SOMATIC HYBRIDIZATION

The availability of protoplast culture systems coupled with cell fusion technologies promises the creation of novel hybrids from sexually incompatible parents. Protoplasts from diverse origins can be induced to fuse by several chemical or electrical processes. Polyethylene glycol is the most commonly used fusogen and very high fusion frequencies can result. The protoplast isolation and fusion process is often easily accomplished. Difficulties usually center around the lack of good protoplast culture systems and the lack of adequate selection systems. The most successful research programs incorporate a strong visual or preferably chemical-based selection system that will distinguish or preferentially favor the growth of the somatic hybrid cells in a background of numerous unfused or self-fused parental cells. Much research continues.

Applications of cell fusion technologies have been few and mostly limited to members of the Solanaceae. Research has demonstrated that the creation of hybrid plants from sexually incompatible parents is possible. However, wide crosses, such as the famed tomato + potato somatic hybrid (13), have resulted in sterility problems and limited possibilities for further plant improvement. Although morphological oddities can be desirable in the ornamental horticulture industry, much current research has concentrated on less wide crosses where transfer of less dramatic, more practical characteristics are desired. For example, Solanum brevidens is a potential donor of disease and physiological resistances for the cultivated potato. A rigid crossability barrier exists between the two species that was overcome through protoplast fusion. The synthesis of a somatic hybrid between S. brevidens and the cultivated potato facilitates a realistic evaluation of the potential of somatic hybridization in potato breeding (1). Other applications have included the transfer of disease resistance in tobacco, herbicide resistance in potato, and higher alkaloid production in Datura.

Companies are also considering the potential of protoplast fusion techniques to facilitate the transfer of cytoplasmically inherited characteristics. Genes coding for cytoplasmic male sterility have been transferred by cell fusions in petunia (8) and tobacco (2), and offers the possibility of transferring other cytoplasmic characteristics such as certain herbicide resistances and ornamental foliage variegation. Somatic hybridization will continue to be an important tool to generate sexually incompatible crosses and transfer polygenic characteristics.

11. GENE TRANSFER

Plant breeders mostly desire adding only a few improved characteristics into their crop rather than incorporation of entire genomes, which typically happens during sexual and somatic hybridizations. Following such hybridizations, the plant breeder typically spends much effort trying to rid the often many undesirable characteristics that were initially transferred along with the few desirable ones.

A system that will enable the transfer of only desirable genes would be a tremendous help in plant improvement programs and would probably revolutionize plant breeding. It is not surprising that recently developed "genetic engineering" techniques have attracted much attention, especially since basic research has shown that genes can be successfully cloned, transferred and expressed in recipient plant cells (4, 10). However, much further research is needed before practical applications are widespread.

Several problems exist that are currently limiting commercially valuable applications of genetic engineering technologies. These include inadequate gene transfer or vector systems, and lack of an ability to recognize commercially useful genes that are worthwhile transferring. Much effort is currently being spent solving these problems.

Successful gene transfers have been accomplished by using Agrobacterium's Ti-plasmid as the vector. However, the effect of the Ti-plasmid is mostly limited to dicotyledonous plants. Other transfer systems such as microinjection and use of viruses as vectors also have limitations. Viruses are often very specific to particular recipient cells, and microinjection is tedious to perform and therefore, relatively few cells can be transformed. Recently, a newly developed technique called electroporation has proven to be a very effective system to mediate gene transfer. It involves generating "pores" in the protoplast's membrane by high voltage electrical pulses. These pores then enable naked DNA or plasmid DNA to readily pass into the recipient cells. Key advantages of this new system is that it is not limited to particular plants and large numbers of cells can be transformed.

Much advancement is being made in the development of suitable vector systems. However, the problem of identifying the commercially useful genes that are worthwhile transferring by genetic engineering techniques is still causing problems. Traits that are polygene controlled currently cannot be transferred by recombinant DNA techniques. Single-gene controlled characteristics are ideal candidates but how to identify the one desirable gene amongst a background of millions of other genes will probably continue being a problem for a long time to come. Effort should be made on the construction of DNA libraries of commercially important crops to help in the identification of valuable genes.

12. CONCLUSION

The previous list of tissue culture technologies, which are being applied to plant improvement and production, are arranged mostly in order of quick value. They can be classified into projects whose goals are short-term (products or applications within 2 years), medium-term (between 2 to 5 years), and long-term (after 5 years) (Table 2). Financiers along with

TABLE 2. *In vitro* technologies, their application rate and commercial examples.

Technologies	Project term[1]	Examples
Micropropagation	Short-term	Orchids, woody and herbaceous ornamentals, plantation crops, and fruits and vegetables
Disease detection and elimination	Short-term	Chrysanthemums, geraniums, strawberries and raspberries
Germplasm preservation	Short-term	Chrysanthemums, geraniums and double petunias
Polyploidization	Short-term	Geraniums, day lilies, marigolds and watermelon
Mutant induction and selection	Short-term	Chrysanthemums, petunias, tomatoes and potatoes
Overcoming breeding incompatibilities	Mid-term	Lilies, impatiens and cucurbits
Haploidization	Mid-term	Tobacco and petunia
Protoclonal variation	Mid-term	Petunia and potato
Somatic embryogenesis	Mid-term	F_1 hybrid celery
Somatic hybridization	Long-term	Potato and petunia
Gene transfer	Long-term	--

[1] Short-term defined as within 2 years, Mid-term is between 2 and 5 years, and Long-term is over 5 years.

researchers will need to decide how soon new products are needed and which technology should be applied to reach their goal.

Tissue culture applications are greatly dependent on improvements in basic technologies, which are developed by universities, companies and other research laboratories. However, the ability to raise internal or external funding for technology development seems to be increasingly dependent on previous successful applications. It appears in the future that much of the tissue culture basic research funding will come from increasingly applied programs.

There continues to be a need for improved micropropagation systems especially the solving of recalcitrant systems towards high shoot regeneration. However, we believe that the long-term future of commercial micropropagation will revolve around somatic embryogenesis, particularly as a method to increase production efficiency and reduce costs. We wish to encourage applied research programs in this area. A commitment to tissue culture mechanization, materials handling and robotics research also should be made, particularly by industry.

The most significant future contributions of tissue culture to plant improvement will come from long-term protoplast manipulation programs including cell fusion and gene splicing. Applications continue to require successful protoplast culture systems including good plant regeneration. Adequate protoplast culture systems have been developed for too few species and we wish to encourage further research in this area. Applications of tissue culture technology, particularly as it relates to plant improvement, should be considered only as one stage in the overall scheme from product idea to market. Horticultural innovation is dependent on many disciplines working together, including plant breeding, genetics, cell biology and physiology, biochemistry and molecular biology, market research and marketing, and engineering.

The United States is fortunate that it presently has a strong leadership position in horticulture and plant biotechnology. However, we firmly believe that the maintenance of our country's strong position will become increasingly more dependent on cooperation among our industry's public research institutions, large corporations and research venture companies. Our international competitors, particularly the Dutch and Japanese, cooperate within their country's industry much better than we, and share the effort of technology development. Tissue culture technologies are becoming increasingly diverse and complicated. So much so, that success on a national scale will require joint national efforts.

REFERENCES

1. Barsby TL, JF Shepard, RJ Kemble, R Wong: 1984. Somatic hybridization in the genus Solanum: S. tuberosum and S. brevidens. Plant Cell Reports 3:165-167.

2. Belliard G, G Pelletier, M Ferault: 1977. Fusion de protoplasts de Nicotiana tabacum a differents: étude des hybrides cytoplasmiques neo-formes. C.R. Acad. Sci. Paris, Ser. D. 284:749-752.

3. Broertjes C, S Roest, GS Bokelmann: 1976. Mutation breeding of Chrysanthemum morifolum Ram. using in vivo and in vitro adventitious bud techniques. Euphytica 25:11-19.

4. Fraley RT, SG Rogers, RB Horsch, PR Sanders, JS Flick, SP Adams, ML Bittner, LA Brand, CL Fink, JS Fry, GR Galluppi, SB Goldberg, NL Hoffmann, SC Woo: 1983. Expression of bacterial genes in plant cells. Proc. Natl. Acad. Sci. U.S.A. 80:4803-4807.

5. Fujimoto M, H Yamagata: 1982. Studies on the utility of artificial mutations in plant breeding. XIII. Mutagenicity of several alkylating agents in rice. Jap. J. Breeding 32:17-25.

6. George EF, PD Sherrington: 1984. Plant propagation by tissue culture; handbook and directory of commercial laboratories. Exegetics, Ltd., Eversley, Basingstoke, Hants, U.K.

7. Hopter JE, PD Ascher, SJ Peloquin: 1967. Inactivation of self incompatibility following temperature pretreatments of styles in Lilium longiflorum. Euphytica 16:215-220.

8. Izhar S, JB Power: 1979. Somatic hybridization in petunia male sterile cytoplasmic hybrid. Plant Sci. Lett. 14:49-56.

9. Kartha KK: 1985. Cryopreservation of plant cells and organs. CRC Press, Inc., Boca Raton, FL.

10. Klee HJ, MF Yanofsky, EW Nester: 1985. Vectors for transformation of higher plants. Bio/Technology 3:637-642.

11. Larkin PJ, WR Scowcroft: 1981. Somaclonal variation--a novel source of variability from cell cultures for plant improvement. Theor. Appl. Genet. 60:197-214.

12. Lawson RH, SD Hearon: 1973. Symptomatology of Cattleya mericlones infected with Cymbidium mosaic virus. Amer. Orchid Soc. Bull. 42:1071-1074.

13. Melchers G, MD Sacristan, AA Holder: 1978. Somatic hybrid plant of potato and tomato regenerated from fused protoplasts. Carlesberg Res. Commun. 43:203-218.

14. Niima Y: 1970. *In vitro* fertilization in the self incompatible plant, *Petunia hybrida*. J. Jap. Soc. Hort. Sci. 39:345-352.

15. Shepard JF: 1982. The regeneration of potato plants from leaf-cell protoplasts. Sci. Amer. 246(5):154-166.

16. Simpkins I, DGA Walkey, HA Neely: 1981. Chemical suppression of virus in cultured plant tissues. Ann. Appl. Biol. 99:161-169.

17. Zubkova M, V Sladky: 1975. The possibility of obtaining seeds following placental pollination *in vitro*. Biol. Plant. 17:276-280.

NEW TECHNOLOGIES IN PLANT TISSUE CULTURE

W. JAN ROWE

Agrigenetics Corporation, 3375 Mitchell Lane, Boulder, CO 80301, U.S.A.

1. INTRODUCTION

This review paper addresses recent technical advances that could be incorporated into current plant culture systems to improve overall laboratory efficiency. Plant biological systems and their impact upon the micropropagation industry will also be discussed.

Current production techniques for the in vitro propagation of plants have allowed for strong and continued growth within the micropropagation industry. Most notable has been the success of commercial laboratories in the production of orchids, foliage and ornamental plants. Currently, demand for most of these products exceeds the capability of the commercial laboratories to produce sufficient numbers of these plants. This demand stimulates growth within the industry. Some companies are producing five to eight million units per year and three major labs are considering expanding to as many as 20 million units per year. This kind of production will saturate certain markets. Since little product diversity exists, competition for well defined markets will force commercial labs to sell their products with lower margins and reduced prices, making it difficult for those with limited cash resources to stay in business.

One of the major difficulties in expanding propagation techniques to other crop areas is production costs. In order to diversify into new areas such as plantation, forest and vegetable crops, one must be able to reduce costs significantly. Vegetable transplants currently sell for approximately 3 cents each compared to about 22 cents and higher for a tissue cultured plant. Another problem will be developing efficient systems to deal with extremely large numbers of plants. Most of these crops are planted seasonally and have heavy peak demand. Reduced labor costs and improved in vitro storage techniques will be critical factors for the expansion of the technology.

Zimmerman, R.H. et al. (eds), Tissue Culture as a Plant Production System for Horticultural Crops.
 ISBN 90-247-3378-2.

Cost studies indicate that labor inputs are a significant percentage of the total production cost (3, 39, 50). As mentioned earlier, plantlets are sold for 20 cents or more with rooted liners selling for 40 to 60 cents each. Workers can only handle 700 to 1,400 plants per day depending upon the species. This is equivalent to the production of 200,000 plants per year per employee. Based on these figures, direct labor cost would be estimated at $15,000 per year or approximately 8 cents per plant. To illustrate how much of a difference can exist in labor costs, Vietnamese farmers are producing tissue cultured potato plants at a total cost of 0.1 cents each. In this case, labor costs are 120 dollars per year per person (59). Given technical support and low transportation costs, future production centers will transfer to countries where labor is less expensive.

Due to the labor-intensive nature of micropropagation today, it is not clear whether economy of scale exists in this industry. While increasing the number of units produced, one must also increase the number of employees and supervisors proportionally and larger capital investments are required. Consequently, both labor and overhead costs escalate. As commercial labs grow, organizational problems also become critical. Therefore the management structure must be well defined. Supporting informational systems for management are required such as inventory control, production scheduling and space utilization. Personnel training and quality control are also critical components for successful large-scale production and add significantly to the management responsibilities. But reducing labor costs and learning how to handle large numbers of plants are only a few issues facing the micropropagation business today. This means that issues such as improvement of working environments, development of techniques for reducing contamination and automation of the tissue culture system itself need to be examined.

As the industry grows and matures, it must recognize and solve problems that affect cost, quality and performance of the plants produced. Research and production labs continue to be excited about micropropagation and plant tissue culture as a tool for crop improvement, and the applications for this kind of technology continue to expand. Commercial labs have been slow to change the techniques developed over the past two decades at university, government and research laboratories. This is in part due to the fact that very little money has been available for research dealing with the commercial development of the plant production process.

It is a critical time for micropropagators to explore new technologies and determine how they might impact the productivity of the tissue culture systems and their facilities.

2. ASEPTIC CULTURE TECHNIQUES

Murashige (35, 37) first described the concept of specific pathogen-free stock cultures. Knauss (26, 27) also stressed the importance of disease indexing in a Dieffenbachia production scheme. Unfortunately, most laboratories have been unable to solve this quality control problem. A good indexing program is necessary during the culture initiation phase but precautions also need to be taken throughout the life of the plant in culture.

At Agrigenetics, all materials and tools that are used in the transfer area are autoclaved prior to use. Forceps and scalpels are packaged in metal canisters. Ten large Potts-Smith forceps can be packaged in each canister while 25 to 30 scalpels can be sterilized in a separate canister. Each canister is sterilized in a sealable package that allows for gas or steam penetration. These are commonly used in the medical and pharmaceutical industry. Dissecting trays (5-1/2" x 8-1/2") with disposable chipboard inserts are also autoclaved in the same packages. The system is safe and frees the operator of worry about whether a tool has been properly sterilized. It also reduces heat load in the transfer hood. The initial costs for the system are high, but this can be easily justified due to reduced contamination rates and increased productivity.

Another interesting possibility is the use of clean room environments for culture transfer. In a clean room, there is no congestion of materials as in a laminar flow hood. Gowning procedures for class 100 provides the benefit of reducing sources of contamination from clothing, skin and hair. When working in a clean room, flame sterilization should not be used because rapid heating of instruments causes particulates to be suspended in the air. A laboratory design could be optimized by connecting a double door autoclave from the media preparation facility to the inside of the clean room. This would reduce the probability of contaminants entering the work area from supplies and containers. Although vertical flow class 100 clean rooms provide a comfortable and sterile work environment, the capital and operational costs are extremely high. As an alternative, isolated vertical flow tunnels have the benefits of a clean room but are less expensive. This type of sterile work area is essential if subculture procedures become automated due to the size of the machinery involved.

3. BIOLOGICAL SYSTEMS

The objective of any micropropagation system is to asexually reproduce an individual plant selection using in vitro techniques. Axillary bud multiplication is probably the most commonly used method of propagation (28, 35, 36) and the most genetically conservative. It is generally a low yielding system and, as alternatives, adventitious shoot induction and somatic embryogenesis have received considerable attention as methods that could potentially revolutionize the micropropagation industry through automation (15, 62). Unfortunately, it may not be possible to apply these processes to all crops. Although over 100 species have been reported to regenerate by somatic embryogenesis, few are being used commercially due to low and inconsistent yields (1, 2, 47, 58, 60).

3.1. Somatic embryogenesis

Somatic embryogenesis was first recognized in carrot cultures in the late 1950's, and the process by which embryoids form on proembryonic masses has been described in detail for some time (23, 34, 56). Ammirato (1, 2) recently reviewed the subject, and Browers and Orton (6) have outlined a scheme for the production of celery using somatic embryos.

The yields of somatic embryos from cell suspensions of carrot are tremendous. Lutz et al. (33) reported that 1 ml packed cell volume containing approximately 2×10^6 cells would produce over 5,000 uniform embryos after 14 days on 2,4-D free medium. Styer (57) has recently reviewed the design and application of bioreactors for large-scale handling of embryogenic cell suspensions. Using these devices, culture conditions can be optimized for growth of the plant cells, and very high yields can be expected. Batch cultures of 1 to 5 liters could easily produce hundreds of thousands of plants. Another benefit of this kind of system is that the embryos are already singulated from the suspension culture. It is therefore easy to separate them by size, making them amenable to precision handling.

Development of the embryos proceeds normally if nurtured properly. In our lab, the addition of 10% by volume coconut water in the third and fourth weeks of regeneration has dramatically improved survivorship of the embryos during the hardening-off period. We are now able to achieve limited success when transferring somatic embryos directly to soil in a growth chamber environment (JR Wong, personal communication).

Suspension culture systems are not without their problems. The first is that very few species, produce a regenerable cell suspension and once established, the embryogenic potential of the cultures can decrease over time (51). It is still the exception rather than the rule. Even within species, not all genotypes respond in a similar manner (7). In carrot, some of this variation can be handled with minor changes in hormone levels, but in celery this has not always worked (33, 40). Variation, either genotypic or phenotypic, in plants regenerated from suspension cultures is well documented and may be a limiting factor in the commercialization of these techniques. The potential for variation seems to increase with the amount of time cells and tissues remain in culture (30, 40). Therefore quality control will be extremely critical. As much as 1 year may be required to test the performance of a cultivar before it would be appropriate to increase stock cultures. During production it could take as long as 20 weeks to obtain additional feedback from field trials. This means that one is committed to 5 months of production without information about the quality of plant material being produced. This is suggested in Figure 1, where the low energy inputs of conventional celery production are compared to a costly, lengthy embryogenic production system.

Substantial progress has been made with some plantation crops such as oil palm, date palm, coffee and cocoa using somatic embryogenesis as a micropropagation technique (14, 22, 41, 42, 46, 53). In oil palm, high yielding mature palms with a predetermined oil composition are being propagated via somatic embryogenesis from zygotic embryo, root, petiole and meristematic tissues (43, 52, 54). These explants are induced to form callus on medium supplemented with 2,4-dichlorophenoxyacetic acid levels from 9-45μM (2-10 mg/liter) with or without low levels of benzyladenine (BA) or kinetin (52). It has been reported that phenolic compounds often leach into the medium from the wounded tissues and this appears to be promoted by the addition of cytokinins. Activated charcoal is now commonly included in the medium to deal with this problem, but hormone additives may need to be increased significantly, due to the adsorptive properties of charcoal (45, 46). Callus growth is slow, often with a doubling time of 40 days (22). Embryo differentiation is induced by omission or reduction of

Seed (Conventional)

Automated double seeding of open pollinated cultivar into plugs or flats in greenhouses.

↓

Germination 80%.

↓

Select vigorous seedlings. Thin all doubles. Grow in greenhouses (60-70 days).

↓

Harden outdoors (7 days).

↓

Transfer to field using automatic transplant system; 3-4 cents per transplant in field.

↓

Harvest after 120 days.

Somatic Embryo (Non-Conventional)

Field plantings; Select elite genotypes.

↓

Initiate callus cultures from petiole sections (30-60 days).

↓

Initiate cell culture in agitated liquid medium. → Regenerate plants via somatic embryogenesis on a small scale to test protocols; 1000-5000 plants. ↓ Conduct field trials. Look for good survivorship and uniformity for selected horticultural traits. → (back to Scale up)

↓

Scale up - increase cell cultures to 5000 ml packed cell volume for weekly production runs (100-150 days)

↓

Separate 700 ml packed cell volume and induce embryogenesis by removing the auxin (30 days).

↓

Embryos have developed and may now be size and density separated to obtain uniform populations. Yield of 1,500,000 embryos.

↓

Transfer to flat or plug transplant system in greenhouse (85-90 days).

↓

Harden and transfer to field using mechanized transplant system. → Harvest first crop after 120 days.

FIGURE 1. Scheme for celery production based on 10% market share in California. Approximately 40.6 acres per week.

auxin in the medium and development of the embryo and the plant may also require several months (41, 45). A major problem with this procedure is that it is difficult to obtain explants. In order to excise axillary buds, roots or petioles, one must use machetes or chainsaws (46). It is also difficult to surface sterilize these tissues. Lengthy soaks in mercuric chloride or chlorine are required (41). In addition to these complications, yields are low and considerable labor is still required to singulate and plant somatic embryos. Researchers in Malaysia (9) report that clonal palms grow and set fruit normally, and produce viable seeds and pollen. They also found that variability within clonal stands is considerably less than within seedling stands.

3.1.1. Somatic embryo delivery systems. Various delivery systems have been discussed since naked embryos are in no way equivalent to a seed. Heavy losses would be expected if somatic embryos were planted directly into the field. Fluid drilling, encapsulation and seed tapes have all been suggested as potential delivery systems (13, 36, 44).

Plant Genetics, Inc., recently reported the successful harvest of hybrid celery grown from encapsulated somatic embryos (17). It is not surprising that this early success would be with a plant that is in the Umbelliferae. Indeed it is a logical extension of work by Drew (13) and others at the National Vegetable Research Station in Wellesbourne, U.K., who have conducted research on delivery systems, particularly fluid drilling of somatic embryos.

3.1.1.1. Fluid drilling. Fluid drilling technology was developed for the planting of pre-germinated seeds (11, 12, 18) to provide more uniform germination and reduce time to emergence in adverse field conditions. The concept of fluid drilling is to suspend the seeds in a carrier gel, supplemented with additives, that is then pumped into the soil. Drilling devices have been developed that separate and precision drill pre-germinated seeds with high efficiency (18). As a delivery system, fluid drilling should work well if the embryos have sufficient vigor to become autotrophic. It has been reported, however, that carrot somatic embryos small enough to be handled in bulk and fluid drill require a carbon source under conditions of sterility and high humidity for an extended period of time in order to become autotrophic (13, 33). It may be beneficial to include growth supplements in these gels for nutrition during early stages of field establishment. At this time, few feel that this is a viable delivery system for somatic embryos.

3.1.1.2. Encapsulation. The encapsulation of somatic embryos was first reported by Redenbaugh et al. (44). Encapsulation involves the formation of a gel matrix around individual embryos. The benefit as a delivery system to the field or greenhouse would be that conventional seeding machines could be used, although this has not yet been proven. This technique may also prove useful as an interface between tissue culture systems and high density field transplanters. This is particularly true of vegetable crops that are routinely transplanted. This concept of encapsulation is attractive in that nutrients and agricultural chemicals can be delivered simultaneously. The technique may be more useful in a system for primed seed than for embryo delivery.

3.2. Adventitious shoot systems

The ability of many ornamental plants such as *Begonia rex*, *Sedum*, African violet (*Saintpaulia*) and lily to develop secondary meristems and adventitious shoots has been exploited in many micropropagation strategies (5, 19, 24). In African violets, for example, new shoots arise from epidermal cells (38). This means that many tissues of African violets have the potential for regeneration. Start and Cumming (55) first described this technique for the *in vitro* propagation of *Saintpaulia* using small leaf sections placed on a modified MS medium supplemented with BA, adenine sulfate, and naphthaleneacetic acid. As many as 80 shoots were produced per explant, which is considerably higher than can be obtained from axillary shoot systems.

Many vegetable seed companies are developing techniques for the *in vitro* propagation of parental lines for F_1 hybrid seed production. Since large numbers of plants are required for seed production, highly efficient cloning protocols are desired. Genotypic uniformity is essential to the success of this strategy. In *Brassica*, for example, tissue culture methods can be used to maintain self-incompatible inbreds instead of using conventional bud pollination. Anderson (4) first described the propagation of broccoli plants from floral buds using a modified MS medium supplemented with 2-isopentenyladenine, indoleacetic acid, adenine sulfate and sodium phosphate. Crisp and Walkey (10) also described similar procedures for the production of cauliflower plants. More recently, Klimaszewska and Keller

(25) reported shoot formation in Brassica napus from epidermal and subepidermal layers of stems and petioles indicating a potential for greater yields in Brassica species. Results from Agrigenetics Corporation indicate that large numbers of adventitious shoots can be generated from stem cross sections. Our data also indicate that the time required to size up plantlets for further sectioning actually reduces total yield from the cultures.

These in vitro adventitious shoot systems are attractive because plants do not need to be handled in an exact and precise manner. It then becomes possible for workers to bulk handle cultures. It is not difficult to imagine a system where plants or shoots are mechanically chopped and placed back on nutrient medium for further multiplication. An extreme example of this approach was first published by Cooke (8) for the propagation of Davallia and Platycerium ferns. Stock cultures were added to a sterile blender containing non-solidified agar medium and homogenized for a few seconds. The plant fragments were then suspended in agar solution and dispensed into containers. The yields from this kind of culture system are tremendous. The technique is applicable to many species of ferns and works extremely well with gametophytic tissues. More recently Lindemann (32) was awarded a patent for a culture system apparatus. Boston ferns were produced from plant fragments using blender techniques similar to those described by Cooke. The homogenate was dispensed into aluminum trays which were then enclosed in sterile bags and transferred to a temperature-controlled growth module. Lindemann claims broad applications to potatoes, lilies, crocus, philodendron, raspberry, grapes and other plants.

Interestingly, the blender techniques are not widely used within the industry, even for the production of ferns, although the benefits of the system are clear. The procedures are easily learned and the yields are consistent. Productivity can increase as much as tenfold. In fact, the markets could be easily saturated using these methods. Nevertheless, a number of problems limit the application of this technology: 1) very few plant tissues can survive this treatment, 2) considerable cellular debris is produced during the process which may have toxic effects, 3) there is a limited number of plants which can produce adventitious shoots, 4) there is a lack of uniformity in plants produced, 5) long culture periods are required to obtain large plantlets, 6) plants often need to be singulated, and 7) sterility must be absolute.

Alternative methods to blending need to be explored. Food processor designs could be easily adapted to produce uniform cuts that would reduce debris and increase the number of plant tissues that could survive. Plants such as sugar cane, Brassica spp. and redwood may be amenable to this kind of procedure.

Smith (49) suggested the development of a system in which explants remain on a physical support while liquid nutrient medium is changed. Ideally, this kind of handling system would allow plantlets or embryos to fit a configuration comparable to a flat or tray. Such a system would also allow for plant tissues to be grown at high density and spaced to final density later. Uniformity of growth is critical to meet current horticultural standards. Grading for size and quality is an important aspect for pricing. It will be critical to determine what stage of plant growth is optimal for grading.

3.3. Axillary bud systems

The technique of stimulating axillary branching or culturing nodal sections *in vitro* are probably the most commonly used procedures in micropropagation (31). Typically shoot tips or axillary cuttings are placed on a basal medium containing cytokinins which suppress apical dominance. These methods have now been used on a wide variety of species from foliage plants to woody ornamentals (29). Since the multiplication and growth of the plant proceeds normally, it is also genetically conservative. The uniformity of plant material produced is excellent. There is also the added benefit that plants produced tend to be more compact and full in habit. These traits are highly valued by the nursery trade.

One of the best known applications of this technique is in the production of disease-free stock for potato. The conventional method of vegetative propagation using virus-tested stem cuttings results in the production of 800-900 plants over the first 3 years, whereas *in vitro* propagation using axillary buds can produce thousands of plants per year without compromising the disease-free status. Potatoes are particularly exciting in that they produce "minitubers" *in vitro* (48). Each nodal section has the potential for producing tubers. Hu and Wang (21) reported high rates of tuberization by transferring axillary shoots onto MS medium with 22-44 μM BA and

0.23 M sucrose and using 8-hour photoperiods. Minitubers could be easily transferred to field or greenhouse using mechanical transplanters.

The propagation of plants using nodal sections is probably the simplest technique and it may have the most potential for the future if automation is to become a reality. Many vegetable, forest and plantation crops could be integrated into this kind of system. We can begin to think about making plant growth conform to certain specifications that make it amenable to handling by machines or robots.

Although there has been considerable discussion among researchers, until recently there have been few published reports of the application of robotics to the field of plant tissue culture. Hartney and Kabay (20) have reported that the Centre for Research on Intelligent Systems at Deakin University in Australia is developing a microprocessor controlled machine to mechanize the subculture process. Apparently a prototype is being developed that could potentially produce plants at a rate of one plant per second. Even if a plant is transferred once every 5 seconds, this machine could produce nearly 5,000,000 plants per year, which is comparable to the current production levels of many labs.

For robotics to be successful, some sort of vision system may be required to differentiate between leaves, stems, apical shoots and nodes in order to select the desired explant. A robot could be programmed to recognize nodes by the angle of petiole to the stem, or by means of the leaf around the node. End effectors attached to the robot arm could then cut and transfer shoots to new containers. Modules could be developed to handle either culture vessels or rooting trays entering or leaving the clean room work cell where the robot is stationed (see Figure 2). Programming costs for the robot vision system may be extremely high. Programs would have to be broadly applicable and easily modified to accommodate minor variation in plant form and growth. If this application of robotics could realize its full potential, it would surely revolutionize the micropropagation industry and new market opportunities would certainly develop.

There is a definite potential for automation of plant transfer from the rooting stage to a finished liner. The CSIRO in Australia is currently evaluating the use of a seed chain (63) for root formation, hardening off, and automated transfer to larger pots. Alcoa Corporation in Australia has apparently developed a vacuum-operated device to enable the rapid transfer of rooted plants from agar to a seedling tray (20).

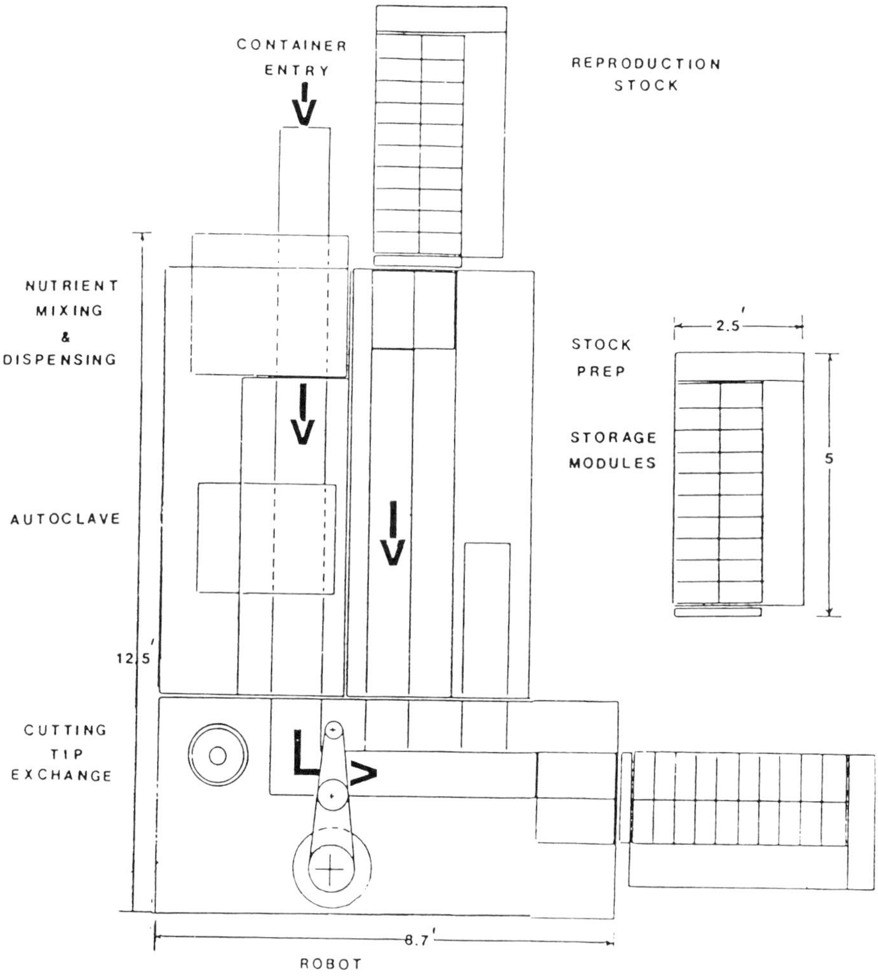

FIGURE 2. Mechanized tissue culture laboratory design.

There is increased interest about the possibility of interfacing laboratory techniques with high efficiency transplant systems that are currently being developed. For instance, polyurethane and phenolic based foams are now being used as a substrate for rooting, although toxic effects on plants have been reported (61). Gebhardt (16) successfully used wide pore phenol resin foam from Smithers-Oasis to root red raspberry shoots (Rubus idaeus cv. 'Gigant'). He concluded that root growth in foam was superior to that in agar, and that the primary advantages were decreased root friability, increased root hairs, and increased shoot length.

Zimmerman and Fordham (64) have successfully used the Techniculture® plug system for rooting micropropagated apple cuttings after root primordia were induced in vitro. They have demonstrated that acclimatization of cuttings, which occurred during root elongation, was simplified and accelerated using the Techniculture® plug system. They have also shown that plant survivorship increased and that transplanting was facilitated since there was little root disturbance. They think that these combined advantages make this technique more attractive for commercial utilization than other previous methods.

The future of the micropropagation industry clearly lies in the ability to research new technologies such as the application of computer mechanized handling systems. Traditionally, universities have not done this kind of research. Until recently, commercial laboratories have not had sufficient capital to develop process technology. Collectively, this could be achieved by forming an association funded by micropropagation companies, seed companies and biotechnology firms. This group could sponsor research within universities and other organizations to establish standards and develop specific products for the industry.

If new technologies do not impact the industry soon, commercial laboratories in the United States and Europe will certainly face competition from other countries that have lower labor costs and as a result could lose significant market shares. It is time to question the fundamental methodologies that drive the industry today.

REFERENCES

1. Ammirato PV: 1983. Embryogenesis, pp. 82-123. In: DA Evans, WR Sharp, PV Ammirato, Y Yamada (eds). Handbook of plant cell culture, Vol. 1: Techniques for propagation and breeding. Macmillan, New York.

2. Ammirato PV: 1983. The regulation of somatic embryo development in plant cell cultures: suspension culture techniques and hormone requirements. Bio/Technology 3:68-74.

3. Anderson WC, GW Meagher, AG Nelson: 1977. Cost of propagating broccoli plants through tissue culture. HortScience 12:543-544.

4. Anderson WC, JB Carstens: 1977. Tissue culture propagation of broccoli, Brassica oleracea (Italica group), for use in F_1 hybrid seed production. J. Amer. Soc. Hort. Sci. 102:69-73.

5. Bilkey PC, BH McCown, AC Hildebrandt: 1978. Micropropagation of African violet from petiole cross-sections. HortScience 13:37-38.

6. Browers MA, TJ Orton: 1984. Cell and tissue culture of celery (Apium graveolens L.) and potential applications. In: YPS Bajaj (ed). In vitro improvement of crops. Springer-Verlag, Berlin.

7. Brown DCW, A Atanassov: 1985. Role of genetic background in somatic embryogenesis in Medicago. Plant Cell Tissue Organ Culture 4:111-122.

8. Cooke RC: 1979. Homogenization as an aid in tissue culture propagation of Platycerium and Davallia. HortScience 14:21-22.

9. Corley RHV, KC Wooi, CY Wong: 1979. Progress with vegetative propagation of oil palm. Planter 55:377-380.

10. Crisp P, DGA Walkey: 1974. The use of aseptic meristem culture in cauliflower breeding. Euphytica 23:305-313.

11. Currah IE, D Gray, TH Thomas: 1973. The sowing of germinated vegetable seeds using a fluid drill. Ann. Appl. Biol. 76:311-318.

12. Currah IE: 1977. Fluid drilling research. National Vegetable Research Station Report. Wellesbourne, Warwick, England.

13. Drew RKL: 1979. The development of carrot (Daucus carota L.) embryoids (derived from cell suspension culture) into plantlets on a sugarfree basal medium. Hort. Res. 19:79-84.

14. Dublin P: 1984. Cacao. pp. 541-563. In: PV Ammirato, DA Evans, WR Sharp, Y Yamada (eds). Handbook of plant cell culture. Vol. 3, Crop Species. Macmillan, New York.

15. Evans DA, WR Sharp, CE Flick: 1981. Plant regeneration from cell cultures. Hort. Rev. 3:214-314.

16. Gebhardt K: 1985. Development of a sterile cultivation system for rooting of shoot tip cultures (red raspberries) in duraplast foam. Plant Sci. 39:141-148.

17. Gebhart F: 1984. Plant Genetics, Inc. reports successful vegetable harvest from "synthetic seeds." Genet. Eng. News 4(1):31.

18. Gray D: 1981. Fluid drilling of vegetable seeds. Hort. Rev. 3:1-27.

19. Hartmann HT, DE Kester: 1975. Plant propagation principles and practices. 3d ed. Prentice-Hall, New Jersey.

20. Hartney VJ, ED Kabay: 1984. From tissue culture to forest trees. Comb. Proc. Intl. Plant Prop. Soc. 34:93-99.

21. Hu CY, PJ Wang: 1983. Meristem, shoot tip and bud cultures, pp. 177-227. In: DA Evans, WR Sharp, PV Ammirato, Y Yamada (eds). Handbook of plant cell culture. Vol. 1, Techniques for propagation and breeding, Macmillan, New York.

22. Jones LH: 1974. Propagation of clonal oil palms by tissue culture. Oil Palm News 17:1-8.

23. Jones LH: 1974. Factors influencing embryogenesis in carrot cultures Daucus carota L. Ann. Bot. 38:1077-1088.

24. Jungnickel F: 1976. Regenerative properties of isolated petioles of Saintpaulia ionantha influenced by 6-benzylaminopurine or alpha-naphthaleneacetic acid. Bio. Chem. Physiol. 170:457-459.

25. Klimaszewska K, WA Keller: 1985. High frequency plant regeneration from thin cell layer explants of Brassica napus. Plant Cell Tissue Organ Culture 4:183-197.

26. Knauss JF: 1976. A tissue culture method for producing Dieffenbachia picta cv. 'Perfection' free of fungi and bacteria. Proc. Florida State Hort. Soc. 89:293-296.

27. Knauss JF, JW Miller: 1978. A contaminant, Erwinia carotovora, affecting commercial plant tissue cultures. In Vitro 14:754-756.

28. Krikorian AD: 1982. Cloning higher plants from aseptically cultured tissues and cells. Biol. Rev. 57:151-218.

29. Lane WD: 1982. Propagation in vitro and tissue culture of woody, deciduous fruits and nuts, pp. 20-36. In: DT Tomes, BE Ellis, PM Harney, KJ Kasha, RL Peterson (eds). Applications of plant cell and tissue culture to agriculture and industry. Univ. of Guelph Press, Guelph, Ontario.

30. Larkin PJ, WR Scowcroft: 1981. Somaclonal variation: a novel source of variability from cell culture for plant improvement. Theor. Appl. Genet. 60:197-214.

31. Lawrence RH, Jr: 1981. In vitro plant cloning system. Env. Expt. Bot. 21:289-300.

32. Lindemann E: 1984. System and apparatus for plant tissue culture propagation. U.S. Government Patent: 1-8. (Patent #4,463,522.).

33. Lutz JD, JR Wong, J Rowe, DM Tricoli, RH Lawrence, Jr: 1985. Somatic embryogenesis for mass cloning of crop plants, pp. 105-116. In: RR Henke, KW Hughes, MJ Constantin, A Hollaender (eds). Tissue culture in forestry and agriculture. Plenum Press, New York.

34. McWilliam AA, SW Smith, HE Street: 1974. The origin and development of embryoids in suspension cultures of carrot (Daucus carota). Ann. Bot. 38:243-250.

35. Murashige T: 1974. Plant propagation through tissue cultures. Annu. Rev. Plant Physiol. 25:135-166.

36. Murashige T: 1978. The impact of plant tissue culture on agriculture, pp. 15-26. In: TA Thorpe (ed). Frontiers of plant tissue culture. Calgary Univ. Press, Calgary.

37. Murashige T: 1980. Plant growth substances in commercial uses of tissue culture, pp. 426-433. In: F Skoog (ed). Plant growth substances. 1979, Springer-Verlag, Berlin.

38. Naylor EE, B Johnson: 1937. Histological study of vegetative reproduction in Saintpaulia ionantha. Amer. J. Bot. 24:673-678.

39. Oglesby RP: 1978. Tissue cultures of ornamentals and flowers: problems and perspectives, pp. 59-61. In: KW Hughes, R Henke, M Constantin (eds). Propagation of higher plants through tissue culture: a bridge between research and application. Technical Information Center, U.S. Department of Energy, Oak Ridge.

40. Orton TJ: 1986. Case histories of genetic variability in vitro: celery. In: IK Vasil (ed). Cell culture and somatic cell genetics of plants. Vol. 3, Academic Press, New York.

41. Paranjothy K: 1979. Oil Palm, pp. 591-605. In: PV Ammirato, DA Evans, WR Sharp, Y Yamada (eds). Handbook of plant cell culture. Vol. 3, Crop species, Macmillan, New York.

42. Pence VC, PM Hasegawa, J Janick: 1979. Asexual embryogenesis in Theobroma cacao L. J. Amer. Soc. Hort. Sci. 104:145-148.

43. Rabechault H, S Cas: 1974. Recherches sur la culture in vitro des embryons de palmier a huile (Elaeis guineensis Jacq. var. dura Becc.). X. Culture des segments d'embryons. Oleagineux 29:73-78.

44. Redenbaugh K, J Nichol, ME Kossler, B Paasch: 1984. Encapsulation of somatic embryos for artificial seed production. In Vitro 20:256.

45. Reynolds JF, T Murashige: 1979. Asexual embryogenesis in callus cultures of palms. In Vitro 15:383-387.

46. Reynolds JF: 1982. Vegetative propagation of palm trees, pp. 182-207. In: JM Bonga, DJ Durzan (eds). Tissue culture in forestry. Martinus Nijhoff/Dr. W. Junk, Boston.

47. Sharp WR, MR Sondahl, L Caldas, SB Maraffa: 1980. The physiology of in vitro asexual embryogenesis. Hort. Rev. 2:268-310.

48. Sluis C, C Rivera: 1984. Microtubers. Spudman 1:14, 15, 36 and 38.

49. Smith DR: 1984. Tissue culture for clonal forestry: present and future technologies. Forest Industries and Biotechnology. Proc. 16th New Zeal. Biotech. Conf.: 9-20.

50. Smith DR: 1985. Radiata pine (Pinus radiata D. Don), pp. 274-291. In: YPS Bajaj (ed). Biotechnology in agriculture and forestry 1: trees I. Springer Verlag, Berlin.

51. Smith SM, HE Street: 1974. The decline of embryogenic potential as callus and suspension cultures of carrot (Daucus carota L.) are serially subcultured. Ann. Bot. 38:223-241.

52. Smith WK, JA Thomas: 1973. The isolation and in vitro cultivation of cells of Elaeis guineensis. Oleagineux 123-127.

53. Sondahl MR, T Nakamura, HP Medina-Filho, A Carvalho, LC Fazuoli, WM Costa: 1983. Coffee, pp. 564-590. In: PV Ammirato, DA Evans, WR Sharp, Y Yamada (eds). Handbook of plant cell culture. Vol. 3, Crop species, Macmillan, New York.

54. Staritsky G: 1970. Tissue culture of the oil palm (Elaeis guineensis) as a tool for its vegetative propagation. Euphytica 19:288-292.

55. Start ND, BG Cummings: 1976. In vitro propagation of Saintpaulia ionantha Wendl. HortScience 11:204-206.

56. Steward FC: 1958. Interpretations of the growth from free cells to carrot plants. Amer. J. Bot. 45:709-713.

57. Styer DJ: 1985. Bioreactor technology for plant propagation. Basic Life Sci. 32:117-130.

58. Tisserat B, EB Esan, T Murashige: 1979. Somatic embryogenesis in angiosperms. Hort. Rev. 1:1-78.

59. Uyen N Van, P Vander Zaag: 1983. Vietnamese farmers use tissue culture for commercial potato production. Amer. Potato J. 60:873-879.

60. Wetherell DF: 1978. In vitro embryoid formation in cells derived from somatic plant tissues, pp. 102-124. In: KW Hughes, R Henke, M Constantin (eds). Propagation of higher plants through tissue culture, a bridge between research and application. Technical Information Center, U.S. Dept. of Energy, Oak Ridge.

61. Wheeler RM, SH Schwartzkopf, TW Tibbitts, RW Langhans: 1985. Elimination of toxicity from polyurethane foam plugs used for plant culture. HortScience 20:448-449.

62. Williames G: 1982. A revolution in the nursery. (Aust.) Pty. Ltd., 61 Victoria Street, Warragul, Victoria.

63. Williams EG, G Maheswaran: 1985. Embryo propagation by direct somatic embryogenesis and multiple shoot formation, pp. 366-367. In: RR Henke, KW Hughes, MJ Constantin, A Hollaender (eds). Tissue culture in forestry and agriculture. Plenum Press, New York.

64. Zimmerman RH, I Fordham: 1985. Simplified method for rooting apple cultivars in vitro. J. Amer. Soc. Hort. Sci. 110:34-38.

PLUG SYSTEMS FOR MICROPROPAGULES

D.D. MCCOWN

Knight Hollow Nursery, Inc., 236 East Sunset Court, Madison, WI 53705, U.S.A.

1. INTRODUCTION

Transplant plugs have become widely utilized in the horticulture and forestry industries where large numbers of vegetable, bedding plant, and tree seedlings must be handled. Much of the technology has become quite refined and is now being rapidly adapted to automation. Since the response and growth of micropropagules is very analogous to that of seedlings, a logical approach is the marriage of the two technologies of transplant plug systems and micropropagation. This paper will discuss some of the current attempts at consumating this marriage and the problems and potentials that have resulted. We view plug systems as a major component of the micropropagation scheme. Ideally, plugs should serve as the rooting and acclimation environment and as the medium for the first stage of growth of the crop.

2. CRITERIA OF AN IDEAL PLUG SYSTEM

The usefulness of any plug system for micropropagation depends on how closely it approaches the characteristics that are generally required for such a specialized application. The more of the ideal characteristics a plug system contains, the more generally applicable the system becomes. It is my hope that within the not so distant future, a widely-useful plug system will emerge that will do much to increase the efficiency of transition from laboratory to the growth environment as well as contribute to the standardization of the micropropagated product.

The plug medium should be uniform, highly reproducible, and stable. The physical characteristics must provide an excellent moisture/air relationship (probably close to 50/50 by volume). This should be achievable with a minimum of set-up time and maintenance. The medium need not necessarily approach the qualities of a soil as such chemical factors

Zimmerman, R.H. et al. (eds), Tissue Culture as a Plant Production System for Horticultural Crops.
© 1986. Martinus Nijhoff Publishers, Dordrecht. Printed in The Netherlands. ISBN 90-247-3378-2.

as pH and nutrient charge are readily-controlled in the intensive management environment common to microculture facilities. There is a trade-off between "buffering capacity" and the ease with which the chemical characteristics can be changed; a highly buffered system will need a minimum of maintenance. However, it may be difficult to tailor the medium for specific requirements of a crop.

The plug medium must be non-toxic. This sounds relatively simple; however, because microcuttings appear to be uniquely sensitive to toxins, such low toxicity is not easily achieved and has been a major problem in the adoption of existing plug designs to microculture.

The plugs and their carrier must be certifiably free of pests and pathogens. The small size and succulence of microcuttings makes them particularly susceptible to pathogenic attack, especially since they are held in humid and stagnant environments for considerable periods during rooting and early acclimation.

The ease of inserting microcuttings is a factor since at present it is a hand-operation and thus labor (and cost) intensive. It would seem essential that a viable plug system have at least the potential for mechanization of this step.

Efficient space utilization is important since microcuttings generally require sophisticated and expensive physical space for rooting and acclimation. One unit of culture space will require 2 to 3 units of rooting space; thus the sheer numbers of plants generated by even small laboratories make space efficiency a significant consideration. Space efficiency and weight are also factors in shipping, especially considering that many propagules will be shipped by air throughout the U.S. and to international markets. It is important to have a plug system adaptable to a containerized shipping unit.

One major advantage of any plug system is the ease of transferring a single plant to a new container. Unlike bed-grown material, there is less root injury and subsequent transplant shock. To be applicable to woody plants, there must be some mechanism for root-pruning and/or preventing root circling. Mechanized transplanting of plugs to field or larger-sized containers has been accomplished for some plug systems. It is a requirement for any system that will be applicable to large industries (e.g., forestry, vegetables, agronomic).

3. CURRENTLY-USED PLUG SYSTEMS

We know of no currently available plug system that meets all of these criteria, especially with woody plant microcuttings. There are three different types of plug systems in commercial use. Probably the most widely used is an individually designed and filled plug to meet a particular laboratory/nursery's needs. Such systems are now in common use in the bedding plant, nursery, vegetable, and forestry industries. A rigid tube-like container is filled with a premixed peat, sand, perlite, and/or vermiculite mixture with appropriate amendments. This type of a system has a number of disadvantages. The plug material is not stabilized (i.e., it is loose) and thus relies on the rooting of the plant to form a stable plug suitable for transplanting. This requirement places rather narrow restraints on the timing of when the crop can be transplanted or shipped. Insufficient rooting results in plugs that are easily knocked-out of the container or are destroyed during planting. Rooting heavy enough to form a stable plug often occurs only when the plant is large and thus has a short storage life in the plug. Many microcuttings, especially those of woody plants, require a plug with significant depth. A small diameter (space efficient), deep plug is difficult to fill uniformly. These plugs are also often difficult to remove from the container. While this type of plug is appropriate for some grower/propagation operations, it is not a desirable alternative for operations that are exclusively micropropagation facilities. Micropropagation facilities need a standardized and pre-made plug system that can be used directly in the laboratory, shipped readily, and easily used by the receiver.

Two commercial plug systems designed to accommodate microcuttings are available and they are Techniculture plugs (Castle & Cooke, Techniculture, Inc., Salinas, CA) and Oasis Horticubes (Smithers-Oasis, Kent, OH). The Techniculture plug is a stable mixture of peat moss and a hydrophilic latex binder. The medium will not break-up and separate from plant roots. Plugs come in a 33 x 33 cm styrofoam tray with either 400 small (1.27 cm diameter) or 144 large plugs (1.9 cm diameter). Both diameter plugs are offered in two depths (4.45 or 6.35 cm). While there are no nutrients in the Techniculture plug, they do have considerable cation ion exchange capacity. The pH is adjusted to 6.3.

The Oasis Horticubes are a polyurethane, open-cell foam cube that come 288 or 144 small cubes (1.59 cm/side) or 72 large (3.49 cm/side) per plastic tray. The material is essentially inert and does not contain nutrients but also has no cation-ion exchange or buffering capacity. The Horticube material is hydrophobic and water is held in the cube by capillarity which imposes a limitation on the system in regards to cell size.

Both of these commercial plug systems have a number of advantages. Both media are uniform and stable. Horticubes are shipped dry and are very light weight while Techniculture plugs, because of the peat-base, are shipped moist and thus drying should be prevented. Both are pasteurized and as long as sanitary techniques are practiced, disease problems should be minimal. Space efficiency is best with the Techniculture plugs (3874 plugs/m^2 versus 1808 Horticubes/m^2), but rooted cuttings must be transplanted while still relatively small. Planting ease with Techniculture plugs is also superior as plugs have a planting cavity specially designed for a microcutting. Techniculture offers a number of different cavity sizes so unusual sized microcuttings (e.g., rosettes) can be accommodated. Oasis Horticubes have a minimal depression in the center and insertion of microcuttings relies on forceps/microcutting pressure (or the use of a scalpel) to produce the appropriate-sized cavity. While this may slow the planting operation, contact of the microcutting with the rooting material will be enhanced. Techniculture plugs have addressed the need to root prune crops (particularly important for woody plants) as the tapered flat has an open bottom facilitating air pruning. The Horticubes were designed for herbaceous crops and have a limited mechanism to prevent root circling or to prune roots.

4. PLANTING PLUG SYSTEMS

If the crop to be planted is very fast growing, it may be appropriate not to fertilize prior to planting. Data on potato microcuttings (B.H. McCown, personal communication) showed plants prefertilized had weak, elongated stems and poorer root development than flats that were fertilized after root development. In addition, prefertilized flats of rooted potato microcuttings were difficult to acclimate. Most crops, however, benefit from fertilization at planting.

The quality of the microcuttings is critical to rooting success. With most crops, there is a direct correlation between cutting size and speed and success of rooting. Good leaf development on the microcutting is especially important. Microcuttings should be in active growth at the time of sticking and remain in active growth throughout the rooting period. With several woody species shoots that cease growth during the rooting phase often do not readily resume growth and can enter a protracted quiescent period.

5. ROOTING ENVIRONMENT

Humidity is the critical rooting factor and it is imperative microcuttings not dry-out. Three basic systems can be used. Flats or trays of plugs can be put into individual chambers, either clear for controlled environment rooting or translucent for greenhouse production. Capillary mat (the specific type has not been an important factor) is used beneath each flat to provide even moisture. This system provides enormous flexibility when a large number of crops are handled in relatively small numbers. The next upscaling is essentially a sweat-box that might enclose an individual greenhouse bench. The third alternative is a fog system that would maintain high humidity throughout an entire greenhouse range. Mist systems are not generally satisfactory because the on/off cycling allows microcuttings to dry out or keeps the plug system too wet for good rooting. Regardless of the system used, once the crop is rooted, humidity must be reduced, light levels increased and plants acclimated.

We have not found the actual rooting temperature to be a critical factor and temperatures are generally in the range of 21-27°C. What is important is that the temperature of the root (plug) be higher than ambient air as this increases the humidity around the microcuttings. Controlled environment light levels should be as high as can be tolerated by the plants and cooling system. Knight Hollow Nursery (KHN) uses 24 hour light (controlled environment rooting) while other labs use a cyclic light regime (16 hour minimum).

KHN has had mixed results rooting microcuttings. We presently have approximately 100 clones of woody plants in production. These clones span 12 plant families but around 50% are Ericaceae (Rhododendron, Kalmia, Vaccinium). KHN has considerable experience with the Techniculture plug system and finds that for some clones, it approaches the "ideal" system for such woody plants. However, clones do not respond equally. With Ericaceous plants, the pH of Techniculture plugs (6.3) is too high. We routinely soak our flats overnight prior to planting with a dilute acidified fertilizer solution (0.4 g 20-20-20 + 0.42 ml 85% H_3PO_4/1, pH 5.5). Even with this pretreatment, the pH of the plugs gradually increases and flats need to be resoaked periodically. When the pH is held as constant as possible, there is still an individual clonal response. Microcuttings of some clones root very easily and uniformly, others will not root at all, while still others give non-uniform rooting (Fig. 1).

Figure 1. Three cultivars of Exbury Azalea (Rhododendron) microcuttings planted in the small Techniculture plugs. The two cultivars on the left ('Red Letter' and 'Honeysuckle') showed nearly 100% rooting while the cultivar on the right ('Renne') failed to root. Flats were rinsed with tap water prior to planting and pretreated with a dilute fertilizer solution at pH 5.5. The flat was resoaked at weekly intervals with the pH 5.5 solution. Plants were photographed at the end of 12 weeks controlled environment rooting.

Kalmia often produce a ball of roots around the base of the cutting but these roots may not penetrate into the plug material, suggesting the presence of an inhibitor. When these cuttings with non-penetrating roots are replanted in 100% peat moss, rooting continues and high quality plants are produced. Other clones of microcuttings die from the base upwards, which indicates a toxin to which these clones are especially sensitive. The inhibitor or toxin is unknown but is most likely a component of the binder mixed with the peat during plug formation.

We have observed other problems with non-Ericaceous crops. Amelanchier produce roots readily in Techniculture plugs but as roots reach the bottom of the plug, they begin to deteriorate. Plants were difficult to acclimate and survival was less than 50%. Roots grew mainly on the outside of the plug rather than through the plug medium and could be peeled off the surface. Amelanchier microcuttings rooted in peat have at least a 75% rooting/survival rate. We cannot distinguish as to whether the problem is a toxin in the plug or the physical characteristics of the plug. The problems observed with Amelanchier have not been observed with Viburnum, Magnolia, or a number of other woody clones.

6. STORAGE OF ROOTED MICROCUTTINGS IN PLUGS

A limited amount of work has been done on the storage of micropropagules rooted and acclimated in plugs. For example, rooted Rhododendron micropropagules have been stored at 5°C under low light for at least 30 weeks without detrimental effects. In fact, because root growth continued during the storage period, the propagules actually improved in quality. For most types of plants, lighted storage is preferable since it allows for further development and also prevents the onset of deep dormancy in deciduous materials.

The ability to store ready-to-ship micropropagules would be of great benefit for commercial labs since this technique can be used to accumulate an inventory during times when the laboratory is not as full capacity. Thus the normal "ups and downs" of production of commercial micropropagation labs can be smoothed and economic efficiency improved.

As in the rooting of microcuttings, storage utilizes expensive space and thus compactness of the propagule carrier system is paramount. Plug systems appear to be logical choices to meet this requirement.

7. RESPONSE TO DIRECT-PLANTED PLUGGED MICROPROPAGULES

Where the use of plugs has been perfected for micropropagation (as with Gerbera in Europe and Mentha in Wisconsin), the acceptance of the receiving growers has been enthusiastic. In some cases, orders have been made contingent on the use of plugs, primarily because of the convenience to the grower that plugs offer.

The use of plugged propagules for direct planting in containers is well-established, especially in the bedding-plant industry (e.g., the Speedling system). Most of these propagules are seedlings; however, since micropropagules act like seedlings, the application of this technology to micropropagation is direct (as with Gerbera and foliage plants). Information is accumulating on the use of plugs in the establishment of crops directly in the field. Studies at the University of Wisconsin have evaluated the performance of Betula, Ulmus, Amelanchier, Thuja, Solanum and Mentha in field plantings that were established by direct planting of micropropagules established in Techniculture plugs. Survival for all the crops was high (90% or greater) when the micropropagules were of excellent quality, the field properly prepared so that good contact between the soil and the plug was established, and irrigation was available during the critical first 7 to 10 days. When the plants were in active growth at the time of planting, root penetration of soil could be observed within several days of planting. As has been noted with most plug systems, it is critical to have the top of the plugs buried in order to prevent wicking and drying of the plug.

Such transplants are smaller than those often used for field planting and thus are more susceptible to stress in the field environment. Uniform irrigation is essential to prevent these peat-based Techniculture plugs from drying and thus shrinking and losing contact with the soil. In addition, some of the potato plantings showed high sensitivity to herbicide carryover, a condition that was not observed in companion plantings using tubers.

8. CONCLUSIONS

At present, there is no commercially-available plug system on the market that meets all the characteristics of the "ideal" transplant plug system useful for micropropagation. However, the tremendous advantages of combining plug technology with microculture technology would seem to be incentive enough for the development of such systems. The only recommendation that can be safely made now is to evaluate the appropriateness of plugs as they are presently configured with each crop of interest. My only hope is that the frustration that this may create will not result in the eventual divorce of the two technologies!

TOTIPOTENT SUSPENSIONS AS A MEANS OF MULTIPLICATION

A.D. KRIKORIAN, R.P. KANN, STEFANIA A. O'CONNOR AND MINDY S. FITTER

Department of Biochemistry, State University of New York at Stony Brook, Stony Brook, NY 11794, U.S.A.

1. INTRODUCTION

It was disclosed in the mid- to late-1950's that cultured higher plant cells could develop into plantlets via a series of developmental events closely resembling zygotic embryogenesis. As the information base grew, it was recognized that this route of multiplication offered potential for mass cloning of elite specimens. Even so, it became apparent rather quickly that not nearly enough was known about the factors controlling somatic embryogenesis so that these ideas could be applied with any degree of confidence to very many species or cultivars (6).

Since then, a wide range of sometimes disparate investigations has culminated in the horticulturist's ability to multiply plants using a variety of aseptic culture techniques (Table 1). Free cells grown in suspension have not been used to any great extent for propagation. Most of the more commonly used strategies for multiplication appropriately fall under the category of micropropagation because the starting explants or cuttings and the resultant plantlets are small. The term micropropagation also seems particularly accurate especially if one considers that most aseptic work aimed at rapid multiplication really does not, in the strict sense, involve "tissue" culture (5). Many propagators who utilize micropropagation procedures capitalize on the tendency for many plants to form precocious axillary buds and shoots when stimulated by exogenous growth regulators under aseptic culture conditions. Axillary shoots so stimulated, in turn, can produce still other axillary branches and so on. As the system is divided or subcultured, one may have an unlimited potential for new shoot production. But more often than not, roots have to be induced upon these shoots by means of an additional step. This rooting is a purely horticultural operation using small axillary bud or shoot tip explants and is carried out under sterile conditions in nutrient media

Zimmerman, R.H. et al. (eds), Tissue Culture as a Plant Production System for Horticultural Crops.
 ISBN 90-247-3378-2.

TABLE 1. Strategies for multiplication of higher plants in vitro.

- Shoots from terminal, axillary or lateral buds
  - . shoot apical meristems (no leaf primordia present)
  - . shoot tips (leaf primordia or young leaves present)
  - . buds
  - . nodes
  - . shoot buds on roots
- Direct organogenesis
  - . adventitious shoot and/or root formation on an organ or tissue explant without an intervening callus
- Indirect organogenesis
  - . adventitious shoot and/or root formation on a callus
- Somatic embryogenesis
  - . direct formation on a primary explant
  - . indirect formation from cells grown in suspension or on semi-solid media
- Direct plantlet formation via an organ of perennation formed in vitro
- Micrografting
- Ovule culture
- Embryo rescue
- Mega and microspore culture
- Infection with a crown gall plasmid genetically altered to give teratoma-like tumors.

containing auxins that may dramatically foster a growth response that is inherent in the plant. In some cases, the rooting can be performed without asepsis in a soilless rooting medium or even soil. This strategy does not involve either callus or cells grown in suspension (5, 6).

The indirect organogenesis route (Table 1) for multiplication had its origin in the 1950's using the Wisconsin 38 strain of a 'Havana' cultivar of tobacco, *Nicotiana tabacum*. Unorganized, primarily parenchymatous, growths were proliferated as callus cultures under a simple protocol and then were induced to undergo organogenesis with varying levels of added kinetin and indole-3-acetic acid (IAA). In tobacco, and many other plants, controlling the ratio of cytokinin:auxin in large measure determined whether shoots or roots were to be produced adventitiously from an otherwise unorganized callus mass. Many such systems are responsive enough to permit retrieval of rooted plantlets that go on to full maturity but the indirect organogenesis route is usually not discrete or "clean." The presence of varying amounts of callus frequently renders it difficult to manage the system effectively (6). Perhaps more seriously it appears that genotypic changes may become apparent at the end of the culture cycle and this, of course, negates any hope to retrieve uniform or clonal specimens. Most propagators tend to avoid callus systems for this reason (12).

Although each of the strategies outlined in Table 1 has its place and value in research and industry, it is becoming increasingly apparent to us that the really dramatic advances are yet to come and they inevitably will be tied to culture techniques involving cells. That is not to say that cell suspension cultures and somatic embryogenesis and related developmental pathways will replace the other more widely used strategies. But the potential advantages afforded by suspensions are so numerous that it is time for researchers interested in horticulturally important plants to direct more attention to the problem. In cloning on a cellular level, enormous numbers of propagules capable of being reared into mature plants are theoretically feasible since each and every individual cell can, at least in theory, yield a plant. Large numbers of clonal specimens are nominally directly achievable upon manipulation of such totipotent cells (6).

2. STRATEGIES INVOLVING MULTIPLICATION FROM CELLS

Since all organisms are comprised of cells, one could envision breaking down a plant part or organ into its component cells and rearing each of the cells so released into an organized form or somatic embryo. Unfortunately, this simplistic strategy has not been taken to the level where it can be made to work predictably even for research. Pectinases etc. can indeed be used to free cells in some cases but the problem has been to induce cells so released to divide and to continue on a pathway of organized growth. Disaggregation of an organ or plant part into component protoplasts has likewise been achieved for a great many plants (3). But preparation of protoplasts capable of regenerating their walls and undergoing cell division, ultimately leading to a propagule capable of growing into a mature plant, has not been an easy accomplishment. Indeed, those species most worked with such as tobacco go on to make a callus mass. In turn these yield adventitious shoots and roots and then a plantlet (6). This is little different from adventitious organogenesis (Table 1) except for the initial reduction of the system to protoplasts from intact cells from an organ.

A system that would minimize the period in aseptic culture and thus more closely represent the original status of the cells of the parental genotype would be most desirable. This would be so not only from a theoretical but from a practical perspective. There is some evidence that the very act of in vitro culture negatively impacts the status of cells and this is responsible, in part at least, for the deviation from the "parental type" encountered in many culture-derived systems. It stands to reason then that the shorter the period in culture, the better (9).

The above implies, however, that all cells in a higher plant are homogeneous if not identical and that if one merely had the means to reduce them to the least regenerable unit--say a cell or protoplast--and to be able to initiate the organized development of these units, then one would recover a strictly clonal population of plants. There is much evidence that higher plants are far from being comprised of homogeneous cell types (9). They may include cells that differ in their level of differentiation, ploidy and other aspects of their genotype and certainly their cells vary in their phenological and developmental programming and sensitivities (their determination, if you will). On this view then, the a priori argument that working with cells to achieve mass multiplication of select genotypes or specimens may seem to be on shaky ground unless special precautions are taken. But the definitive events that ensue in the course of development from cells to organized plantlets entail a number of stages that provide opportunities for a substantial measure of control to be imposed on the system.

Figure 1 draws attention to several of the features that are of critical importance to any effort aimed at multiplying plants via cells. Since it does not yet seem practical to initiate organized forms directly from a primary explant via isolated cells or protoplasts, the strategy involves first inducing proliferative growth from the primary explant. If one is aiming for uniformity, this means that efforts must be made to avoid chimeras (13). Assuming that all due precautions are exercised to achieve as uniform an explant as possible, then the problem becomes one of determining how to initiate the proliferative growth desired. In the case of the now classic systems of carrot and tobacco, the cell division requirements seem, in retrospect at least, rather simple to fulfill, but there are many examples where the requisite callus growth is still not easy to achieve. Monocotyledons in general still provide challenges to in vitro culture specialists and the reasons why this should be so are still not particularly obvious, but the generally accepted (and to us spurious)

Definitive Events	Status of Cultures	Status of Units Present
	isolated explants	mature, attached, quiescent cells
proliferative growth induced	↓	↓
	cultured explants	randomly proliferating, attached, somatic cells
liberation of cells and small units	↓	↓
	suspension culture growth unorganized	free cells and small derived clusters multiplied and propagated in suspension
organized growth stimulated	↓	↓
	proembryonic stock culture	cells and aggregates with varying capability for orderly organized growth maintained by frequent subculture
embryogenesis	↓	↓
	highly organized embryonic cultures	units in various stages of somatic embryonic development
plantlets develop	↓	↓
	propagated crop of embryos and plantlets	plantlets in profusion

FIGURE 1. Propagation of plants from aseptically cultured cells: the sequence and definitive events. See text for details.

reason, is that they lack cambia. But in many dicotyledonous systems, all of which have cambia, the first callus growth is not necessarily cambium-derived. Be that as it may, it is frequently not a routine task to initiate the right kind of growth. Fortunately, an empirical approach utilizing a good understanding of growth regulatory substances and nutrition can usually yield a permissive medium. A strategy used by most has been to initiate growth on a semi-solid medium (5) and then to transfer to liquid to release cells and small units but it is also possible in many cases to initiate a culture directly in liquid.

It is tacitly understood by workers in the field that culture work involves substantial amounts of selection and husbandry. This is due in part to the fact that one can obviously only subculture that which does grow. Parts of the tissue explant that do not grow provide indirect evidence that all the component cells are not responsive for whatever reason to the growth induction stimulus. The second feature of this sort of work is that one learns through experience that growths that have a "certain look" are morphogenetically responsive whereas others that "look different" are not. It cannot be denied that much of the "art form" inherent in plant tissue culture technology involves subjective recognition of culture types that can perform under further manipulative protocols. Presumably as our understanding of cell biology improves, more objective and quantitative tools can be brought to bear on identification of cells that are responsive. This early stage of the culture process, however achieved, permits the building up of a stock of cells that are rather uniform.

The next stage of a cycle dealing with propagation from cells involves further multiplication of the cells in suspension as unorganized but competent units. This involves a maintenance medium management protocol--frequently utilizing sieving or screening of cells through stainless steel mesh or equivalent filters--to collect more or less equivalent units that can be propagated and multiplied. This stage is important because it provides one with an unlimited supply or volume of material. It also affords a system easy to handle since there is no organization. The next stage involves further stimulation of organized growth potential and the proliferation of totipotent cells as proembryos (variously also referred to as proembryonic masses, PEMs). In some systems like carrot and other umbellifers and certain cereal grasses, these proembryonic suspensions can be serially subcultured in a medium containing a growth regulator like

2,4-dichlorophenoxyacetic acid (2,4-D). Upon reduction of the level of 2,4-D, or elimination altogether, somatic embryogenesis proceeds and a large number of somatic embryos are produced (1). Since the ease with which such cultures can be synchronized varies considerably, one usually encounters a range of sizes and stages in somatic embryo development. The further rearing of these propagules to maturity is generally rather straightforward (6).

There are at least two key features in successful manipulation of morphogenetically competent cell suspensions. The first is that unorganized units (cells or cell clusters) must be controllably propagatable for substantial periods without concomitant organization into plantlets. This keeps the system potentially open-ended. The second is that it must be possible to obtain organized propagules (somatic embryos etc.) when some component(s) or feature(s) of the culture environment in which the cells are normally maintained is (are) altered. This allows one to turn the system "on" or "off." If these features are not realized, one has the disadvantage of a system that may not be open-ended. This is the case in the so-called direct embryogenesis route (Table 1) in which somatic embryos form directly on a primary explant. Here once competent cells have expressed their somatic embryo-producing potential, the system is exhausted of its totipotency. It should be pointed out, however, that some in vitro systems are obtainable where crops of somatic embryos are maintainable by virtue of a budding that occurs on the hypocotyl, etc. Also some embryonic cell clusters that can behave like a "stem-line" or proliferative source for continual production of proembryos can yield somatic embryos for sustained periods. But even here there is a tendency to "run downhill" and the flexibility achievable with cell suspensions that can be maintained and managed open-ended should generally be viewed as preferable by virtue of their greater control (6).

3. DAYLILY AS A MODEL SYSTEM FOR GENERATION OF PLANTLETS FROM COMPETENT SUSPENSIONS

At Stony Brook, morphogenetically competent cultures of Hemerocallis (daylily) have been initiated and routinely maintained for several years (7). These cultures are competent when grown on semi-solid media or in liquid, and, even when grown in liquid and distributed onto agar, plantlets are produced in profusion (6, 7, 8). The daylily system as we now use it in our research studies is of interest for a number of reasons: 1) it is a monocotyledon and callus and cell cultures of these are difficult to work with; 2) daylilies are temperate perennials but share many features in common with tropical monocotyledons. As such have they potential to serve as a model for perennial monocotyledons including such crops as palms and bananas-plantains; 3) daylily hybridizers and enthusiasts are regularly producing large numbers of seedlings and plants that are slow to multiply and as such research efforts could have a practical benefit for multiplication; 4) daylily chromosomes are large enough to be conveniently studied and this provides a significant means whereby one can correlate karyology and phenotype among culture-derived plants and their progenitors; 5) daylilies are relatively easy to grow under field conditions without recourse to excessively sophisticated horticultural practices. This provides the avenue through which populations can be reared to maturity in the field for evaluation.

A major disadvantage of daylilies that we recognize, but which further justifies our utilization of the genus, is that the genetics is poorly known. Many clones are virtually seedless and as such, Hemerocallis can act as a model for those plants that have no or reduced fertility and for which, therefore, little or no genetics exists.

The details of the process have been published (7, 8, 10), but a developmental sequence is provided in Fig. 2 a-f. Callus is initiated from either ovaries of unopened flower buds or from excised shoot apices. The callus when grown in liquid is not very friable but it can be reduced to small units by cutting with a scalpel. This leads to a coarse, compact growth mode that we refer to as nubbin-like. These may be cultured in erlenmeyer flasks or nipple flasks (8). In Fig. 2a, an aliquot derived from a #20 sieve passage is shown in a petri dish. After several successive subcultures the chopping becomes unnecessary and is eliminated. Figure 2b shows the appearance of a suspension maintained by subculturing through a #40 sieve. For photography an aliquot of the fine granular suspension was poured into a petri dish but cultures are primarily grown in a nipple flask in 220 ml of medium and rotated around a horizontal axis at 1 rpm (Fig. 2c).

FIGURE 2. Morphogenetically competent suspension cultures of the diploid daylily 'Autumn Blaze' and their development into organized plantlets. See text for details. a, b and d, X .63; c, X .39; e, X 1.72; f, X .55.

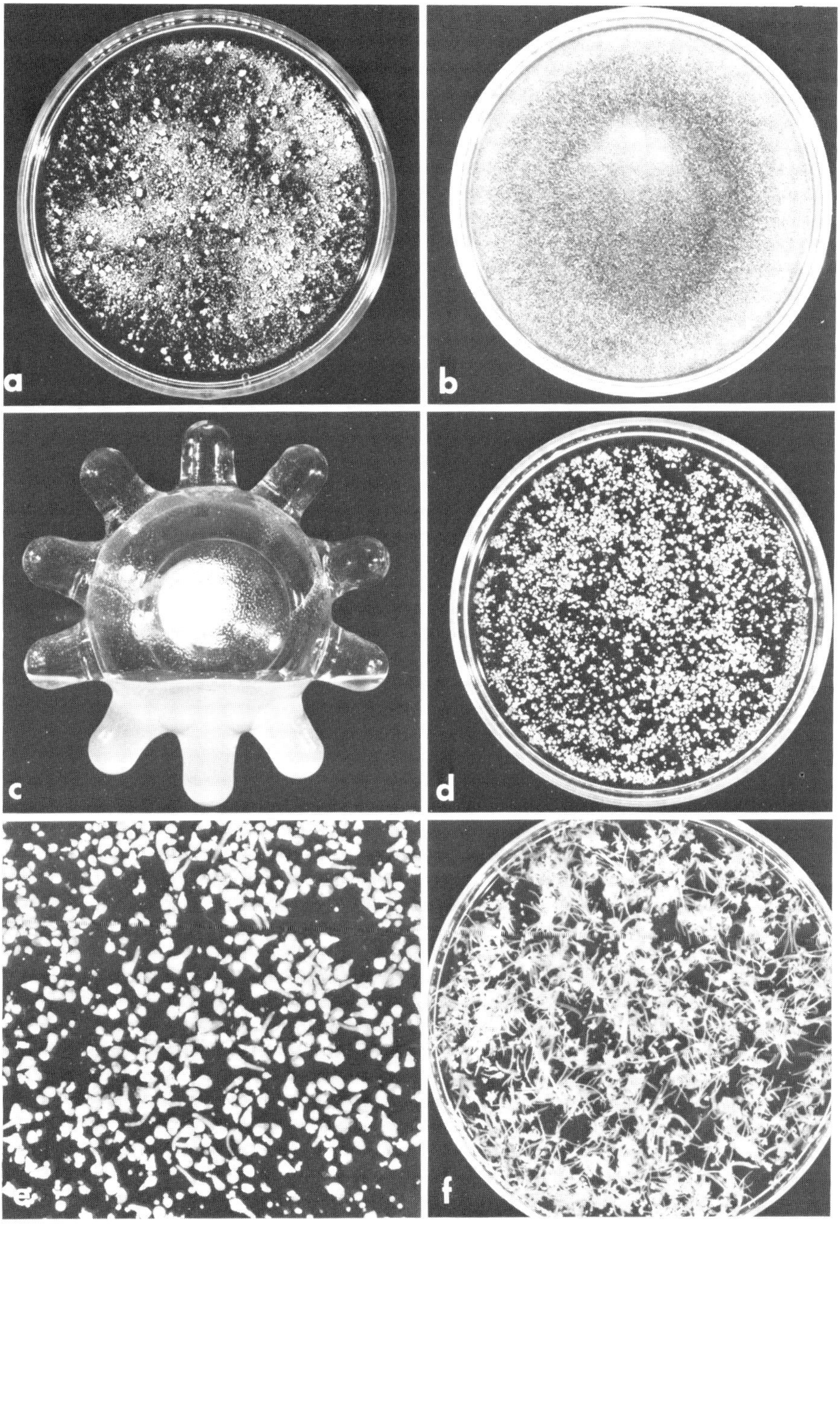
a
b
c
d
e
f

Three to 4 weeks following its previous subculture, a suspension is sieved through a #30 or #40 mesh sieve and is concentrated. An aliquot of cells is then transferred to a nipple flask containing MS basal medium and 10% v/v coconut water (the normal suspension culture maintenance medium also includes 2,4-D). Withdrawal of 2,4-D sets in train a series of events that results in the onset of organized growth (8). Figure 2d shows an aliquot of such units and we refer to them as in the "pre-shoot" stage. Here there is a definite and recognizable polarity in the organized forms but the shoot and root apical regions are not yet fully discrete. Figures 2e and 2f show later stages of development into embryonic plantlets. Their formation is fostered by subculturing of units such as those at 2d into a medium free of coconut water. Also a change in composition of the basal salt medium may be helpful (8), since different clones have different requirements. Continued development to plantlets ready for field planting is more or less routine (10).

We have evaluated the field performance of the diploid clone 'Autumn Blaze' generated by a number of means (11) including protoplast culture (2). A feature of particular importance for phenotypic (and presumably genotypic) stability is that suspension cultures must not be maintained for extended periods. Protoplast-derived plants show high levels of deviation from the prototype clone (4). This emphasizes that frequent re-initiation and monitoring of cultures by such tests as karyotype analysis must be undertaken (6, 9). On the other hand, generation of variant forms and the obtaining of tetraploids from diploids may be viewed as desirable under some circumstances (4). No pretext is made here to suggest that the daylily is a simple system to generate or manipulate. Quite the opposite (11). Being able to control the system is crucial and this will only emerge after extended investigations. One can be encouraged, however, that our ability to deal with morphogenetically competent suspensions for the purpose of clonal multiplication is improving and one can perhaps look forward to inclusion of cell use of suspensions on a practical scale in select propagation strategies.

ACKNOWLEDGEMENTS

The research described was supported from grants from the National Aeronautics and Space Administration. A broadly based research effort entitled "Cells, Embryos and Development in Space" NASA NSG 7270 has permitted a range of aseptic test systems to be developed with the view of exposing them at varying levels of prior organization to the space environment.

REFERENCES

1. Ammirato PV: 1984. Induction, maintenance, and manipulation of development in embryogenic cell suspension cultures, pp. 139-151. In: IK Vasil (ed). Cell culture and somatic cell genetics of plants. Vol. 1, Laboratory procedures and their applications. Academic Press, Orlando.

2. Fitter MS, AD Krikorian: 1981. Recovery of totipotent cells and plantlet production from daylily protoplasts. Ann. Bot. 48:591-597.

3. Fitter MS, AD Krikorian: 1983. Plant protoplasts: Some guidelines for their preparation and manipulation in culture. CalBiochem-Behring, Division of American Hoechst Corporation, LaJolla.

4. Fitter MS, AD Krikorian: 1986. Variation among plants regenerated from protoplasts of diploid daylily. In: FA Valentine (ed). Progress and prospects in forest and crop biotechnology. Springer-Verlag, New York.

5. George EF, PD Sherrington: 1984. Plant propagation by tissue culture. Handbook and directory of commercial laboratories. Exegetics Ltd., Eversley, Basingstoke, England.

6. Krikorian AD: 1982. Cloning higher plants from aseptically cultured tissues and cells. Biol. Rev. 57:151-218.

7. Krikorian AD, RP Kann: 1979. Clonal micropropagation of daylilies, pp. 835-836. In: WR Sharp, PO Larsen, EF Paddock, V Raghavan (eds). Plant cell and tissue culture: Principles and applications. Ohio State University Press, Columbus.

8. Krikorian AD, RP Kann: 1981. Plantlet production from morphogenetically competent cell suspensions of daylily. Ann. Bot. 47:679-686.

9. Krikorian AD, SA O'Connor, MS Fitter: 1983. Chromosomal stability in cultured plant cells and tissues, pp. 541-581. In: WR Sharp, DA Evans, PV Ammirato, Y Yamada (eds). Handbook of plant cell culture. Vol. 1. Techniques for propagation and breeding. Macmillan Publications, New York.

10. Krikorian AD, SA Staicu, RP Kann: 1981. Karyotype analysis of a daylily clone reared from aseptically cultured tissues. Ann. Bot. 47:121-131.

11. Krikorian AD, RP Kann, SA O'Connor, MS Fitter, SS Cronauer, DL Smith: 1986. The range of morphogenetic responsiveness in aseptically cultured daylily tissues and cells. In: FA Valentine (ed). Progress and prospects in forest and crop biotechnology. Springer-Verlag, New York.

12. Mullins MG: 1982. Tissue culture and plant propagation: Coming down to earth. Comb. Proc. Intl. Plant Prop. Soc. 31:162-165.

13. Stewart RN: 1978. Ontogeny of the primary body in chimeral forms of higher plants, pp. 131-160. In: S Subtelny, IM Sussex (eds). The clonal basis of development. Academic Press, New York.

PHENOTYPIC AND GENOTYPIC STABILITY OF TISSUE CULTURED PLANTS

D. A. EVANS and J. E. BRAVO

DNA Plant Technology Corporation, 2611 Branch Pike, Cinnaminson, NJ 08077, U.S.A.

1. INTRODUCTION

The ability to recover intact plants from cultured cells was originally viewed as an efficient method of producing large numbers of clones, an expectation that has been realized. Many commercial laboratories now use tissue culture of shoot tips to propagate a wide range of ornamental plants including orchids, ferns, lilies, etc. and some crop plants, such as oil palm. As predicted, tissue culture propagation has been shown to be economically competitive with conventional propagation and has resulted to a large extent in clonal fidelity (33). However, it was recognized in early experiments that, if cultures were established from explants that did not contain a pre-organized meristem, or if cultures were maintained as callus prior to plant regeneration, the regenerated plants were quite variable. In early reports most of the variation was attributed to the readily detected chromosome instability of cultured plant cells. In many of these studies, the degree of chromosome instability was reported to be proportional to the length of time the cells remained in culture. Recognition of the spontaneous variation inherent in long-term cultures led to the use of cell culture for mutagenesis and selection of genetic variants and for direct recovery of novel genotypes from cell cultures via somaclonal variation. In this paper, we intend to outline the types of genetic variants that have been recovered from cell cultures via somaclonal variation and to discuss the impact of this variation for crop improvement and recovery of new ornamentals.

2. TERMINOLOGY

As genetic variation has been detected in plants for several years, it is not surprising that both general cell culture and specific genetic terminology have evolved that describe the variation recovered from

Zimmerman, R.H. et al. (eds), Tissue Culture as a Plant Production System for Horticultural Crops.
 ISBN 90-247-3378-2.

cultured cells. Alterations have been referred to as phenotypic or genotypic changes. The genotype refers to the sum total of the genetic information, while the phenotype, or appearance of a plant or cell, is recognized to be a combination of genetic (genotypic) and environmental factors. These phenotypic changes that are not the result of genetic alterations are termed epigenetic changes. Hence, in most cases, it is appropriate to characterize variation in the plant or plant cell culture phenotype as a genetic or an epigenetic change. The distinction between these two types of changes is only conclusively demonstrated by detailed genetic evaluation often requiring several sexual generations. Because of this requirement, very few reports of cell culture-induced variation have been conclusively shown to have a genetic basis.

Early variant plants regenerated from cell cultures of geranium were termed "callicones" (45), while plants regenerated from protoplasts of potato were termed "protoclones" (41). Larkin and Scowcroft (28) promoted the use of the more general term "somaclonal variation" for the variation detected in plants derived from any form of cell culture. However, the type of genetic variation recovered in regenerated plants is to a large extent dictated by the genetic constitution of the particular cell population that is regenerated. It is necessary for genetic reasons to distinguish between plants regenerated from somatic and gametic tissue. To this end we use the term somaclones to refer to plants regenerated from cell cultures originating from somatic tissue and gametoclones to refer to plants regenerated from cell cultures originating from gametic tissue. Similarly, mixed terminology has evolved to refer to sexual progeny of plants regenerated from cell cultures. In this paper, as in our earlier publications (13, 15) we use the terminology proposed by Chaleff (9, pp. 94-95) as follows: Plants regenerated from cell culture, irrespective of tissue or origin, are referred to as R or R_o plants. The self-fertilized progeny of R plants are referred to as R_1 plants. Subsequent generations produced by self-fertilization are termed R_2, R_3, R_4, etc.

3. REVIEW OF LITERATURE

Although cloning of identical plants was originally believed to be the principal use for plant cell culture, it has become increasingly clear that under the appropriate cultural conditions a great deal of genetic variability can be recovered in regenerated plants (Table 1). Results from

TABLE 1. Documented cases of somaclonal variation in ornamental species.

Species	Altered characteristics	Inheritance examined	Reference
Chrysanthemum x morifolium	flower size and color, petal shape	no	4
Chrysanthemum x morifolium	plant vigor, leaf morphology and hairiness, lateral bud growth, phyllotaxy	no	47
Dendrobium	flower size and color, petal shape	no	48
Nicotiana alata	time to flower, flower and leaf morphology, pollen viability, plant height	yes	this chapter
Pelargonium spp.	plant and organ size, leaf and flower morphology, oil constituents, fasciation, pubescence, anthocyanin pigmentation	no	45
Pelargonium graveolens	chromosome number, leaf shape and type	no	24
Saintpaulia ionantha	RuBCase activity	no	6

early experiments documented genetic variability in plants regenerated from protoplasts (40). Growth of unorganized callus was thought to be necessary for induction of variability (45), but recent results suggest that genetic variability is even present in populations of plants regenerated directly from leaf explants in the absence of callus growth. Phenotypic variation has been reported in a number of plant species regenerated via organogenesis or embryogenesis.

Genetic variation was first detected as altered chromosome numbers in cultured plant cells. The onset of chromosome instability has been well characterized in many systems, such as Daucus carota (46) and Haplopappus gracilis (43). The most frequently reported variation has been polyploidy (44), attributed to selective growth of normally non-dividing polyploid cells pre-existing in the original explant (11). It has also been reported that the frequency of polyploid cells is dependent on the concentration and type of cytokinin used in the culture medium. Bennici et al. (5) reported that the frequency of polyploid cells in cultures of H. gracilis was dependent on the ratio of kinetin to naphthaleneacetic acid (NAA). Polyploid plants have been recovered in many commercially important plant species including Pelargonium zonale (45), Nicotiana tabacum (8), N. alata (Bravo and Evans, this chapter), Lycopersicon esculentum (13) and Medicago sativa (36). Aneuploid changes, typically involving the gain or loss of a few chromosomes, have also been frequently reported in plant cell cultures. The accumulation of aneuploid cells has been attributed to aging of cultures (11), and in many cases, older cell lines are incapable of plant regeneration. However, aneuploid plants have been regenerated from cultures of N. tabacum (38) and Saccharum spp. (21). Aneuploidy is often associated with sterility, but would certainly be tolerated in plant species that are propagated asexually. In addition to these numerical chromosome changes, observation of anaphase bridges and fragments also suggests that the structure of plant chromosomes is modified in cell culture (43). Chromosome rearrangements have been detected in clones of potatoes (Solanum tuberosum) regenerated from mesophyll protoplasts (40).

Evidence of recovery of single gene mutations produced via somaclonal variation has been presented. The progeny of tomato plants regenerated from leaf-derived callus were examined (13) and thirteen distinct single gene mutations were recovered among 230 regenerated tomato plants. This frequency of visual somaclonal mutations (ca. 1 in 18 regenerated plants) is substantially greater than the cell mutagenesis rate from several cell selection experiments (29). Several of these single gene mutants of tomato have been well characterized and mapped to specific chromosomes.

Cytoplasmic genetic variation has been detected less frequently following plant regeneration from cell culture, although the importance of this class of variation has been emphasized (42). The most compelling evidence for cytoplasmic genetic variation among regenerated plants was presented by Gengenbach et al. (17) who used restriction enzyme analysis of isolated mitochondrial DNA to detect spontaneous variant corn plants regenerated from cell culture. These authors monitored variation in cytoplasmically controlled male sterility, a mitochondrially encoded character. Hence, somaclonal variation has resulted in numerical and structural chromosomal changes, in nuclear genetic modifications and in cytoplasmic genetic variation. This broad spectrum of variation suggests that by using appropriate selection methods all classes of genetic variation could be recovered and used for crop improvement.

No detailed systematic studies have been devised to ascertain the source of somaclonal variation; however, it is possible by examining published literature to gain information regarding the sources of variation. Skirvin and Janick (45) systematically compared plants regenerated from callus of five cultivars of *Pelargonium spp.* Plants obtained from geranium stem cuttings *in vivo* were uniform, whereas plants from *in vivo* root and petiole cuttings and plants regenerated from callus were quite variable. This suggests that some variability is correlated with donor explant and pre-exists in the tissue used to establish cell cultures. On the other hand, evidence from several laboratories suggests that variability is dependent on hormone concentration of the culture medium. Most likely variability is the result of both pre-existing genetic instability and genetic alterations induced during the process of cell culture and subsequent plant regeneration.

Recent indications of somaclonal variation in several crop plants have stimulated interest in application of this method for crop improvement. Initial studies with sugarcane suggested that clones with disease resistance could be regenerated from callus induced from sensitive plants. Krishnamurthi and Tlaskal (27) isolated clones of regenerated sugarcane with resistance to Fiji disease virus and downy mildew. These selected resistant clones did not show reduced sucrose yield. Similarly variants of sugarcane have been isolated with resistance to eyespot disease (28). Most sugarcane variants have been attributed to changes in chromosome number. Researchers in several laboratories have examined plants regenerated from protoplasts of potato following the recovery of somaclones resistant to late blight (30). Potato variants have been isolated with resistance to early blight and with altered growth habit, tuber shape and color and maturity date. These somaclonal variants of potato were attributed to changes in chromosome number (26) and chromosome structure (40). While these variants can be stably propagated asexually, the genetic inheritance of this variation in sugarcane and potato has not been determined. Variation with potential agricultural application has been detected in several plant species that are sexually propagated. These include yield improvement in tobacco, TMV resistance in tobacco, southern leaf blight (Helminthosporium maydis) resistance in corn (Zea mays L.) and leaf shape in geranium (28). Somaclonal variants have also been described in rapeseed for blackleg disease susceptibility (37) and for growth habit, fruit color, and male sterility in tomato (13). All of these variants demonstrate the potential of somaclonal variation for production of new breeding lines.

As detailed genetic analyses have been completed for only a few somaclonal variants, it is not surprising that this method has not been exploited. Most variants reported to date have been chromosomal. While chromosomal changes can be tolerated in asexually propagated crops such as potato and sugarcane, other types of single nuclear gene and cytoplasmic variation may be much more important for crop improvement of sexually propagated crops.

4. CASE STUDY: TOMATO

Evans and Sharp (13) and Evans et al. (15) published a description of experiments designed to generate somaclones of tomato and to ascertain the genetic basis of somaclonal variation.

Tomato seeds of a well-characterized open-pollinated tomato variety, UC82B, were sown in a greenhouse. Young, fully expanded leaves were removed from plants, sterilized, and placed onto a culture medium known to permit regeneration of plants from cultured leaf explants (35). Callus that developed from the cultured leaf explants regenerated shoots within 4 weeks. Regenerated shoots were rooted on medium containing NAA, then transferred to the greenhouse for maturation and fruit collection. Self-fertilized seed was collected from mature fruits on R_0 plants. To evaluate the R_1 generation, seed was sown in the greenhouse where seedling characters, such as chlorophyll deficiency, anthocyanin content, and leaf shape were monitored. R_1 seedlings were transplanted to the field to classify mature plant characteristics such as pedicel type, fruit shape, and fruit and flower color. In addition, preliminary data on agronomic characters were collected in the R_1 field evaluation.

Variation in chromosome number, particularly tetraploidy, 2n=48, was detected. Sterile aneuploid lines were also detected. However, since emphasis was directed to analysis of R_1 progeny of regenerated plants, most aneuploid lines were discarded because little or no R_1 seed was collected. Several R_1 progenies were observed to segregate for morphological characters. These include recessive mutations for male sterility, jointless pedicel, tangerine virescent leaf, flower and fruit color, lethal chlorophyll deficiency, virescence, and mottled leaf appearance and dominant mutations controlling fruit ripening and growth habit. Genetic analysis was completed by evaluating self-fertilized R_2 progeny of selected R_1 plants and, in some cases, by crossing with known mutant lines. Extensive genetic analysis has been completed for several of the variants.

Genetic analysis was first completed for the tangerine-virescent (tv-tcl) character. This character is a single recessive allele that results in orange flowers and fruit and yellow virescent leaves. The R_1 segregation data observed in the field (30 red to 6 tangerine fruit) first suggested control of the trait by a single recessive allele. Fruit was collected from eight individual self-fertilized plants. Six of these R_1 plants had red fruit while two had tangerine fruit. The two tangerine fruit contained seed that bred true for the tangerine-virescent phenotype. Of the six red-fruited plants, three bred true for red fruit and three segregated for red vs. tangerine-virescent. The pooled progeny of the three segregating plants fit a 3:1 ratio (58 red to 17 tangerine). In addition, among the segregating R_2 progeny, the flower, fruit, and leaf color defects cosegregated suggesting control of two pigments, carotenoids (flower and fruit) and chlorophyll (leaf), by a single pleiotropic gene. Such pleiotropic genes controlling carotenoids and chlorophyll, two compounds both found in plastids, have been reported previously. A single mutant R_1 plant was crossed to a known fruit variant, tangerine, and all hybrid progeny had tangerine fruit. As these two mutants do not complement (t-gene), it can be concluded that the new somaclone (tv-tcl) was a mutant for a new allele in a previously known gene at position 95 on the long arm of chromosome 10. This has been the most precise genetic characterization of a new somaclonal variant to date. In addition, by evaluating the self fertilized progeny of the hybrids between the tv-tcl and the earlier tangerine mutant for the virescence character (13), it has been possible to conclude that the locus contains two elements, one controlling chlorophyll synthesis and one controlling carotenoid synthesis, that mutate independently. The new tv allele recovered by somaclonal variation is recessive to the t allele for the virescence phenotype.

Jointless pedicel is a trait that is a well-characterized recessive mutant of tomato. Plants with this character are desirable for mechanical harvesting as the harvested fruit has no stem attached. Two mutants with jointless pedicel were identified in somaclones of UC82B. The first was as expected, i.e., the regenerated plant was normal, and the recessive mutant was detected in about 25% of the R_1 progeny. However, the original regenerated plant of the second variant (j-tc2) already expressed the mutant trait. This jointless R_o plant bred true in both the R_1 and R_2 generations. Based on crosses with a known jointless mutant and with a normal jointed tomato, we have been able to ascertain that the original R_o plant was homozygous recessive for the jointless mutant. Presumably this somaclone originated by mutation, followed by mitotic recombination and subsequent shoot regeneration. This new mutation complements the known j mutation so that it is not encoded in the same gene. However, it does not complement the known j-2 mutant and is, therefore, allelic with this somaclonal mutant and maps at position 0 on the short arm of chromosome 11.

The mottled mutation (m-tcl) uncovered among somaclones of UC82B is a chlorophyll deficient mutant that was identified in the R_1 generation in the greenhouse. Selected mottled R_1 plants breed true for the mottled appearance. The variegation is somewhat similar to the previously reported plastome mutants. When chlorophyll deficient sectors of plastome mutants are placed in vitro, it is possible to establish shoot cultures of pure chlorophyll deficient tissue (19). However, the mottled phenotype of m-tcl is quite stable and is maintained in shoots regenerated from leaf explants of mottled plants. In addition, it is also possible to discern orange-red mottling on fruits of homozygous (m-tcl/m-tcl) plants. Once again, as with tv-tcl, the m-tcl mutant has an effect on both the chlorophyll and carotenoid pigments. This mutant appears to be distinct from any of the hundreds of previously reported tomato mutants.

Several other single gene mutations have now been identified in R_1 progeny tests of regenerated tomato plants including a semi-dominant allele controlling an electrophoretic variant of alcohol dehydrogenase (Medina-Filho et al., unpublished data) and a dominant allele conferring resistance to *Fusarium oxysporum* Race 2 (32). The *Fusarium* resistant mutant is allelic to the previously reported I-2 gene. In addition, several somaclones have been identified with agriculturally important traits that do not appear to be controlled by single genes. For example, new lines have been developed with higher pigment and higher solids.

Morphological variants have also been detected among regenerated plants of UC82B that breed true in the R_1 generation. In many cases, the altered somaclones have larger leaf size, darker leaf color, or reduced fruit set. In some cases, these R_1 plants have been evaluated and preliminary evidence obtained by restriction enzyme analysis of isolated chloroplast DNA suggests that genetic changes have occurred in chloroplast DNA (Flick et al., personal communication). Chloroplast genes are inherited maternally in tomato, hence, one would expect the variant to breed true in R_1 progeny. It is not surprising that chloroplast DNA-variants are uncovered as the number of plastids in a developing shoot apex is much smaller than in a mature cell (cf. 3). Hence, if a mutation occurs in chloroplast DNA, it is more likely to become the dominant plastid type during sorting out if the mutant occurs in 1 of 10 plastids than in 1 of 100 plastids for example.

In addition, several other tomato varieties have been regenerated using the procedure developed for UC82B. Single gene mutations have been recovered in several of these varieties including several male sterile and chlorophyll deficiency mutations. A recessive mutation for chlorophyll deficiency was identified in one tomato breeding line. This new mutant appears distinct from other known chlorophyll deficient mutations. New fruit color mutants have also been detected including a dominant orange mutation in C38 and a recessive yellow mutation in C40. Both C38 and C40 are Campbell Soup Company commercial tomato varieties. The orange mutation appears to be pleiotropic as it is associated with bushy foliage and altered flower color. The yellow mutation, on the other hand, affects only fruit color.

Based on our work, the following has been concluded regarding somaclonal variation in tomato: 1) Chromosome number variation can be recovered in regenerated plants; 2) Several single gene mutations have been recovered in several different tomato varieties; 3) Somaclones include dominant, semi-dominant, and recessive nuclear mutations; 4) The frequency of single gene mutation using our procedure is in the neighborhood of 1 mutant in every 20-25 regenerated plants; 5) Some evidence suggests that new single gene mutants not previously reported using conventional mutagenesis have been recovered using somaclonal variation; 6) The occurrence of 3:1 ratios for single gene mutants in R_1 plants suggests that mutants are of clonal origin and that the mutation occurred prior to shoot regeneration (i.e., no mosaics are detected); 7) Evidence suggests that mitotic recombination (reciprocal or non-reciprocal) may also account for some somaclonal variation; 8) Evidence suggests that mutations in chloroplast DNA (detected by both maternal inheritance and restriction enzyme analysis) can also be recovered; and 9) Agriculturally useful variants leading to development of new breeding lines have been recovered via somaclonal variation.

5. CASE STUDY: ORNAMENTAL TOBACCO

As demonstrated by the work on tomato, it is possible to use somaclonal variation to uncover new, useful genetic variants. In addition to those variants already described, many morphological variants, with novel leaf, flower, and fruit color were detected. These unique variants, though not useful for a commercial horticultural crop, are the type of variants often used to release new, ornamental varieties. To evaluate the potential value of somaclonal variation in the development of new ornamental varieties, we regenerated plants of *Nicotiana alata* 'Nicki Red'.

Nicotiana alata has a diploid chromosome number of 18. It is a member of the section Alatae and is self-incompatible. Young, fully expanded leaves of *Nicotiana alata* cv. Nicki Red (Park Seed Company, Greenwood, South Carolina) were surface sterilized in 8% Clorox for 8 minutes, then rinsed three times in sterile distilled water. Leaves were dried under sterile conditions and then cut with a sterile scalpel into 2.5 cm x 2.5 cm sections and placed onto agar-solidified media composed of Murashige-Skoog macro and micronutrients and B5 vitamins. This modified Murashige and Skoog (MMS) medium was supplemented with 0, 0.5, 1.0, 5.0, or 10.0 uM of 6-benzyladenine (BA) to determine capacity for plant regeneration.

Leaf explants were cultured under 16 hours of light with a mean temperature of 29.4°C (85°F). When regenerated shoots had reached a height of 3 cm, they were excised and placed onto rooting medium: MMS with 0.1 mM 3-aminopyridine (14). Following root development, shoots were transplanted to Jiffy pots and subsequently to 4-inch pots when secondary leaves developed. Plants were transferred to the greenhouse where they were raised to maturity.

Shoots were regenerated from explants cultured on either 1.0 μM or 5.0 μM BA within 28 days. After 56 days in culture, none of the explants on hormone-free medium had regenerated shoots. In medium containing 0.5 μM BA, a small amount of callus was formed from which only one small shoot was regenerated from the eight explants at this treatment. Large shoots with normal morphology were obtained using 1.0 μM BA, but the frequency of shoot formation and number of shoots recovered was less than on the medium containing either 5.0 μM or 10.0 μM BA. Five μM BA was optimum as it resulted in the highest number of shoots in the shortest time period.

The regenerated plants were phenotypically distinct from seed-derived plants. One characteristic frequently observed in these regenerated plants was not heritable. Seventy-two percent of the plants regenerated in the presence of 10 μM BA exhibited first and second leaves that were wider than young leaves of seed grown N. alata (Figure 1). However, subsequent leaves of these regenerated plants developed normally. Hence, this trait is epigenetic. On the other hand, a small percentage of plants regenerated on MMS containing 5.0 to 10.0 μM BA flowered as soon as they were transferred to 4-inch pots, at a height of only 3 inches (Figure 2). This

FIGURE 1. Left: normal leaf shape for N. alata. Right: wide leaf shape observed on many regenerated N. alata plants.

FIGURE 2. Dwarf N. alata plants regenerated from leaf explant culture. Several regenerated plants flowered as soon as they were removed from culture jars.

type of dwarfness was sexually transmitted to the R_1 self-fertilized progeny of these plants (Table 2). Plant height behaved as a quantitatively inherited character as a range of intermediate heights were observed among the R_1 progeny. While the population size was insufficient to ascertain the mode of inheritance, it is nonetheless clear that this trait expressed as mean plant height, is stably transmitted to the R_1 generation. Evaluation of the R_1 generation will permit a more detailed genetic analysis of this trait. Changes in gross chromosome number can be excluded from consideration as all regenerated dwarf plants had 2n=18.

TABLE 2. Statistical comparison of dwarf *Nicotiana alata* self-fertilized progeny for flowering heights. Leaf explants cultured on MS medium containing 5 or 10 μM BA.

Plant material	Mean height (cm)	t-test	Probability
Control (seed-derived)	52.0	--	--
10 μM BA Regenerate, #3	29.5	2.52	0.01
10 μM BA Regenerate, #5	38.7	2.45	0.025
10 μM BA Regenerate, #10	37.2	1.97	0.05
Control (seed-derived)	42.3	--	--
5 μM BA Regenerate, #21	28.7	2.53	0.01
5 μM BA Regenerate #23	20.6	4.19	0.01

Chromosome number was ascertained for each plant regenerated in the presence of 5.0 μM and 10.0 μM BA. At both of these cytokinin concentrations 25% tetraploidy was observed in the root tips of the regenerated plants (Table 3). These tetraploid regenerates were much

TABLE 3. Chromosome number of *Nicotiana alata* plants regenerated from leaf explants, cultured on M or S medium containing 5 or 10 μM BA.

Plant type	2n=18	4n=36
Control (seed-derived)	25	0
5.0 μM BA Regenerate	29	8
10.0 μM BA Regenerate	26	10

slower to flower than comparable diploid regenerates. After 6 months from the date of culture, all of the diploid regenerates, but less than 50% of the tetraploid regenerates, had flowered.

Some tetraploid regenerates were crossed with diploid regenerates. Tetraploid plants were used both as female and male. Crosses were completed using regenerates that had already been observed to self-pollinate and set seed. It was presumed that these plants would have a higher probability of successful crossing than regenerates that had been unable to set seed. Several of these crosses were successful; however, triploid progeny were recovered only when the tetraploid was used as the female parent.

Plants regenerated on MMS containing 5.0 or 10.0 μM BA exhibited decreased pollen variability compared to seed derived plants when stained with acetocarmine (Table 4). Plants regenerated on 10 μM BA were also observed to produce less pollen than normal *N. alata*. The pollen viability

TABLE 4. Pollen viability of *Nicotiana alata* plants regenerated from leaf explants determined by staining with acetocarmine.

Plant type	Pollen viability (%)
Seed-derived control	92.0
5.0 μM BA regenerate	81.6
10.0 μM BA regenerate	48.1

remained low even after plants were 7 months old. These plants eventually produced a small amount of self-pollinated seed after 10-12 months in the greenhouse.

The flower morphology of several regenerated plants was also altered (Figure 3). This trait has not been monitored in subsequent sexual generations as these plants had greatly reduced pollen viability.

Hence, a great deal of variability was observed in these N. alata plants regenerated from leaf explants. This somaclonal variability included variation for morphological traits (flower shape, leaf shape, plant height), pollen viability and chromosome number.

Some of the variation induced may have value for improvement of these ornamental plants. Currently, one of the most undesirable features for using ornamental types of Nicotiana as bedding plants is that they flower at heights of 30 cm or taller. While N. alata plants may be commercially treated with growth retardants such as daminozide, it is desirable to produce dwarf plants that will reach maturity rapidly and flower at a short height. Dwarf plants were recovered among the population of plants regenerated from leaf and petal explants. This population of dwarf

FIGURE 3. Flower shape variant derived from tissue culture. Left: normal N. alata flower. Right: variant flower.

regenerated plants had concomitant decrease in pollen viability. Such reduction in pollen viability is a hinderance to integration of these dwarf somaclonal variants into a breeding program. However, for N. alata, we have also uncovered an alternative approach for the production of dwarfs. Some individual diploid and tetraploid regenerated plants that retained high pollen viability were used in crosses to obtain triploid progeny. Triploids flower at low plant heights compared to seed-derived controls and behaved in many respects as dwarfs. In addition to the dwarf characteristics, triploids have the added advantage of reduced seed set.

6. CASE STUDY: PLANTS FROM PROTOPLASTS AND PROTOPLAST FUSION PRODUCTS

Somaclonal variability has been documented in plants regenerated from protoplasts. This variability was first reported for the autotetraploid potato 'Russet Burbank'. Although only partial cytogenetic analysis has been reported, it is apparent that regenerated clones are highly variable. Over 10,000 clones have been screened and lines have been identified that are resistant to diseases to which 'Russet Burbank' is susceptible (39). Regenerates included clones that were resistant to Alternaria solani toxin and to Phytophthora infestans. These variant lines have retained the resistance phenotype through a number of vegetative generations (41). Although it has been recognized in work on other crops that it is not necessary to use protoplasts to recover somaclonal variants, it should be recognized that all populations of plants regenerated from protoplasts probably contain somaclonal variants.

Variability is also observed in somatic hybrids produced through protoplast fusion. Interspecific or intraspecific sexual hybrids are usually genetically uniform. In most instances, particularly if hybrids are unilaterally incompatible, only certain nuclear-cytoplasmic combinations are possible. As both agronomic and non-agronomic characteristics are variables, traits important for crop improvement may be expressed in somatic hybrids and not in comparable sexual hybrids. This phenomenon emphasizes the potential value of somatic hybrids even when produced between sexually compatible species. Evans et al. (12) have compared somatic and sexual hybrids between Nicotiana nesophila and N. tabacum and reported increased phenotypic variability in the somatic hybrids.

The phenotypic variability observed in somatic hybrid plants is a reflection of genetic phenomena that occur prior to plant regeneration. Four potential sources of variability have been identified:

i) Nuclear genetic instability in wide species combinations may result in chromosome elimination. Directional chromosome elimination has been documented in several intergeneric plant cell hybrids (25) and this instability has resulted in recovery of aneuploid plants resembling one or the other parental type for all intergeneric hybrid plants (22). Aneuploid plants have also been recovered for more closely related interspecific somatic hybrids (7), although some of this chromosome variation may be due to the use of protoplasts isolated from long-term aneuploid cell cultures.

ii) Intergenomic mitotic recombination (translocation) results in unique mixtures of genetic information. Recombination could be observed in several somatic hybrids (14) when the fate of genetically controlled spot formation was monitored on leaves of somatic hybrids. If mitotic recombination occurs prior to shoot regeneration, this nuclear genetic variation can be detected when regenerated clones are compared. It is likely that chemical or physical agents could be used to increase the frequency of intergenomic genetic exchange.

iii) Somaclonal variability has been observed in plants regenerated from mesophyll protoplasts (40). This phenomenon is the result of either pre-existing genetic variation in leaf cells or is induced by undefined components of the protoplast, callus or plant regeneration media. It is as yet impossible to measure the effects of somaclonal variation on interplant variation of somatic hybrids, however, it should be noted that leaf mesophyll protoplasts, used to recover somaclonal variability in potato, have been used to produce most somatic hybrid plants.

iv) Following fusion, the organelles of the two parent lines are mixed and then recombine or segregate prior to regeneration resulting in substantial differences between clones of somatic hybrids. This fourth source of variability in somatic hybrids represents one of the important differences between somatic and sexual hybrids. During fertilization, the male cytoplasm is preferentially excluded from the developing embryo. Thus, conventional sexual hybrids contain and express only maternally derived cytoplasmic genes. As protoplast fusion combines both parent cytoplasms into a single protoplast, fusion can be used to produce nuclear-cytoplasmic combinations not possible using conventional breeding (19).

Evidence from a number of somatic hybrids suggests that although two types of cytoplasms are initially mixed during protoplast fusion, resulting in heteroplasmons, eventually one parental type cytoplasm predominates resulting in cytoplasmic segregation. Following segregation, which may not occur until after meiosis, genetic information from this second cytoplasm is irreversibly lost. This phenomenon was first reported by examination of the fate of the large subunit of fraction 1-protein in Nicotiana glauca + N. langsdorffii somatic hybrids (10). Random chloroplast segregation resulted in fixed langsdorffii or glauca-type fraction 1- protein in each clone prior to meiosis (10 langsdorffii-type and 7 glauca-type), except for one clone in which segregation followed self-fertilization. These two different types of cytoplasms then segregate after meiosis. Cytoplasmic segregation has been observed for other cytoplasmically controlled traits. Cytoplasmic hybrids (cybrids) containing mixed cytoplasms of cytoplasmic male sterile (cms) Petunia hybrida and normal P. axillaris segregated to male sterile and male fertile cytoplasm during the first and second meiotic cycle (23). The male fertile character behaves as a dominant cytoplasmic trait as male fertility was expressed in putative cybrids. Similar segregation of male sterility versus male fertility has been reported in other cybrids (20). Following analysis of four cytoplasmic characters (large subunit fraction 1-protein, cms, tentoxin sensitivity and restriction enzyme patterns of cytoplasmic DNA), cosegregation was observed for all four traits (1). Similarly, we have found that fraction 1-protein and tentoxin sensitivity cosegregated in N. glauca + N. tabacum somatic hybrids (16). Cosegregation implies that following cytoplasmic segregation, the cytoplasm contains all characters from one parent or the other, and therefore, precludes widespread cytoplasmic recombination of chloroplast genes.

Despite reports of cosegregation of cytoplasmic genes, biochemical evidence using restriction analysis of mitochondrial DNAs of some cybrids has suggested that recombination may occur between parental mitochondrial DNA prior to cytoplasmic segregation in somatic hybrids between different male sterile and fertile parent plants (2). In these reports, mitochondrial recombination following protoplast fusion produced a wide range of morphological phenotypes controlling male sterility in the hybrid plants. Similarly, Nagy et al. (34) detected recombination in mitochondrial DNA of *N. tabacum* + *N. knightiana* somatic hybrids resulting in novel mitochondrial DNA restriction enzyme patterns. Recombination of chloroplast DNA has recently been reported in higher plants (31). Most likely these intriguing results of cytoplasmic segregation and recombination predict an era of extensive research on cytoplasmic genetics in higher plants. Cytoplasmic markers for disease resistance (18) and antibiotic resistance (29) should facilitate research in this area. The production of unique nuclear-cytoplasmic combinations using protoplasts will aid in the development of novel germplasm using conventional methods.

7. IMPLICATIONS FOR BREEDING

Based on the results presented for tomato and ornamental *Nicotiana*, an overall breeding strategy is evident for use of somaclonal variation to develop new cultivars of horticultural crops and ornamentals.

As single gene mutations and organelle gene mutations have been produced by somaclonal variation, an obvious strategy is to introduce the best available cultivars into cell culture to select for improvement of a specific character. Hence, somaclonal variation could be used to uncover new variants that retain all the favorable qualities of an existing cultivar while adding one additional trait, such as disease resistance or herbicide resistance. Work with sugarcane and tomato has already suggested that this approach is feasible. Once new R_1 variants are identified, these should be field tested in replicated plots to ascertain genetic stability. Seed should be increased at the same time to permit rapid cultivar development of promising lines. Reciprocal crosses between desirable R_1s and seed-derived controls can be used to gain an understanding of the genetic basis of the somaclonal variants. New promising breeding lines can be introduced into cell culture to add an additional character or to improve agronomic performance or ornamental appearance of a selected somaclonal variant. By using this approach, it is possible to produce new breeding lines with desirable traits in a short period of time.

8. SUMMARY

While tissue culture cloning has found widespread application in commercial horticultural operations for micropropagation, somaclonal variation has not been so widely used. It is apparent with the unique variants that are generated, the case of technology, and the case of integration into a conventional breeding program, that this biotechnology tool will have increasing usefulness in horticulture during the next several years. The value of somaclonal variation for producing incremental changes can be equally exploited for production of horticultural crops.

REFERENCES

1. Aviv D, R Fluhr, M Edelman, E Galun: 1980. Progeny analysis of the interspecific somatic hybrids: Nicotiana tabacum (cms) + Nicotiana sylvestris with respect to nuclear and chloroplast markers. Theor. Appl. Genet. 56:145-150.

2. Belliard G, F Vedel, G Pelletier: 1979. Mitochondrial recombination in cytoplasmic hybrids of Nicotiana tabacum by protoplast fusion. Nature 281:401-403.

3. Bendich AJ, LP Gauriloff: 1984. Morphometric analysis of cucurbit mitochondria: the relationship between chondriome volume and DNA content. Protoplasma 119:1-7.

4. Ben-Jaacov J, RW Langhans: 1972. Rapid multiplication of Chrysanthemum plants by stem tip proliferation. HortScience 7:289-290.

5. Bennici A, M Buiatti, F D'Amato, M Pagliai: 1971. Nuclear behavior in Haplopappus gracilis callus grown in vitro on different culture media. Colloq. Internation. CNRS 193:245-250.

6. Bhaskaran S, RH Smith, JJ Finer: 1983. Ribulose bisphosphate carboxylase activity in anther-derived plant of Saintpaulia ionantha Wendl. Shag. Plant Physiol. 73:639-642.

7. Bravo JE, DA Evans: 1985. Protoplast fusion for crop improvement. Plant Breed. Rev. 3:193-218.

8. Brossard D: 1976. Influence of kinetin on formation and ploidy level of buds arising from Nicotiana tabacum pith tissue grown in vitro. Z. Pflanzenphysiol. 78:323-333.

9. Chaleff RS: 1981. Genetics of higher plants. Cambridge University Press, Cambridge.

10. Chen K, SG Wildman, HH Smith: 1977. Chloroplast DNA distribution in parasexual hybrids as shown by polypeptide composition of fraction-1 protein. Proc. Natl. Acad. Sci. USA 74:5109-5112.

11. D'Amato F: 1977. Cytogenetics of differentiation in tissue and cell cultures, pp. 343-357. In: J Reinert, YPS Bajaj (eds). Applied and fundamental aspects of plant cell, tissue and organ culture. Springer-Verlag, Berlin.

12. Evans DA, CE Flick, SA Kut, SM Reed: 1982. Comparison of Nicotiana tabacum and Nicotiana nesophila hybrids produced by ovule culture and protoplast fusion. Theor. Appl. Genet. 62:193-198.

13. Evans DA, WR Sharp: 1983. Single gene mutations in tomato plants regenerated from tissue culture. Science 221:949-951.

14. Evans DA, LR Wetter, OL Gamborg: 1980. Somatic hybrid plants of Nicotiana glauca and Nicotiana tabacum obtained by protoplast fusion. Physiol. Plant. 48:225-230.

15. Evans DA, WR Sharp, HP Medina-Filho: 1984. Somaclonal and gametoclonal variation. Amer. J. Bot. 71:759-774.

16. Flick CE, DA Evans: 1982. Evaluation of cytoplasmic segregation in somatic hybrids in the genus Nicotiana: tentoxin sensitivity. J. Hered. 73:264-266.

17. Gengenbach BG, JA Connelly, DR Pring, MF Conde: 1981. Mitochondrial DNA variation in maize plants regenerated during tissue culture selection. Theor. Appl. Genet. 59:161-167.

18. Gengenbach BG, CE Green, CM Donovan: 1977. Inheritance of selected pathotoxin resistance in maize plants regenerated from cell cultures. Proc. Natl. Acad. Sci. USA 74:5113-5117.

19. Gleba YY: 1979. Nonchromosomal inheritance in higher plants as studied by somatic hybridization, pp. 775-788. In: WR Sharp, PO Larsen, EF Paddock, V Raghavan (eds). Plant cell and tissue culture principles and applications. Ohio State Univ. Press, Columbus.

20. Gleba YY, DA Evans: 1984. Genetic analysis of somatic hybrid plants, pp. 322-357. In: DA Evans, WR Sharp, PV Ammirato, Y Yamada (eds). Handbook of plant cell culture. Vol. 1. Techniques for propagation and breeding. Macmillan, New York.

21. Heinz DJ, WP Mee: 1971. Morphologic, cytogenetic, and enzymatic variation in Saccharum species hybrid clones derived from callus tissue. Amer. J. Bot. 58:257-263.

22. Hoffmann F, T Adachi: 1981. "Arabidobrassica": Chromosomal recombination and morphogenesis in asymmetric intergeneric hybrid cells. Planta 153:586-593.

23. Izhar S, Y Tabib: 1980. Somatic hybridization in Petunia. Part 2: Heteroplasmic state in somatic hybrids followed by cytoplasmic segregation into male sterile and female fertile lines. Theor. Appl. Genet. 57:241-245.

24. Janick J, RM Skirvin, RB Janders: 1977. Comparison of in vitro and in vivo tissue culture systems in scented geranium. J. Hered. 68:62-64.

25. Kao KN: 1977. Chromosomal behavior in somatic hybrids of soybean - Nicotiana glauca. Molec. Gen. Genet. 150:225-230.

26. Karp A, RS Nelson, E Thomas, SWJ Bright: 1982. Chromosome variation in protoplast-derived potato plants. Theor. Appl. Genet. 63:265-272.

27. Krishnamurthi M, J Tlaskal: 1974. Fiji disease resistant Saccharum officinarum var. Pindar subclones from tissue cultures. Proc. Int. Soc. Sugar Cane Technol. 15:130-137.

28. Larkin PJ, WR Scowcroft: 1981. Somaclonal variation--a novel source of variability from cell cultures. Theor. Appl. Genet. 60:197-214.

29. Maliga P: 1980. The need and the search for genetic markers in plant cell cultures, pp. 107-114. In: F Sala, B Parisi, R Cella, and O Ciferri (eds). Plant cell cultures: results and perspectives. Elsevier/North Holland.

30. Matern U, G Strobel, JF Shepard: 1978. Reaction to phytotoxins in a potato population derived from mesophyll protoplasts. Proc. Natl. Acad. Sci. USA 75:4935-4939.

31. Medgyesy P, E Fejes, P Maliga: 1985. Interspecific chloroplast recombination in a Nicotiana somatic hybrid. Proc. Natl. Acad. Sci. USA 82:6960-6964.

32. Miller SA, GR Williams, HP Medina-Filho, DA Evans: 1985. A somaclonal variant of tomato resistant to race 2 of Fusarium oxysporum f. sp. lycopersici. Phytopathology 75:1354. (Abstract)

33. Murashige T: 1974. Plant propagation through tissue culture. Annu. Rev. Plant Physiol. 25:135-166.

34. Nagy F, I Torok, P Maliga: 1981. Extensive rearrangements in the mitochondrial DNA in somatic hybrids of Nicotiana tabacum and Nicotiana knightiana. Molec. Gen. Genet. 183:437-439.

35. Padmanabhan V, EF Paddock, WR Sharp: 1972. Plantlet formation from Lycopersicon esculentum leaf callus. Can. J. Bot. 52:1429-1432.

36. Reisch B, ET Bingham: 1981. Plants from ethionine-resistant alfalfa tissue cultures: variation in growth and morphological characteristics. Crop Sci. 21:783-788.

37. Sacristan MD: 1982. Resistance response to Phoma lingam of plants regenerated from selected cell embryogenic cultures of haploid Brassica napus. Theor. Appl. Genet. 61:193-200.

38. Sacristan MD, G Melchers: 1969. The caryological analysis of plants regenerated from tumorous and other callus cultures of tobacco. Molec. Gen. Genet. 105:317-333.

39. Shepard JF: 1980. Mutant selection and plant regeneration from potato mesophyll protoplasts, pp. 185-219. In: I Rubenstein, B Gengenbach, RL Phillips, CE Green (eds). Genetic improvement of crops. University of Minnesota Press, Minn.

40. Shepard JF: 1982. The regeneration of potato plants from leaf-cell protoplasts. Sci. Amer. 246:154-166.

41. Shepard JF, D Bidney, E Shahin: 1980. Potato protoplasts in crop improvement. Science 28:17-24.

42. Sibi M: 1978. Multiplication conforme, non conforme. Le Selectionneur Francais 26:9-18.

43. Singh BD, BL Harvey, KN Kao, RA Miller: 1975. Karyotypic changes and selection pressure in Haplopappus gracilis suspension cultures. Can. J. Genet. Cytol. 17:109-116.

44. Skirvin RM: 1978. Natural and induced variation in tissue culture. Euphytica 27:241-266.

45. Skirvin RM, J Janick: 1976. Tissue culture-induced variation in scented Pelargonium spp. J. Amer. Soc. Hort. Sci. 101:281-290.

46. Smith SM, HE Street: 1974. The decline of embryogenic potential as callus and suspension cultures of carrot (Daucus carota L.) are serially subcultured. Ann. Bot. 38:233-241.

47. Sutter E, RW Langhans: 1981. Abnormalities in Chrysanthemum regenerated from long-term cultures. Ann. Bot. 48:559-568.

48. Vajrabhaya T: 1977. Variations in clonal propagation. pp. 179-201. In: J Arditti (ed). Orchid biology: reviews and perspectives, Columbia Univ. Press, NY.

SUMMARY OF PANEL DISCUSSION ON PHENOTYPIC AND GENOTYPIC STABILITY OF TISSUE CULTURED PLANTS

D. A. EVANS (moderator), I. Y. E. CHU, R. D. HARTMAN AND H. J. SWARTZ

DNA Plant Technology Corp., Cinnaminson, NJ 08033; Geo. J. Ball Inc., West Chicago, IL 60185; Hartman's Plants, Inc., Palmdale, FL 33944; and Department of Horticulture, University of Maryland, College Park, MD 20742, U.S.A., respectively.

Variation in phenotype of plants produced in tissue culture can be characterized as either transient or permanent. Transient variation includes such characteristics as increased branching of plants, greater susceptibility to disease, and lack of uniform response to flower induction stimuli. These variations disappear within a few weeks to a few years, depending upon the crop. Some of these variations are useful. For example, increased runnering of tissue cultured strawberry plants is advantageous for nursery production. Increased branching of ornamental foliage plants results in a fuller pot plant, but the effect is lost in subsequent generations in the greenhouse when cuttings are taken from the tissue cultured plants.

Permanent variation can be either a hindrance or a benefit depending upon the situation. Variants exhibiting obvious differences, such as yellow or white sectors in leaves of strawberry, can be rogued whenever seen and before plants are distributed to the ultimate user. Such variants may also occur with chimeral plants in which the variant reverts to the less desirable original form. An example is the reversion of heavily branched Dieffenbachia cultivars ('Camille', 'Compacta') to the more vigorous, less branched 'Perfection', which is no longer in demand. On the other hand, permanent (somaclonal) variation is sought for variants that can be selected and used directly as new cultivars or that can be

Zimmerman, R.H. et al. (eds), Tissue Culture as a Plant Production System for Horticultural Crops.
© 1986. Martinus Nijhoff Publishers, Dordrecht. Printed in The Netherlands. ISBN 90-247-3378-2.

incorporated into conventional plant breeding programs. Both of these pathways have already been used successfully with horticultural crops.

Proliferation from axillary shoots is preferred when it is desired to minimize variation; shoots produced from adventitious buds are more likely to produce variants, particularly when the desired characteristics of the plant being propagated result from a chimera. Composition of the medium plays an important role because lower cytokinin levels give fewer reversions (variants) and more of the desirable plant types. One can exert some control over somaclonal variation produced by controlling the composition of the medium. It is undesirable to have too high a percentage of somaclonal variants among the plants produced because many different changes are occurring simultaneously. With a lower rate of variation, it is more likely that fewer characteristics have changed so that the variant plant resembles the parent in most respects.

A variant condition often found in cultures is the phenomenon known as vitrification. This is characterized by water-soaked appearance of the tissue, abnormal development of stomates, and low epicuticular wax formation resulting in plants that are difficult to acclimatize and transplant. This is a problem with many plants and has been noted especially in members of the Caryophyllaceae; it varies among species and cultivars. The condition can be minimized or controlled by increasing the agar and/or sucrose concentrations, improving air exchange in the culture vessels, and reducing cytokinin concentration.

PATHOGEN DETECTION AND ELIMINATION

ROGER H. LAWSON

U. S. Department of Agriculture, Agricultural Research Service,
Florist and Nursery Crops Laboratory, Beltsville, MD 20705, U.S.A.

1. INTRODUCTION

Detection of a pathogen is an essential step in plant disease diagnosis. Diagnosis of biotic pathogens requires a knowledge of biotic pathogens and factors in the environment that are involved in disease development. Disease diagnosis is often an art based on experience with a particular crop that allows a series of observations of the plant in both a diseased and healthy state and from knowing the often subtle differences in symptoms and their variability over time (13). It is also essential to observe and interpret the significance and implications of changing patterns of disease distribution that occur over time.

It is important to recognize that the "one cause-one disease" phenomenon is rare and procedures are needed to clarify situations where several individual factors may contribute significantly to disease development (49). In instances where disease can be correctly diagnosed as a result of a specific cause, the one factor predominates over all other causal factors. All the factors in the plant's environment, including water, nutrients, insects, weeds, as well as fungi, bacteria, nematodes, viruses and viroids, may be involved in the etiology of disease.

Recognizing the complex nature of disease requires an awareness that diagnosis often depends on an increased understanding of the influence of two or more pathogens on the same plant, relationships among the pathogens, and their host plants and the environment.

From the foregoing, it should be apparent that diagnoses based on symptoms alone, or where potential causal factors are studied as single

Zimmerman, R.H. et al. (eds), Tissue Culture as a Plant Production System for Horticultural Crops.
 ISBN 90-247-3378-2.

agents or independently, may give misleading results because more complex interactions occur. It may be more important to consider the complex etiology of some diseases to be as much an ecological as a plant pathological problem. The ecological approach may involve evaluating insect populations, assessing soil properties, or measuring components of weather (49).

Detection of a specific biotic agent in a disease complex is only one aspect of disease diagnosis. It is important to recognize this fact since one or more of the other factors that influence disease development and affect diagnosis may also be important in influencing detection of a specific pathogen.

Prior to the 1940's, some important diseases of major floral, nursery, and fruit crops were controlled by the application of chemicals and the choice of resistant cultivars. During the 1940's and 1950's a combination of factors led to improved crop health. In the chyrsanthemum industry, for example, cultured cutting techniques allowed the selection of cuttings free of pathogenic fungi and bacteria and the production of a much greater range of susceptible chyrsanthemum cultivars. Soil fumigation came into use along with the development of pathogen-free cuttings and provided a method to control soilborne fungal pathogens. Coincident with these developments was the discovery of the transmissible nature of the stunt disease and the associated development of a certification procedure for production of clean stock. The successful introduction of pathogen-free stock into commercial culture depends on the adoption of sanitary procedures and control of reinfection of the clean stock.

Disease control in high value asexually propagated genera of floral and fruit crops has depended on the development of clean stock procedures for producing large numbers of uniform plants that are disease free. Indexing procedures to detect viral, bacterial, and fungal pathogens in protected nuclear stocks has a long tradition in the industry.

There are many factors to be considered in adopting certification procedures for clean stock production in horticultural crops. Although the ideal objective is to eliminate all pathogens, there are both technical limitations as well as economic considerations that may impose limitations on achieving this objective.

2. CROP VALUE

What is the benefit to a crop when one or more pathogens are eliminated from the plant? In those crops that are produced from nuclear stock plants that are grown in a protected environment, the investment in developing and protecting the nuclear stock plants is often very great. For example, in chrysanthemum and carnation culture the principal propagators in the United States have for many years applied a variety of techniques to maintain small populations of protected plants that are propagated in large numbers and serve as the principal source of supply for the industry. The nuclear stock concept has been the basis for development of improved health of many fruit crops including decidious fruit trees, small fruits, and cane berries. Much of the effort in developing these programs has been directed toward producing virus-free nuclear stocks.

2.1. Greenhouse crops

Many greenhouse crops are high value crops and crop quality is of great importance in determining the value of the plant material. In some instances the crop is produced entirely in the protected environment from propagation to final sale. This procedure provides the advantage that many different kinds of pathogens can be controlled in this protected environment.

In other instances, crops that are initially propagated in a protected environment are grown in an exposed field environment in the build-up stage of secondary propagation. Vegetative propagules from these plants are then returned to the protected environment to finish the crop in the final stage of production. The obvious limitation of this procedure is sometimes not adequately appreciated. Clean stock transported to the less secure environment provides a greater opportunity for pathogens from outside the crop to be transmitted by wind, rain, or natural insect vectors into the crop. Although disease may not be obvious when propagules are returned to the protected environment, the disease subsequently appears and may cause serious loss.

When is a plant really free of a specific biotic agent? This question is often difficult to answer because of many factors. Symptoms associated with infection may be a primary reason. Symptoms are often influenced by many other biotic and non-biotic factors that have been mentioned previously. Because visual observations of symptoms on the intact plant are often inadequate for specific diagnosis, it is necessary to develop techniques for detecting biotic agents in plants growing in conditions where they are subjected to a wide variety of environmental influences. In some instances these techniques are available because the biotic agent occurs in such high concentration that it can be detected by procedures that have limited sensitivity. In other instances the procedures used in detection are highly sensitive and very low concentrations of the disease agents can be detected.

2.2. Management of stock plants used in tissue culture

Stock plants used as a source for propagation in tissue culture should be grown in protected environments to avoid unnecessary exposure to biotic agents. Proper precautions should be taken to avoid transmission of fungal bacterial or viral agents if they are introduced into the clean nuclear stock. This may be accomplished by observing basic rules of sanitation, controlling insects, and avoiding inadvertent contact of plants. The most important vector of these biotic agents may be man. Great care should be taken to prevent pathogen spread while taking cuttings, propagating by budding or grafting, or through other harvesting processes that create wounds and allow transfer of plant sap or plant parts containing the biotic agent.

Plants should be grown in environments that provide optimal conditions for development of the biotic agent that is being tested. For example, temperature will have a direct effect on the concentration of virus in a host plant. High temperatures suppress the development of many viruses and nuclear stock should be tested during cooler seasons of the year. The cooler season is usually spring when the outset of growth results in an increased concentration of the disease agent corresponding to the break in winter dormancy.

3. METHODS OF PATHOGEN DETECTION

The first step in conducting tests in the laboratory or greenhouse should be the collection of affected plants in all stages of disease development and with a complete range of symptoms. Samples may include above ground parts as well as roots, if a soil problem is possibly involved. Tests of soil for bulk density, pH, salt concentration, nematode infestation, and for other pathogenic organisms may be required (13).

The primary question that should be answered is whether an infectious agent is involved. The first step in this process should include a thorough study of the surface and sections of small pieces of affected tissues with low and high powered microscopes to determine whether bacteria, fungi, or nematodes are present and what tissues are affected. This type of examination should precede attempts to isolate a pathogen since its nature and location in the affected tissue provides information on which tissue to select for isolation and methods of surface sterilization and the selection of culture media. Because many media and surface sterilization procedures are often selective, it is important to wash some of the material in tap water and place it in a moist chamber or in water or on water agar in a petri dish to promote sporulation (13). Specific methods useful in diagnosis of fungi and bacteria are beyond the scope of this article but some references are included (13, 42, 46). Although the consistent association of bacteria and/or fungi with a particular disease may suggest a causal relationship, isolation of the organism and tests for pathogenicity are important in establishing the nature of disease.

3.1. Viruses

Bioassay procedures have been the traditional methods for detection and diagnosis of diseases caused by viruses and viroids. Traditional reliance on these methods was required because alternative procedures had not been developed that could be used to detect low levels of disease agents or various strains of disease agents in the field. Ideally, detection procedures should allow processing of a large number of samples in a short time.

Inoculation of indicator plants with an extract from a diseased plant often produces a symptom that in a few days, weeks, or months, provides proof of the presence of a pathogen. Although this procedure is still widely used to detect some viruses and viroids, the method is often not suitable for processing a large number of samples in a short time and has the disadvantage that different strains of a virus may produce markedly different symptoms on the same host. In addition, some viruses are transmitted by sap and are also transmitted by insect, nematode and fungal vectors. Other viruses are only transmitted by a vector.

Immunological diagnosis has been used for many years and relies on the use of antibodies prepared to viral coat protein. The initial preparation of antibodies to viral coat protein is the first step in immunological diagnosis. Several general review articles have been published recently (2, 6, 47).

Among the early immunological test procedures that found widespread application in plant virology were the chloroplast agglutination and microprecipitin tests. These tests relied on visualization of the antigen-antibody reactions in a liquid as clumped chloroplasts or complex antigen-antibody precipitates seen in the light microscope.

More recently, the use of a solid phase to adsorb an antibody or antigen with subsequent attachment of an antigen or antibody and the use of an antibody-enzyme conjugate to detect the antigen has become an important procedure in plant virus detection (6). Two variations of this method are most commonly used, but the principle is the same in both cases. The method referred to as enzyme-linked immunosorbent assay (ELISA) utilizes the wells of a plastic microtiter plate made of polystyrene or polyvinylchloride. The two most commonly used procedures are the double antibody sandwich method (7) and the indirect double antibody method (43). The wells are filled with the appropriate antivirus gamma globulin (IgG) and the IgG is irreversibly adsorbed to the surface. After the plate is rinsed, a separate plant extract is added to each well. Viral antigen present in the extract is specifically trapped by the bound antibody. The way in which the antigen-antibody complex is detected varies, but the final step involves the addition of an antibody-enzyme conjugate that gives a colored product in the presence of an appropriate substrate. The intensity of this color reaction is proportional to the amount of antibody-enzyme conjugate bound. The formation of a colored product with an intensity significantly above the background of the control indicates the presence of a particular virus in the plant extract.

The double antibody sandwich procedure reportedly has a narrow strain specificity (3, 6, 22). This is a useful test for separating and distinguishing strains but is less suitable for detection of a broad range of virus isolates with a single antiserum. Because the double antibody sandwich method is very strain specific, the enzyme-IgG conjugate prepared with antibodies against one virus strain often does not react with related strains (23, 47).

In a typical protocol for the double antibody sandwich procedure, each well of a plastic microtiter plate is charged with 200 μl of a pH 9.6 solution of the appropriate antivirus IgG that is usually prepared by precipitating antiserum with 2M ammonium sulfate and separation of the immunoglobulin on a DEAE-cellulose column (7). The antivirus IgG is incubated for 0.5 to 4 hours, during which time the antivirus IgG is immoblized. The microtiter plate is then washed and the plant extract (200 to 300 μl) in phosphate-buffered saline in the presence of nonionic detergent are added to each well in the plate. The plates are incubated overnight at 6°C to allow trapping of any viral antigen by the antibody that is adsorbed to the surface of the plastic well. The plate is again rinsed with buffer and antivirus IgG that is covalently coupled to an enzyme is added and incubated at 30°C for several hours to allow binding to any viral antigen previously complexed with immobilized antibody. The plate is then washed a final time and the appropriate substrate is added to each well and the plates are incubated for 0.5 to 3 hours at room temperature. The colored enzyme reaction product is then measured either visually or optically with a spectrophotometer.

Alkaline phosphatase is the enzyme most commonly used. The enzyme is coupled to IgG with the fixative glutaraldehyde. Although this enzyme will hydrolyze any monophosphate ester, the usual substrate in the procedure is the colorless p-nitrophenyl phosphate, which is converted to yellow p-nitrophenol and inorganic phosphate.

In the indirect double antibody ELISA procedure, wells of the microtiter plate are coated with either goat or chicken antivirus IgG. Plant extracts are incubated in the wells as previously described in the first procedure. A second antivirus IgG specific for the same virus, but produced in a different species of animal than is used in the initial coating IgG, is then incubated in the plate. Rabbit antivirus IgG may, for example, be used as the second antibody. After further rinsing, the plates are incubated with antibodies to total rabbit γ-globulins prepared in an animal such as goat. The goat antivirus rabbit IgG is conjugated with alkaline phosphatase and the bound conjugated IgG is detected with the appropriate substrate.

The indirect double antibody sandwich methods have the advantage that a single enzyme conjugated IgG preparation can be used regardless of the virus tested. In addition, this method has the advantage that the conjugated IgG prepared to total rabbit, goat, mouse, or chicken γ-globulin are available commercially and that a broader range of serologically related viruses can be detected (48). This method has proven more reliable for detection of serologically related strains of the same virus (23, 47) than the double antibody sandwich method. The method has the disadvantage that antibodies to each virus must be prepared in two species of animals.

Nitrocellulose membranes can also be used as solid supports for ELISA tests. The sensitivity of the test is reported to be 2 to 4 times more sensitive than tests in microtiter plates. This procedure has been described as dot-immunobinding, immunoblot, or DOT-ELISA (4, 14, 37). The increased sensitivity of immuno dot blotting may make this procedure useful for virus detection in crude leaf sap (38).

3.2. ELISA applications in diagnosis of plant virus infection

ELISA has become the single most important procedure for plant virus detection and disease diagnosis because the methodology is applicable to such a wide range of viruses and host species. The method is simple, rapid, sensitive, and requires only small amounts of antiserum. Tests can be performed with simple equipment and in some diagnostic situations reliable results can be obtained by visual evaluation without need for a spectrophotometer. Many agricultural crops can now be routinely tested for the presence of one or more virus disease (25).

3.3. Important considerations in reagent and sample preparation

Virus diseased plants may not be uniformly infected. Because of this uneven distribution (1, 8, 16), it is very important to develop reliable sampling procedures. Comparisons of different tissues and a range of tissue ages should be made. It is usually not adequate to sample from only a single position and a single age of tissue in routine assays of field samples.

Antisera prepared in rabbits in the conventional manner may have different properties when obtained from different test animals or from the same animal when immunized over a period of time. It has been recognized for many years, for example, that some animals in a population of the same genotype are much better antibody producers and will develop much higher titered antisera than other animals injected with the same antigenic preparation. In addition, an animal injected with the same antigen over a period of time will produce antibodies that are less specific than those obtained from early bleedings from the same animal following the initial injections of the viral antigen (27).

3.4. Electron microscopy in virus detection

Each plant virus has a characteristic size and shape. This property can be used to assist in identification of a specific pathogen and is also useful in routine detection and disease diagnosis. Virus particles are either rod-shaped, filamentous, or small polyhedra that can be observed in the electron microscope in the presence of an electron-dense stain. The method can usually be used most successfully for detection of rod-shaped or filamentous particles in crude sap. Small polyhedral particles are more difficult to detect because they may be confused with some normal cell components that are similiar in size.

Immune electron microscopy (IEM) is a method of virus detection and identification that is receiving increased attention (30). In the IEM technique, a drop of virus-specific antiserum is applied to a supporting transparent film that is first placed on an electron microscope support grid. A freshly cut surface of a leaf is placed in the drop for 1 or 2 seconds. The antiserum clumps the virus particles and coats the particles with antibody if the antibody in the antiserum is related to the virus. The virus particles are visualized by negatively staining the preparation with an electron-dense stain. Virus particles can be visualized in samples where the virion concentration is below the level of detection by conventional negative staining.

Virus particles may also be attached to antiserum-coated electron microscope support films. With this method, known as the Derrick technique (9), the increase in particle number over untreated control grids was up to 30,000, usually 5,000-10,000 for carnation cryptic virus and 100-fold for tobacco rattle and tomato bushy stunt viruses. A comparison of the number of virus particles attached to a grid treated with cymbidium mosaic antiserum and a grid treated with normal serum shows a large increase in the number of particles attached to the grid treated with the specific antibody. The method has been modified to include a second application of antiserum after attachment of the virions. In this procedure, a "decoration" or coating of the antigenic sites with antibody occurs, and specific attachment can be observed when the homologous or a related antibody is applied (30).

These procedures have been used to detect chrysanthemum virus B, carnation vein mottle and carnation necrotic fleck viruses, orchid tobacco mosaic and cymbidium mosaic viruses, and narcissus latent virus (25). Flexuous rod-shaped viruses can be detected in bulbous iris with direct negative staining of the sap. Because of the improved sensitivity, the IEM method should be applicable for detecting and identifying iris mild and severe mosaic, bean yellow mosaic, and other rod-shaped viruses in iris, gladiolus, tulip, lily, and other bulb species and to many rod-shaped viruses in flower crops.

Small polyhedral viruses can also be detected in IEM. Negative stains other than neutral phosphotungstate must be used, however, to preserve viruses such as cucumber mosaic, prunus necrotic ringspot, and apple mosaic (30). The IEM method was nearly twice as sensitive as ELISA in detecting arabis mosaic, prunus necrotic ringspot, and strawberry latent ringspot virus in extracts from rose (44). The high nonspecific background from healthy roses after overnight incubation of sap at 4°C made detection of arabis mosaic and prunus necrotic ringspot viruses difficult in an ELISA test (44). Both prunus necrotic ringspot and apple mosaic virus may decrease in concentration as leaves mature, since neither was detected in mature leaves.

Clumping and the modified Derrick procedure can trap low concentrations of virus and can be used to detect virus in crude sap preparations. Application of the IEM procedure may be most useful in research diagnosis involving specific virus and strain identification and in routine indexing where a limited number of samples are tested from nuclear stocks or plants from tissue culture where very high sensitivity is required.

3.5. Monoclonal antibodies

Somatic cell hybridization between a mouse myeloma (cancer) cell and mouse spleen cell results in cell hybrids or hybridomas that retain properties of both the myeloma cell and spleen cell. The fused cell hybrid has the ability to produce specific antibodies, contributed by the spleen cell, and the ability to grow in culture, contributed by the myeloma cell. The hybridoma technique provides the opportunity to produce highly specific (monoclonal) antibodies that react to a single antigenic determinant.

Monoclonal antibodies may be very useful in distinguishing closely related viruses (15, 21) in studies of viral epidemiology and distribution in the field. In other experiments it has been possible to select monoclonal antibodies that will react with a single determinant that is common to several different viruses (21).

3.6. Viroids

Viroids are the smallest known disease causing agents. About 10 diseases caused by viroids have been described (10). A wide spectrum of important crop species are infected and many show severe symptoms of disease. Viroids may have a long incubation period and some plants can be viroid infected and may remain symptomless. Indexing for the presence of viroids by symptom development in a suitable indicator plant may take 2 or 3 weeks in tomato for the potato spindle tuber disease but many months for others such as chrysanthemum stund viroid. Since viroids are small ribonucleic acid molecules that are not encapsidated with protein, immunological procedures used to detect the surface antigens of conventional viruses are not applicable for viroid detection.

3.7. Viroid detection

Polyacrylamide gel electrophoresis (PAGE) has been used to detect chrysanthemum stund viroid, potato spindle tuber viroid, and coconut cadang cadang viroid by analyzing for the viroid molecule. The method has been applied as a rapid, efficient, and reliable method for routine assay of chrysanthemum stunt viroid (20) and potato spindle tuber viroid (33). Using this procedure, the stunt viroid could be detected in as little as 50 mg of infected tissue of *Chrysanthemum morifolium*. The stunt viroid may be detected in 5% slab gels by ethidium bromide staining (31) in samples as small as 6 mg. A viroid band could be visualized with 0.01 μg of purified viroid RNA. The stunt viroid is in stem tips, young leaves and old leaves, lamina, veins, bark, and roots, but the lower concentration of viroid in the lower older leaves, veins and bark required 25 mg samples to detect the disease agent.

Healthy chrysanthemum plants contain two small RNAs different from the 9S cell RNA that are detected in the analysis of large samples. Although the 9S RNA can be clearly distinguished from the viroid RNA in a 5% polyacrylamide gel, other small host RNAs migrate to a position similar to the viroid. Because of the low concentration of these healthy RNAs, they are not detected when small samples are analyzed. These small RNAs can be eliminated after lithium chloride treatment of the sample and absorbing the viroid RNA to cellulose in the presence of 30% ethanol. After elution of the viroid with ethanol-free buffer, the viroid is precipitated with ethanol and separated by PAGE. Although this method has been successfully used on 5 to 50 g samples, care must be taken in interpretation of results because of the healthy RNAs. The modified procedure can be completed in 3 days (32).

In the Netherlands (34), a modified method using 5 g of young shoots extracted in a larger volume of buffer and in the absence of the 2-mercaptoethanol used by Morris and Smith (33) increased the reliability of the test. However, the number of bands may vary in the gel and a host nucleic acid band may occur just below the viroid band. Because the front is not always at the same location in the gel, the presence of the host nucleic acid may make the results difficult to interpret.

A new two-dimensional gel electrophoretic technique has been developed for the rapid and sensitive detection of viroids and virusoids (41). Viroid concentrations as low as 60 ng/g of tissue can be detected. The method is rapid, simple, and sensitive. The method combines a silver staining procedure with two-dimensional or bidirectional gel electrophoresis. In the two-dimensional gel, only a single sample can be analyzed on a single gel. In the "bidirectional" gel, several samples can be analyzed in the same gel. After the initial run, a 1 cm-wide rectangular strip is cut from the gel that is known to contain the viroids. The position of the viroid in the strip that is cut is determined by using a reference sample of partially purified viroid. The strip is set across the bottom of the gel chamber that is filled with a denaturing gel. The first gel is run from top to bottom and in the second run from bottom to top.

The general assay method used in "bidirectional" gels does not require the separation of RNA and DNA, polysaccarides or polyphenols. Because of the high resolution of the procedure, the preparative method can be as simple as phenol extraction and ethanol precipitation. The sensitivity of the method is shown to have a lower limit of 600 pg of potato spindle tuber viroid (PSTV). The total amount of tissue used was 0.1 g. Ten percent of this amount was loaded onto the gel. The corresponding sensitivity of 0.06 μg PSTV/g of tissue is similar to the sensitivity reported for hybridization (36) but hybridization may be more sensitive than silver staining. Time required to conduct the complete assay is less than 8 hours from tissue extraction to the final results from the second gel run. A major advance in this technique is the ability to separate and clearly distinguish pathogenic RNA species from healthy plant components.

3.8. Virus detection by hybridization

Serological detection is based on the antigenic nature of the virus coat protein. The coat protein gene contains only about 10 percent of the total viral genome and serological assays therefore detect only 10 percent of the genetic information of the virus. Some viruses that are distantly related, or apparently unrelated serologically, may have genes in common other than the coat protein gene.

A procedure that detects the most conserved portion of the viral genome should have the greatest cross reactivity. Hybridization analysis is a new method that has been applied to the detection of plant viruses and viroids (12). Although the method is more complicated and requires more time to conduct the test than the ELISA assay, the procedure may be useful in special circumstances with viruses and offers an means for detection of viroids that cannot be detected by conventional serological methods.

The principle of hybridization analysis of DNA and RNA molecules is based on biological properties that are determined by the sequence of the four nucleotides that compose the molecules. The nucleotide sequence of a virus or viroid is usually constant and a procedure that can distinguish between nucleotide sequences in nucleic acids is useful in detection of a particular nucleic acid.

The hybridization technique is based on incubation of single-stranded complementary DNA with single-stranded viral RNA under conditions where the two molecules anneal to form a DNA-RNA hybrid. The A, C, G, and T residues in DNA faithfully base pair with the U, G, C, and A residues in RNA, respectively. Complementary DNA (cDNA) is synthesized enzymatically using purified viral RNA as the template.

Hybridization analysis for indexing viruses and viroids requires preparation of highly radioactive cDNA to the viroid or viral nucleic acid (43). The isotope most often used is ^{32}P because of the ease of detection. Preparation of the cDNA is accomplished enzymatically on the viral or viroid nucleic acid so that its sequence is exactly complementary. When ^{32}P-cDNA is incubated with a nucleic acid extract of a test plant under appropriate conditions, it will hybridize to any complementary (or related) viral or viroid sequences present to produce a double-strand ^{32}P-cDNA:RNA hybrid in the case of viroid and RNA viruses. The extent of hybrid formation is a measure of the concentration of viral or viroid sequences in the plant extract (43). The procedure has been used in preparing ^{32}P-cDNA for the indexing of avocado sunblotch disease, coconut cadang cadang disease, citrus exocortis disease, and for the quantitation of the levels of chrysanthemum stunt viroid in chrysanthemum using a liquid hybridization procedure (43). In the liquid hybridization method, the ^{32}P-cDNA:RNA hybrid is acid precipitated and collected on a glass filter where the radioactivity is measured (43).

In a second method the test nucleic acid is immobilized on a nitrocellulose membrane. The method is simpler and more practical for testing a large number of samples (43). Potato spindle tuber viroid can be routinely detected directly in tuber sprouts and foliage using this method. An extract is made from the tissue sample. Following clarification of the extract, a small aliquot is spotted on a nitrocellulose membrane. The nucleic acid is immobilized by baking the membrane and the membrane is exposed to a prehybridization mixture followed by the ^{32}P-cDNA probe. After washing the membrane to remove the unhybridized probe, the membrane is auto-radiographed using X-ray film to detect those dots containing ^{32}P-cDNA hybridized to the bound viral or viroid nucleic acid. Both healthy and positive controls should be included in the test.

Although ^{32}P is a highly sensitive marker to label the cDNA probe, its value is limited because of a short half-life. Nucleotide analogs labeled with biotin can effectively replace radioactive reagents (24). There is a strong interaction between biotin and avidin. Avidin can be coupled to an indicator molecule, such as alkaline phosphatase, that will detect low concentrations of biotin. Biotin-labeled DNA probes can be detected after hybridization to RNA immobilized on nitrocellulose membranes (26). Color development similar to that obtained in ELISA indicates the presence of the specific viral nucleic acid.

3.9. Double-stranded RNA (ds-RNA)

Double stranded RNA (ds-RNA) has been isolated from plants, animals, insects, bacteria and fungi infected with RNA viruses. The potential of ds-RNA for detection and diagnosis of plant viruses has been investigated. Detection of high molecular weight ds-RNA is the subject of a separate chapter in this volume and will not be considered further in this article.

3.10. Bacterial diseases

Bacterial diseases of plants have often been recognized only by the symptoms they produce. During the last several years however, greater emphasis has been placed on utilization of a few tests that will allow the investigator to determine whether a given bacterial culture is likely to be the suspected pathogen and which of the isolated organisms should be selected for further study as possible plant pathogens. The first step in diagnosis should be to examine the diseased plant and record the symptoms.

Two methods are commonly used to isolate bacteria. The first method involves dilution plating. A small piece of diseased tissue is removed aseptically from the edge of a typical lesion at the border of healthy tissue. The tissue is placed in a few drops of sterile water and teased apart with sterile needles or ground with a mortar and pestle. Small quantities of the suspension are streaked on agar plates and individual colonies of bacteria are allowed to develop. After streaking, the dishes are incubated at about 25°C and most pathogens appear after 4-5 days or after 10-14 days for slower growing pathogens. The most frequently occurring organism is subcultured, unless it is recognized as a saprophyte.

The second method of isolation is similar to that used for fungi. Pieces of infected tissue are placed directly on the agar medium and bacteria grow around the tissue. This method is not often used because the saprophytes present usually grow faster than the pathogen, and overgrow the pathogen.

Plant pathogenic bacteria may be divided into two groups. One group is frequently isolated from cultivated plants and grows on standard bacteriological media. The second group consists of those that are uncommon and do not grow on standard media. Bacteria classified in the first group can be differentiated on the basis of colony characteristics on certain agar media and through the use of Gram stain and anaerobic growth tests. Members of the second group are initially identified on the basis of 1) growth on complex media, 2) presence or absence of cell wall, 3) Gram stain, or 4) obligate anaerobiosis (40). After isolation and tentative identification of an organism, pathogenecity tests are usually required to establish final proof of identity.

In addition to those techniques described, serological methods of bacterial identification have been increasingly important (39). Test methods may include tube agglutination and precipitation, Ouchterlony double diffusion procedure, immunofluorescence, and ELISA. Each method of testing has certain advantages. For example, the Ouchterlony double diffusion (ODD) test can be used effectively to differentiate among strains of bacteria (39). In routine identification, the immunofluorescence test with specific antiserum to membrane protein complex or glutaraldehyde-fixed cells is probably the best test available (39). More recently, however, the ELISA test is becoming more widely used for routine detection of bacteria. The form of the antigen, either as intact bacteria or a bacterial component, is very important and may differ between different bacteria and the antiserum used for detection. Test methods must be developed for the specific use and application.

4. PATHOGEN ELIMINATION

Twenty years ago the intensive monoculture of crop plants was recognized as a potential problem in the spread of many diseases and pests (19). Rapid vegetative propagation and international trade contributed to the problem. Production and distribution of virus-tested plant material depends on 1) exploiting erratic virus distribution within the plant, 2) chemotherapy, 3) heat treatment, and 4) meristem-tip culture (19).

Obtaining virus-free plants may be based on regular inspection and selection of plants with symptoms in many crops. Rogueing is the most effective method of eliminating virus-infected plants in species that show noticeable symptoms. Unfortunately, many viruses cause few or no noticeable symptoms and must be detected by laboratory tests.

In instances where all propagules of a vegetatively propagated cultivar are not completely virus-infected, virus-free plants can be selected by testing. Where cultivars are uniformly infected, special methods will be required to obtain virus-free plants.

Heat treatment of some virus-infected plants has been successfully used to eliminate about half of the known viruses in vegetatively propagated horticultural species. The most common form of heat treatment is to expose actively growing plants at 35 to 40°C in air for several weeks (19).

It is not possible to definitely categorize the kinds of virus that can be inactivated by heat treatment and those that cannot. Different strains of a virus may react differently. For example, cucumber mosaic, tobacco ringspot, tomato aspermy viruses and the chrysanthemum stunt viroid each have strains that can, and strains that cannot, tolerate high temperatures. Most yellows type mycoplasma induced diseases and many mechanically transmitted viruses with isometric particles can be eliminated by heat treatment (19). The ability of a virus to withstand high temperatures in the plant is not related to *in vitro* stability. Tomato spotted wilt, a virus that is inactivated in 10 minutes at about 45°C survives in plants at 36°C. Tomato bushy stunt, a virus that is stable after 10 minutes treatment at 80°C *in vitro*, is reduced in concentration in plants treated at 36°C, and most cuttings removed from heat-treated plants 3 weeks after treatment are virus-free (19).

Meristem-tip culture has been used to obtain virus-free dahlias, potatoes, carnations, and chrysanthemums, as well as many other crops. Although this method alone has been successful, it has been combined with heat treatment to obtain virus-free plants where meristem-tip culture alone has failed (19, 35).

Chemotherapy as a means of controlling plant viruses has had limited success because virus synthesis and normal host metabolism depend on the same biosynthetic processes. Many attempts have been made to apply chemicals to the infected host plant to reduce viral replication, prophylactic inhibition of viral transmission or production of virus-free host seed or propagation stock. Applications of ultraviolet light, X-rays, heavy metals, cations, nucleotide analogs, plant growth regulators, antibotics, herbicides, or fungicides can all reduce viral replication (11, 18, 29). These treatments have usually not been successful in eliminating viral infections *in vivo* (45). Complex remission of symptoms caused by apple chlorotic leafspot virus in

Chenopodium quinoa occurred when plants were treated with a soil drench of ribavirin or virazole. It was not determined if the lack of symptoms was due to masking or a loss of virus (17). Later, it was shown that suppression of symptoms caused by the green ring mottle causal agent and lack of graft transmission in Prunus serrulata resulted from the application of ribavirin sprays to infected plants. Sprays of amantadine, ribavirin, methasazone, or formycin applied to leaves of virus-infected cowpeas induced temporary remission of symptoms caused by cowpea chlorotic mottle virus (28). Ribavirin has also been used to eliminate lily symptomless virus and tulip breaking virus from lilies in tissue culture (5).

The potential of chemotherapentants has not been realized in control of plant viruses. Further work will be required to expand the potential use to treat plants in tissue culture and as seed treatments.

REFERENCES

1. Adams AN: 1978. The detection of plum pox virus in Prunus species by enzyme-linked immunosorbent assay (ELISA). Ann. Appl. Biol. 90:215-221.

2. Bar-Joseph M, SM Garnsey: 1980. Enzyme-linked immunosorbent assay (ELISA): Principles and applications for diagnosis of plant viruses, pp. 35-59. In: K Maramorosch, KF Harris (eds). Ecology of plant viruses. Academic Press, New York.

3. Barbara DJ, MF Clark, JM Thresh: 1978. Rapid detection and serotyping or prunus necrotic ringspot virus in perennial crops by enzyme-linked immunosorbent assay. Ann. Appl. Biol. 90:395-399.

4. Berger PH, DW Thornbury, TP Pirone: 1984. Highly sensitive serological detection of potato virus Y. Phytopathology 84:847.

5. Blom-Barnhoorn GJ, J van Aartrijk: 1984. The regeneration of plants free of LSV and TBV from infected Lilium bulb-scale explants in the presence of virazole. Acta Hort. 164:163-168.

6. Clark MF: 1981. Immunosorbent assays in plant pathology. Annu. Rev. Phytopathol. 19:83-106.

7. Clark MF, AN Adams: 1977. Characteristics of the microplate method of enzyme-linked immunosorbent assay (ELISA) for the detection of plant viruses. J. Gen. Virology 34:475-483.

8. Converse RH: 1978. Uneven distribution of tobacco streak virus in Santiam blackberry before and after heat therapy. Phytopathology 68:241-244.

9. Derrick KS: 1973. Quantitative assay for plant viruses using serologically specific electron microscopy. Virology 56:652-653.

10. Diener TO: 1979. Viroids and viroid diseases. Wiley - Interscience, New York.

11. Fraser RSS, RJ Whenhan: 1978. Chemotherapy of plant virus disease with methyl benzimidazol-2yl-carbamate: effects on plant growth and multiplication of tobacco mosaic virus. Physiol. Plant. Pathol. 13:51-64.

12. Gould AR, RH Symons: 1983. A molecular biological approach to relationships among viruses. Annu. Rev. Phytopathol. 21:179-199.

13. Grogan RG: 1981. The science and art of plant-disease diagnosis. Annu. Rev. Phytopathol. 19:333-351.

14. Gumpf DJ, W Kositratana, CY Zheng: 1984. Dot-immunobinding assay for virus detection. Phytopathology 74:847.

15. Halk EL, HT Hsu, J Aebig, J Franke: 1984. Production of monoclonal antibodies against three ilarviruses and alfalfa mosaic virus and their use in serotyping. Phytopathology 74:367-372.

16. Hamdorf G: 1982. The detection of plum pox virus (PPV) by indicator plants and enzyme-linked immunosorbent assay (ELISA). Acta Hort. 130:151-159.

17. Hansen AJ: 1984. Effect of ribavirin on green ring mottle causal agent and necrotic ringspot virus in Prunus species. Plant Dis. 68:216-218.

18. Hirai T, T Shimomura: 1965. Blasticidin S, an effective antibiotic against plant virus multiplication. Phytopathology 55:291-295.

19. Hollings M: 1965. Disease control through virus-free stock. Annu. Rev. Phytopathol. 3:367-396.

20 Horst RK, SO Kawamoto: 1980. Use of polyacrylamide gel electrophoresis for detection of chrysanthemum stunt viroid in infected tissues. Plant Dis. 64:186-188.

21. Hsu HT, J Aebig, WF Rochow: 1984. Differences among monoclonal antibodies to barley yellow dwarf viruses. Phytopathology 74:600-605.

22. Koenig R: 1978. Elisa in the study of homologous and heterologous reactions of plant viruses. J. Gen. Virol. 40:1-10.

23. Koenig R, HL Paul: 1982. Variants of ELISA in plant virus diagnosis. J. Virol. Methods 5:113-125.

24. Langer PR, AA Waldrop, DC Ward: 1981. Enzymatic synthesis of biotin-labelled polynucleotides. Novel nucleic acid affinity probes. Proc. Natl. Acad. Sci. U.S.A. 78:6633-6637.

25. Lawson RH: 1981. Controlling virus diseases in major international flower and bulb crops. Plant Dis. 65:780-786.

26. Leary JJ, DJ Brigati, DC Ward: 1983. Rapid and sensitive colormetric method for visualizing biotin-labelled DNA probes hybridized to DNA or RNA immobilized on nitrocellulose: Bio-blots. Proc. Natl. Acad. Sci. U.S.A. 80:4045-4049.

27. Lister RM, J Hammond, DL Clement: 1983. Comparison of intradermal and intramuscular injection for raising plant virus antisera for use in ELISA. J. Virol. Methods 6:179-182.

28. Mancino LJC, GN Agrios: 1984. Effects of antiviral compounds on symptoms and infectivity of cowpea chlorotic mottle virus. Plant Dis. 68:219-222.

29. Matthews REF: Plant virology. 1981. Academic Press, New York.

30. Milne RG, E Luisoni: 1977. Rapid immune electron microscopy of virus preparations, pp. 265-281. In: K Maramorosch, H Koprowski (eds). Methods in virology, Vol. 6. Academic Press, New York.

31. Monsion M, G Macquaire, JC Bachelier, J Dunez: 1980. Detection of chrysanthemum stunt and chlorotic mottle viroids by slab gel electrophoresis. Acta Hort. 110:321-328.

32. Monsion M, G Macquaire, JC Bachelier, J Dunez: 1981. Recent progress in chrysanthemum viroid detection. Acta Hort. 125:227-232.

33. Morris TJ, EM Smith: 1977. Potato spindle tuber disease: procedures for the detection of viroid RNA and certification of disease-free potato tubers. Phytopathology 67:145-150.

34. Mosch WHM, H Huttinga, FA Hakkaart, JA DeBokx: 1978. Detection of chrysanthemum stunt and potato spindle tuber viroids by polyacrylamide gel electrophoresis. Neth. J. Plant Path. 84:85-93.

35. Murashige T: 1974. Plant propagation through tissue cultures. Annu. Rev. Plant Physiol. 25:135-166.

36. Owens RA, TO Diener: 1981. Sensitive and rapid diagnosis of potato spindle tuber viroid disease by nucleic acid hybridization. Science 213:670-671.

37. Powell CA: 1984. Detection of three plant viruses by an immuno-blot assay. Phytopathology 74:847.

38. Rybicki EP, MB von Wechmar: 1982. Enzyme-assisted immune detection of plant virus proteins electroblotted onto nitrocellulose paper. J. Virol. Methods 5:267-278.

39. Schaad NW: 1979. Serological identification of plant pathogenic bacteria. Annu. Rev. Phytopathol. 17:123-147.

40. Schaad NW: 1980. Laboratory guide for identification of plant pathogenic bacteria. Amer. Phytopathol. Soc., St. Paul, Minn.

41. Schumacher J, JW Randles, D Riesner: 1983. A two-dimensional electrophoretic technique for the detection of circular viroids and virusoids. Anal. Biochem. 135:288-295.

42. Streets RB, Sr: 1969. The diagnosis of plant diseases. Coop. Ext. Serv., Agric. Exp. Stn., Univ. Arizona, Tucson. 134 pp.

43. Symons RH: 1984. Diagnostic approaches for the rapid and specific detection of plant viruses and viroids, pp. 93-124. In: T Kosuge, EW Nester (eds). Plant-microbe interactions. Macmillan Publishing Co., New York.

44. Thomas BJ: 1980. The detection by serological methods of viruses infecting the rose. Ann. Appl. Biol. 94:91-101.

45. Tomlinson JA: 1982. Chemotherapy of plant viruses and virus diseases, pp. 23-44. In: KF Harris, K Maramorosch (eds). Pathogens, vectors and plant diseases: approaches to control. Academic Press, Inc., New York.

46. Tuite J: 1969. Plant pathological methods - fungi and bacteria. Burgess, Minneapolis.

47. Van Regenmortel MHV: 1982. Serology and immunochemistry of plant viruses. Academic Press, New York.

48. Van Regenmorel MHV, J Burckard: 1980. Detection of a wide spectrum of tobacco mosaic virus strains by indirect enzyme-linked immunosorbent assays (ELISA). Virology 106:327-334.

49. Wallace HR: 1978. The diagnosis of plant diseases of complex etiology. Annu. Rev. Phytopathol. 16:379-402.

PRODUCTION OF CULTURE VIRUS-INDEXED GERANIUMS

WENDY OGLEVEE-O'DONOVAN

Phytolab, Inc., 152 Oglevee Lane, Connellsville, PA 15425, U.S.A.

The production of culture virus-indexed (CVI®) geraniums as performed by Phytolab, Inc. and Oglevee Associates, Inc. involves many processes developed to eliminate detectable pathogenic vascular bacteria, fungi and viruses. Once clean, these high quality selected plants are increased through a structured sanitary quality control system in order to reach the consumer. By law, to be an elite grower of CVI® geraniums in Pennsylvania, all plant material must be renewed annually from clean material, and no plants can be held in the system longer than 12 months.

The first process in the CVI® system involves culture indexing, a laboratory method of testing individual cuttings for the presence of vascular bacteria and fungi (2, 6, 8). Cuttings to be tested are handled individually under strict sanitation. Basal portions approximately 3 cm in length are removed from each cutting and surface sterilized. Small, one mm thick cross-sections containing vascular tissue are removed and placed in appropriate nutrient broth. If present, the vascular bacterium and fungus multiply in this rich medium, being readily visible to the eye in 14 days. Any culture indicating bacterial or fungal growth of any kind is discarded along with its corresponding cutting. In this way, the vascular pathogens are detected and then eliminated from the system. Of importance in pelargonium cultivars are the vascular bacteria, *Xanthomonas pelargonii* and *Pseudomonas solanacearum*, and the fungus, *Verticillium albo-atrum*. All can be spread throughout every phase of cutting geranium production, unnoticed during cool, dark weather, only to cause up to 100% loss when the warm sunny weather stresses the infected plants. In comparison, culture indexed plants grown under strict sanitation restrictions (1) have virtually 0% loss due to either *Xanthomonas*, *Pseudomonas*, or *Verticillium* with a reduction of up to 4 months in finishing time. Without culture indexing to remove these three vascular pathogens, geraniums are not economically profitable.

Zimmerman, R.H. et al. (eds), Tissue Culture as a Plant Production System for Horticultural Crops.
 ISBN 90-247-3378-2.

Once plants are culture indexed, they begin their long journey through a system designed to eliminate viruses. This part of the system involves thermotherapy and meristem-tip culture, followed by indexing for detectable viruses (11).

During the thermotherapy plants are grown at 100°F during a 16-hour day and 95°F during an 8-hour night for three weeks to reduce the concentration of spherical viruses in the plant (3). This heat treatment allows for the removal of slightly more tissue than just the few cells of the meristem itself for tissue culture, while increasing the odds of obtaining plants that index negatively for the presence of certain viruses (3).

Once the heat treatment is completed, all meristem-tips of 0.1-0.5 mm in height containing the meristem plus the first leaf primordia (3), are removed under sterile conditions and grown in vitro (4, 11). Even though meristems do contain virus (12), because the amount of virus has been decreased due to the heat treatment, the less tissue excised, less likely a virus would be contained in the explant. However, in the numerous cultivars with which we are forced to deal, the smaller the meristem-tip or explant, the longer the in vitro requirement. This in turn potentially offers minute horticultural variations in the resulting plants (10). In order to limit these intolerable variations, the cultures of the geranium meristem-tips are maintained no longer than 12 weeks, no multiplication is forced in the cultures and callus is not acceptable. In addition, cultures are routinely observed to insure that the growth is from the original meristem-tip; if it originates elsewhere, the culture is discounted. In this way, the maintenance of the horticultural characteristics unique to each selected cultivar, such as fast finishing, early blooming, number of blooms, plant habit, and plant zonation, which are so vitally important in the commercial industry, are insured. Resulting plants are transplanted into a sterilized peat-like medium and uniquely numbered in order that each plant's origin, history and previous performance record can be traced at any time. Plants are then grown in isolation until they are virus indexed; to this point, they are still only culture-indexed plants.

The virus indexing is done several times during the system's cycle before the plant is released. Bioassays are performed by gently swabbing carborundum-dusted leaves of Chenopodium quinoa, and cotyledons of Phaseolus vulgaris, Vigna sinensis, and Cucumis sativus with a slurry of geranium tissue ground in 0.05 M phosphate buffer containing 3%

polyethylene glycol 6000 at a 1:5 dilution (W:V). The leaves and cotyledons are rinsed and observed for symptoms over a two-week period. If any viral symptoms appear on the indexing plants, the corresponding geranium is either discarded or placed back into the system where the heat treatment and meristem-tip process are repeated. The bioassay is designed to detect and eliminate tobacco ringspot, tomato ringspot, pelargonium ringspot, pelargonium ring pattern, pelargonium line pattern, pelargonium leaf curl, cucumber mosaic, tobacco mosaic, tobacco rattle, tomato black ring and arabis mosaic viruses. Additional testing for tobacco and tomato ringspot viruses is done using enzyme-linked immunosorbent assay (ELISA) throughout the following year (7). All negative plants are saved and tested for horticultural quality. Only those plants meeting the highest "stringent" horticultural standards remain and are increased under strict sanitation (1). They are then distributed for sale under license through the system designed by Oglevee Associates, under which one plant can be ultimately responsible for 100,000 plants on the market place in 3 years.

All parts of the system from the laboratory to the licensed producer are monitored and subject to the regulations of the Pennsylvania Department of Agriculture and approved by Oglevee Associates. The pathological methods used to produce the Elite-Nuclear plants and the pathological quality of the plants must also meet the Pennsylvania Department of Agriculture's standards.

Because the quality of the final product is only as good as the sensitivity of the virus-indexing technique, there are constant work and research to improve the methods being used for viral detection and elimination; there will probably always be "just one more" to eliminate. Many of the viruses now being found have only been identified in geranium and specific isolation methods must be developed, fine tuned, and made commercially available. As more pelargonium viruses are recognized and isolated, it is our goal to work cooperatively in order to have ELISA or other state of the art methods made available for their detection.

The horticultural improvements due to the elimination of detectable viruses are the buds and florets that previously aborted due to tobacco and tomato ringspot viruses (5, 9) and that are now more numerous. Plants now exhibit not only larger numbers of bloom but larger blooms. The plants are more vigorous than ever before, yet are self-branching. Many cultivars can be finished pot-to-pot without pinching. The finishing time is reduced and the plants are capable of blooming in low light weather. These CVI® geraniums are the most economical choice for the progressive grower.

The process of producing CVI® geraniums is built upon cooperation between science, business and government. However, the benefits of this time-consuming, redundantly involved, never-ending process are rewarding to all.

REFERENCES

1. Baker KF: 1957. The U.C. System for producing healthy container grown plants. Univ. Calif. Agr. Ext. Serv. Manual 23. Berkeley, CA. Pp. 21-22.

2. Dimock AW: 1962. Obtaining pathogen-free stock by cultured cutting techniques. Phytopathology 52:1239-1241.

3. Hollings M: 1965. Disease control through virus-free stock. Ann. Rev. Phytopatholology 3:367-396.

4. Horst RK, SH Smith, HT Horst, WA Oglevee: 1976. In vitro regeneration of shoot and root growth from meristematic tip of Pelargonium x hortorum Bailey. Acta Hort. 59:131-142.

5. Murdock DJ, PE Nelson, SH Smith: 1976. A histopathological examination of Pelargonium infected with tomato ringspot virus. Phytopathology 66:844-850.

6. Nelson PE, JF Tammen, RR Baker: 1960. Control of vascular wilt diseases of carnation by means of culture-indexing. Phytopathology 50:356-360.

7. Newhart SR, CP Romaine, R Craig: 1980. A rapid method for virus-indexing the florist's geranium. HortScience 15:811-813.

8. Oglevee-O'Donovan WA: 1982. Culture-indexing for vascular wilts, pp. 233-241. In: JW Mastalerz, EJ Holcomb (eds). Geraniums III. PA Flower Growers, University Park, PA.

9. Scarborough BA, SH Smith: 1977. Effects of tobacco- and tomato-ringspot viruses in the reproductive tissues in Pelargonium x hortorum. Phytopathology 67:292-297.

10. Skirvin RM, J Janick: 1976. Tissue culture induced variation in scented Pelargonium spp. J. Amer. Soc. Hort. Sci. 101:281-290.

11. Smith SH, WA Oglevee-O'Donovan: 1977. Meristem-tip culture from virus infected plant material and commercial implications, pp. 453-460. In: WR Sharp, PO Larsen, EF Paddock, V Raghavan (eds). Plant cell and tissue culture. Ohio State University Press, Columbus, Ohio.

12. Walkey DGA, MJW Webb: 1968. Virus in plant apical meristems. J. General Virol. 3:311-313.

DIAGNOSIS OF PLANT VIRUSES USING DOUBLE-STRANDED RNA

RAMON L. JORDAN

U. S. Department of Agriculture, Agricultural Research Service,
Florist and Nursery Crops Laboratory, Beltsville, MD 20705, U.S.A.

ABSTRACT

Double-stranded RNA (dsRNA), an indicator for the presence of viruses in plants, can be isolated rapidly by simple methods from small tissue weights from multiple greenhouse and field collected samples using relatively simple laboratory equipment. The number, size, intensity and complexity of dsRNA detected by gel electrophoresis are distinctive and consistent for different virus groups and can aid in the diagnosis of a virus to group, member and even strain. Techniques for the isolation, detection, analysis and reagent use of dsRNAs from plants infected with various specific viruses and virus-like agents will be presented.

1. INTRODUCTION

A conventional virological approach to resolving the virus etiology of a particular disease is to isolate and purify the virus and produce an antiserum, in addition to characterizing the nucleic acid and determining other physical and biological properties. Success in this area is usually dependent upon transmission of the agent to an herbaceous host and subsequent purification to a homogeneous state. The isolation, analysis and use of viral replicative form double-stranded RNA (dsRNA) is another approach that can be utilized for virus disease diagnosis.

The development and use of dsRNA detection as a diagnostic tool (14) evolved from a premise that virus isolation and successful purification from and detection in tissue extracts is usually very host and/or tissue dependent (because virus stability in host extracts can be affected by

Zimmerman, R.H. et al. (eds), Tissue Culture as a Plant Production System for Horticultural Crops.
 ISBN 90-247-3378-2.

enzymes and oxidative substances), whereas successful nucleic acid extraction and isolation is usually tissue independent (because of the use of organic solvents and protein denaturants during extraction procedures). In fact, Morris and Dodds (14) demonstrated that dsRNA could be simply and readily isolated from virtually any type of tissue.

The detection and description of viral dsRNA for disease diagnosis is based on the premise that only plants infected with RNA viruses or virus-like agents contain homogeneous segments of high molecular weight ($>0.1 \times 10^6$) dsRNA (8, 15). Therefore the presence of dsRNAs, which are the replicative templates from which viral RNA genomes are made, indicates the presence of an RNA virus or virus-like agent. Additional diagnostic information (possible virus group and even virus strain) can be obtained from the pattern of genomic and subgenomic dsRNA species upon gel electrophoretic analysis of the isolated dsRNAs.

Techniques for the isolation, analysis and reagent use of dsRNA will be described and illustrated in this report. A more general discussion on plant viral dsRNA can be found in the review article by Dodds, Morris and Jordan (4).

2. MATERIALS AND METHODS

The following procedures for the isolation and analysis of dsRNA include modifications of the original, as well as subsequent, published reports (3, 10, 11, 14, 15).

2.1. Standard dsRNA extraction

Fresh tissue samples (4.0-40 g for 'column method'; 0.1-1.0 g for 'mini-batch method') are ground to a powder in liquid nitrogen in a pre-chilled mortar and pestle. The frozen powder is brushed into a chilled 50-250 ml centrifuge bottle (column method) or 3 ml centrifuge tube (mini-batch method). Samples can be extracted immediately or stored frozen at -20°C indefinitely.

The frozen powder is extracted with 1 to 5 volumes of 2xSTE (single strength = 0.1M NaCl, 0.05 M Tris-HCl, 1 mM EDTA, pH 7.0) containing 1 to 3% SDS (sodium dodecylsulphate), 0.1-0.5% 2-mercaptoethanol, 0.5 mg/ml fractioned bentonite (or 1% polyvinyl pyrrolidone-10). Once the tissue has thawed, the slurry is extracted with an equal volume of STE-saturated phenol (adjusted to pH 7 with ammonium hydroxide), and 0.5 volume of chloroform: pentanol (24:1, v/v). The emulsion is clarified by centrifugation (10,000 g, 10 min.), and the aqueous phase is adjusted to 15-18% ethanol (v/v).

2.2. Isolation and concentration of dsRNA

2.2.1. Standard column method. The supernatant is subjected to two cycles of CF-11 (Whatman, Clifton, NJ 07014) cellulose chromatography on small columns (14; method 1). CF-11 columns are made by suspending 2.5g CF-11 cellulose in 30 ml of 15-18% ethanol (in 1xSTE), and then pouring the suspension into a 30 ml disposable plastic syringe plugged with filter paper. Each cycle of chromatography consists of loading the column with 20-40 ml of sample in 15-18% ethanol (in 1xSTE), washing the column with 80-120 ml of 15-18% ethanol in STE, and eluting the dsRNA from the column with 14 mls of 1xSTE. DsRNA is concentrated by ethanol precipitation by adjusting the samples to 67% ethanol (v/v) and 50mM sodium acetate, and storing them for at least 4 hr at -20°C.

2.2.2. Mini-batch method. Supernatants from 0.1-1.0 g tissue (already adjusted to 15-18% ethanol) are transferred to a 1.5 ml microfuge tube containing 50 mg CF-11 cellulose pre-equilibrated with 15-18% ethanol in 1xSTE. (CF-11 containing microfuge tubes are prepared by pipetting 1.0 ml of a 50 mg/ml CF-11 suspension in 15-18% ethanol to a microfuge tube, centrifuging to pack the CF-11, and decanting the supernatant). The microfuge tube is vortexed 3 times in 15 min to allow the dsRNA to absorb the cellulose and then centrifuged to pack the cellulose-bound dsRNA. The CF-11 pellet is washed 3-4 times by adding 1.0 ml 15-18% ethanol in STE, vortexing, centrifuging and decanting the supernatant. After the final wash, the dsRNA is eluted by adding 0.5 ml 1xSTE to the cellulose pellet, vortexing, centrifuging, and transferring the dsRNA-containing supernatant to a second microfuge tube. DsRNA is concentrated by ethanol precipitation as above.

2.3. Gel electrophoretic analysis

Samples are prepared for electrophoresis by resuspending centrifuged (12,000 g, 20 min.) ethanol precipitatates in electrophoresis buffer (40 mM tris, 20 mM sodium acetate, 1 mM EDTA, pH 7.8) containing 10% glycerol. DsRNAs are analyzed on 6-8% polyacrylamide (40:1, acrylamide: bisacrylamide) mini-slab gels (0.8 mm thick; Idea Scientific, Corvallis, OR 97339). Samples are electrophoresed for 2-4 hr at 30 mA, 100V. Gels are stained for 10 min in 10-50 ng/ml ethidium bromide in electrophoresis buffer, destained in water, and photographed on Polaroid 57 film (ASA 3000) on an ultraviolet light transilluminator (302 nm transmission) for 2-90 sec with yellow and red filters.

The dsRNAs from three fungal viruses are used as molecular weight markers (Hm9, MW = 5.7×10^6; PcV, MW = 2.15, 1.85×10^6; PsV, MW = 0.89, 0.42×10^6) (e.g., see 10). Prior to electrophoresis, samples can be treated with deoxyribonuclease (DNase I) at 10 μg/ml in 5 mM $MgCl_2$ or ribonuclease at 10 μg/ml in 0.3 M NaCl to enzymatically degrade DNA and single-stranded RNA (ssRNA) contaminants, respectively.

2.4. Hybridization analysis

Hybridization techniques utilizing radioactive end-labeled RNA or virus specific dsRNA as hybridization probes for the rapid detection of virus-related nucleic acid species in gels (9) and in crude or purified nucleic acid preparations spotted on nitrocellulose (1, 6, 16) have been described in detail.

3. RESULTS

3.1. Extraction and purification of dsRNA

Young, succulent, newly infected symptomatic tissue generally yield the best quality and highest concentration of nucleic acids. Leaf midribs or bark tissue give better dsRNA results for phloem restricted viruses. Increasing the proportion of aqueous extractant is usually necessary to reduce chromatographic and gel analysis problems associated with the high concentrations of polysaccharide contaminants present in rosaceous and other woody host tissues. Nuclease treatment of samples prior to and/or after gel electrophoresis (in situ) can eliminate that small amount of DNA and/or ssRNAs that are isolated from certain tissues (even with extensive washings) and ensure accurate diagnosis and conformation of the dsRNA species detected. The optimum ethanol concentration (15-18%) required to selectively bind dsRNA to the CF-11 cellulose needs to be determined for each batch of CF-11 (4).

3.2. Detection and analysis of dsRNA

Examples of the detection and analysis by polyacrylamide gel electrophoresis of dsRNAs isolated by the above procedures from healthy plants, and plants infected with known viruses, or plants suspected to be virus-infected are illustrated in Figures 1 and 2. No homogeneous high molecular weight dsRNAs were detected in healthy tobacco or chinese cabbage plants (Figure 1, B, D and F). The presence of heterogeneous low molecular weight dsRNAs found in some plants (8) does not affect a gel-based detection of viral dsRNAs since these low molecular weight molecules ($<0.1 \times 10^6$) are normally electrophoresed off the end of the gel. Because of the existence of cryptic viruses (temperate viruses with dsRNA genomes) in certain varieties or seed lots, the analysis of healthy control plants is very important (4).

Routine detection of dsRNAs from relatively small amounts of tobacco mosaic virus (TMV)-infected tobacco, turnip yellow mosaic virus (TYMV)-infected cabbage, and cucumber mosaic virus (CMV)-infected tobacco was readily accomplished using the standard column method. The expected dsRNAs, which are twice the size of the respective genomes of TMV, TYMV, and CMV, were readily detected (Figure 1, C, E, G, respectively). For example, analysis of TMV-infected tobacco and TYMV-infected cabbage, which

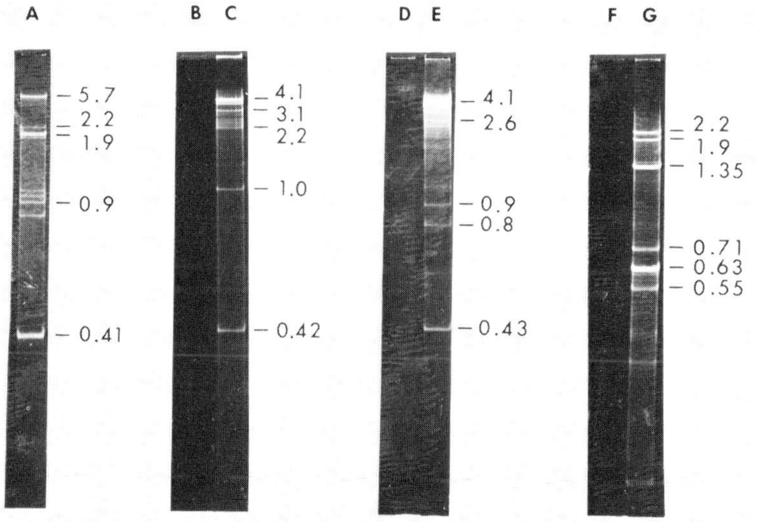

FIGURE 1 - Gel analysis of dsRNAs isolated by the column method from: 0.2 g healthy tobacco (B, F); 0.1 g tobacco mosaic virus-infected tobacco (C); 0.2 g healthy chinese cabbage (D); 0.05 g turnip yellow mosaic virus-infected chinese cabbage (E); 0.1 g cucumber mosaic virus-infected tobacco (G). Molecular weight size markers of dsRNA from three fungal viruses were run in (A). Molecular weights ($x10^6$) are as indicated for samples electrophoresed for 4 hr on a 6% polyacrylamide mini-slab gel.

have ssRNA genomes of 2.0-2.1 x 10^6 MW, revealed the expected 4.1 x 10^6 MW replicative form dsRNAs. Also 4-12 other dsRNAs (MWs approximately 3.1-0.4 x 10^6) were readily detected (Figure 1, C and E). The resolution of these dsRNAs (possibly the replicative form dsRNAs of subgenomic mRNAs; 4, 17) by electrophoresis into complex yet consistent patterns are of obvious diagnostic value for each of the viruses.

An example of using a dsRNA based analysis for the detection of previously unrecognized yet suspected, virus infections of avocado is illustrated in Figure 2. Three major dsRNA patterns were detected in avocado by gel analysis of dsRNA isolated by the column method (Figure 2A) and three different viruses or virus-like agents have been proposed (10). An example of a typical result using the mini-batch method in analyzing sixteen individual avocado plants in a routine indexing program (11, 12) is illustrated in Figure 2B. All three dsRNA patterns could be detected

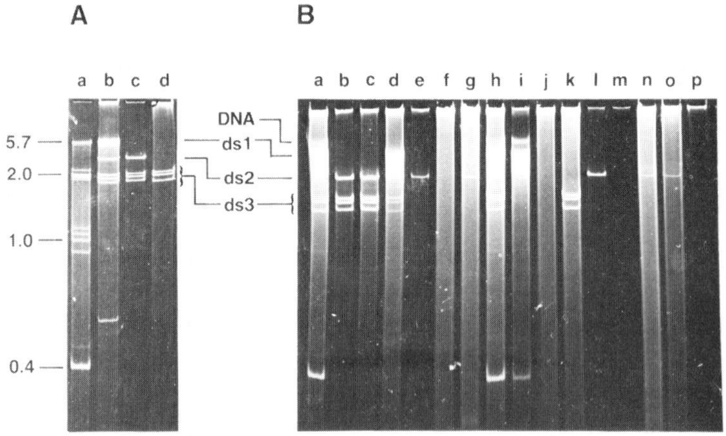

Figure 2 - Six-percent polyacrylamide minislab gel analyis of dsRNAs isolated from presumed healthy and virus-infected avocado plants. (A) DsRNAs isolated by the column method from 0.1 g of tissue from avocado plants known to have: dsRNA patterns 1(MW = 6.0,0.5 x 10^6), 2(MW = 3.0x10^6) and 3(MW =2.0, 1.9, and 1.7 x 10^6) (b); dsRNA patterns 2 and 3 (c); dsRNA pattern 3 (d); and dsRNA from three fungal viruses (a). (B) DsRNAs isolated from 0.3 g of avocado tissue by the mini-batch method from sixteen individual plants. Four samples (f, j, m and p) contain no detectable dsRNA. One, two, or all three dsRNA patterns were detected in the other samples (a-e, g-i, k, l, n, o). Samples were electrophoresed as in Figure 1.

singly and in all possible combinations; healthy (negative dsRNA) samples were also detected.

Not all virus-host combinations, however, yield as much dsRNA per gram starting tissue as those illustrated above. More than adequate amounts of dsRNA for a positive gel diagnosis are usually obtained from 0.1-0.5g of tissue. That is, the amount of dsRNA recovered from 0.1-0.5g of tissue of most virus-infected plants is sufficient to observe dsRNA bands in one gel lane. When diagnosing "unknowns", we usually extract about 4-5g of tissue per sample. If the results of analyzing 1-2g of that tissue are negative (i.e., no detectable dsRNA), we analyze the remaining 3-4g as a total sample.

4. DISCUSSION

The replication of single-stranded RNA plant viruses proceeds through the synthesis in infected cells of a double-stranded RNA structure, termed replicative form (RF), corresponding to twice the size of the genomic ssRNA. Since plants not infected with RNA viruses or virus-like agents do not contain homogeneous segments of high molecular weight dsRNA, the presence of dsRNA in a plant is an indication of an RNA virus or virus-like infection. The dsRNA detected could be the genome of a dsRNA virus or the replicative dsRNAs of a ssRNA virus. In addition to the major RF dsRNA, small dsRNAs have also been detected in many virus-infected plants (Figure 1; 3, 4, 9, 14, 17). The consistent pattern of genomic and 'subgenomic' dsRNA species is diagnostic for a given virus and virus group; that is, the number and size of such dsRNAs detected by gel electrophoresis are diagnostic parameters by which an unknown could be identified to one of the groups of plant viruses.

Some of the methods for the isolation, analysis and reagent use of dsRNA have been presented and illustrated. Other methods to purify dsRNA have been described (see references in 4). For example, precipitation of dsRNA in 4 M lithium chloride and solubility in 2 M lithium chloride has been demonstrated (2) and in fact can be used to confirm the identity of the nucleic acid species isolated by cellulose chromatography. In our experience, cellulose chromatography, whether by the column or mini-batch method, coupled with polyacrylamide gel electrophoretic analysis, has been found to be a rapid, reliable, and sensitive tool for the routine detection of dsRNA from relatively small amounts of greenhouse and field collected tissue of both herbaceous and woody hosts. In fact with the simpler mini-batch procedure we have found that a full-time technician can analyze over 120 samples per week, from sample preparation to final gel result.

The basic techniques presented have been shown to be useful for the detection of dsRNA in known virus-infected plants (Figure 1; 1, 3, 4, 7, 14, 15, 17) and in suspected, yet unproven, virus-infected plants (Figure 2; 5, 10, 11, 13). Mentioned briefly was the potential reagent use of dsRNA as labelled diagnostic probes. Purified dsRNAs can be melted, hydrolyzed and end-labelled with ^{32}P to generate an equimolar population of (+) and (-) sense viral sequences which can be used to detect homologous sequences in sap or gels (6, 9, 16; Dodds, Jordan and Mirkov, unpublished). Advantages of this approach include the ability to produce disease specific probes from known and unknown agents with equal ease, without requiring knowledge about virus particles nor their purification.

In summary, key features of the dsRNA based diagnostic procedure include: (i) the ease with which dsRNA can be routinely detected by simple methods from relatively small amounts of virus-infected tissue from virtually any host, even from naturally infected field plants; (ii) the number, size, intensity and complexity of dsRNAs detected by gel electrophoresis are distinctive for different virus groups and are consistent and independent of host type or tissue type; (iii) gel analysis can be completed within 48 hr of receipt of sample and is one of the best techniques for detecting mixed infections, including satellites; (iv) purified dsRNA can be a source of reagents to produce probes which can detect homologous virus nucleic acids by such methods as dot-blot hybridization; (v) the diagnostic information can be obtained without prior knowledge of the possible identity of the virus nor how to purify the virus particle. A single method purifies dsRNA regardless of which virus or virus-like agent, known or unknown, is causing dsRNA to accumulate in whichever host is being extracted.

REFERENCES

1. Bar-Joseph M, A Rosner, M Moscovitz, R Hull: 1983. A simple procedure for the extraction of double-stranded RNA from viral infected plants. J. Virol. Methods 6:1-8.

2. Diaz-Ruiz JR, JM Kaper: 1978. Isolation of viral double-stranded RNAs using a LiCl fractionation procedure. Prep. Biochem.8:1-17.

3. Dodds JA, M Bar-Joseph: 1983. Double-stranded RNA from plants infected with closteroviruses. Phytopathology 73:419-423.

4. Dodds JA, TJ Morris, RL Jordan: 1984. Plant viral double-stranded RNA. Annu. Rev. Phytopath. 22:151-168.

5. Falk BW, TJ Morris, JE Duffus: 1979. Unstable infectivity and sedimentable dsRNA associated with lettuce speckles mottle virus. Virology 96:239-248.

6. Garger SJ, T Turpen, JC Carrington, TJ Morris, RL Jordan, JW Dodds, LK Grill: 1983. Rapid detection of plant RNA viruses by dot blot hybridization. Plant Mol. Biol. Reptr. 1:21-25.

7. Gildow FE, ME Ballinger, WF Rochow: 1983. Identification of double-stranded RNAs associated with barley yellow dwarf virus infection in oats. Phytopathology 73:1570-1572.

8. Ikegami M, H Fraenkel-Contrat: 1979. Characterization of double-stranded ribonucleic acid in tobacco leaves. Proc. Natl. Acad. Sci. U.S.A. 76:3637-3640.

9. Jordan RL, JA Dodds: 1983. Hybridization of 5'-end labelled RNA to plant viral RNA in agarose and acrylamide gels. Plant Mol. Biol. Reptr. 1:31-37.

10. Jordan RL, JA Dodds, H Ohr: 1983a. Evidence for virus-like agents in avocado. Phytopathology 73:1130-1135.

11. Jordan RL, JA Heick, JA Dodds, H Ohr: 1983b. Rapid detection of sunblotch viroid RNA and virus-like double-stranded RNA in multiple avocado samples. Phytopathology 73:791.

12. Jordan RL, HD Ohr, GA Zentmyer: 1983c. Avocado blackstreak disease: A newly recognized major disease of avocado (Persea americana Miller) in California. Phytopathology 73:970.

13. Mirkov TE, JA Dodds: 1983. Detection of dsRNA in lettuce big vein infected lettuce. Phytopathology 73:961.

14. Morris TJ, JA Dodds: 1979. Isolation and analysis of double-stranded stranded RNA from virus infected plant and fungal tissue. Phytopathology 69:854-858.

15. Morris TJ, JA Dodds, B Hillman, RL Jordan, SA Lommel, SJ Tamaki: 1983. Viral specific dsRNA: diagnostic value for plant virus disease identification. Plant Mol. Biol. Reptr. 1:27-30.

16. Rosner A, M Bar-Joseph, M Moscovitz, M Mavarech: 1983. Diagnosis of specific viral RNA sequences in plant extracts by hybdridization with a polynucleotide kinase mediated, ^{32}P-labeled, double-stranded RNA probe. Phytopathology 73:699-702.

17. Zelcer A, KF Weaber, E Balazs, M Zaitlin: 1981. The detection and characterization of viral related double-stranded RNAs in tobacco mosaic virus-infected plants. Virology 113:417-427.

PATHOGEN INDEXING IN LARGE-SCALE PROPAGATION OF FLORIST CROPS

BOLIGALA C. RAJU AND JANE C. TROLINGER

Plant Pathology Department, Technical Division, Yoder Brothers, Inc., P.O. Box 68, Alva, FL 33920, U.S.A.

ABSTRACT

Plant disease can spread rapidly in vegetatively propagated florist crops simply due to the nature of the propagation process. Hence, it is important to utilize culture and also virus and viroid indexing systems when producing the stock plants. Culture indexing generally signifies the use of liquid culture media for the isolation of systemic bacteria and fungi. Virus-viroid indexing is accomplished using indicator cultivars, herbaceous hosts and serologic techniques. Pathologically certified stock is evaluated for horticultural qualities and the most acceptable material is used for cutting production.

1. INTRODUCTION

Systemic fungal pathogens like Fusarium oxysporum or bacterial pathogens like Erwinia chrysanthemi can cause serious losses in ornamental plants even when a single plant of nucleus stock is infected since vegetative propagation is the means by which many of the florist crops are produced. Similarly, even one plant in the nucleus stock program with a virus or viroid disease can result in serious economic losses in propagation as well as in the finished product. Thus, zero tolerance for the presence of pathogens in the nucleus stock programs of florist crops is demanded.

The devastating effects of Verticillium wilt of chrysanthemum, bacterial wilt of carnation caused by Pseudomonas caryophylli and bacterial blight of geranium caused by Xanthomonas pelargonii led to the development of culture indexing programs for ornamental crops (1, 3, 5). Chrysanthemum stunt disease, originally described in 1947 (2), threatened the commercial survival of chrysanthemum and it was realized that a nucleus stock that was completely free of stunt disease was necessary to eliminate the problem (6). As a result of these studies, virus-viroid indexing systems with

Zimmerman, R.H. et al. (eds), Tissue Culture as a Plant Production System for Horticultural Crops.
 ISBN 90-247-3378-2.

indicator cultivars, herbaceous hosts and serologic techniques were developed and are now used by most commercial propagators of florist crops.

2. CULTURE INDEXING

This is a laboratory procedure performed with each individual cutting of the crop in question to determine if systemic fungi or bacteria are present. The basal portion of a stem about 1.0-2.0 cm long is removed and coded with the same number as the cutting from which the section was removed. The stem portion is then surface-sterilized in 5% sodium hypochlorite solution, rinsed in sterile water, blotted on a sterile absorbent towel to remove the moisture, cut into three to five small pieces with a sterile razor blade and finally placed in a test tube containing either nutrient or potato-dextrose broth. The entire process is carried out under aseptic conditions. Coded test tubes are incubated at room temperature for 10-14 days. During the incubation period, all the cuttings from which the stem sections were removed are individually placed in plastic bags and stored at 32°-33° C.

After the two-week incubation period, test tubes with the plant material are examined for the presence of fungal or bacterial growth. Microbial growth in the medium is indicated by the presence of fungal mycelium or turbid liquid indicating bacterial multiplication. When the medium is clear and there is no sign of microbial growth, it is considered culture-negative and the cutting is used for virus-viroid indexing. A cutting is considered culture-positive when microbial growth is observed in the medium and/or the liquid is turbid. Positive cultures are discarded from the program.

3. VIRUS-VIROID INDEXING

Virus and viroid pathogens require living tissue for replication and symptom expression. Since these pathogens can not be cultured, grafting procedures were developed initially in 1949 to index for chrysanthemum stunt. Selection of a sensitive indicator that will show symptoms of a particular disease under commercial greenhouse conditions is extremely important before a program can be developed. Some of the indicator plants like mistletoe show distinct symptoms of chrysanthemum stunt under controlled conditions, but the symptoms are not evident under the low light conditions of winter in a greenhouse. Our studies have shown that the chrysanthemum cultivar Blanche is the most reliable indicator of stunt throughout the year. Since stunt disease spreads rapidly by mechanical means and only a few infected plants can affect the quality of several thousand stock plants, the support stock must be free of stunt. Hence, a double-grafting program has been developed. Polyacrylamide gel electrophoresis (PAGE) originally developed for the detection of potato spindle tuber viroid has been shown to also detect chrysanthemum stunt (4). However, since our studies have indicated that in some cultivars stunt can not be detected using the PAGE system alone, PAGE can be used as a screening tool for the leaf samples before they enter into the indexing process.

Enzyme linked immunosorbent assay (ELISA) has been used recently by several commercial propagators of florist crops for the detection of viral diseases. ELISA was demonstrated to be highly sensitive, reliable, and economical for the detection of aspermy virus and chrysanthemum virus-B (6). Similarly, ELISA was found to be excellent for the detection of carnation mottle, etched ring, necrotic fleck, and ring spot viruses. Our studies have indicated that spring is the best time to detect carnation viruses in the affected plants.

Throughout the virus-viroid indexing program, each plant is maintained individually, with no cross-handling, in insect-proof greenhouses with positive pressure. Chlorinated water with drip irrigation is used in the greenhouses and proper sanitation procedures are practiced. When the virus-viroid tests are negative, cuttings are harvested from the stock plants, and the meristems are taken for placement in a special tissue culture medium. These meristems are then grown into small rooted plants in tissue culture and supplied to the nucleus program.

4. NUCLEUS PROGRAM AND HORTICULTURAL CERTIFICATION

Pathologically certified plant material from tissue culture tubes is supplied to the nucleus program that should be located in an isolated and restricted area away from the production facilities. Greenhouse and plant health maintenance requirements are the same as explained previously. Each clone from the tissue culture bottle is planted individually in a pot and this plant material is called the A-nucleus.

The first available seven cuttings from each plant from A-Nucleus plants are harvested, planted, and grown separately in another greenhouse, and this plant material is called B-Nucleus. Cuttings from B-Nucleus are then clone flowered separately, horticultural characteristics are compared, and the best clones are selected by the breeder. This material is then used for production stock. The entire process is repeated once a year.

Several studies have demonstrated that the stock from pathological and horticultural certification programs grows and flowers uniformly and produces top quality flowers. However, it must be emphasized that these plants can also get infected just like any other plants with various diseases when planted in infected soil. Hence, it is of utmost importance to have effective scouting programs for plant pests and reliable disease control measures in order to get the complete benefits of the certified stock.

REFERENCES

1. Dimock AW: 1940. Importance of Verticillium as a pathogen of ornamental plants. Phytopathology 30:1054-1055.

2. Dimock AW: 1947. Chrysanthemum stunt. N.Y. State Flower Grow. Bull. 26:2.

3. Gruba EF: 1952. Culturing carnation cuttings. Carnation Craft 21: 4-5.

4. Mosch WHM, H Huttinga, FA Hakkaart, JA DeBokx: 1978. Detection of chrysanthemum stunt and potato spindle tuber viroids by polyacrylamide gel electrophoresis. Neth. J. Pl. Pathol. 84:85-93.

5. Munnecke DE: 1956. Development and production of pathogen-free geranium propagative material. Plant Dis. Reptr. Suppl. 238:93-95.

6. Raju BC, CJ Olson: 1985. Indexing systems for producing clean stock for disease control in commercial floriculture. Plant Disease 69:189-192.

CERTIFICATION OF HORTICULTURAL CROPS--A STATE PERSPECTIVE

L.M. GOFF

Maryland Department of Agriculture, Annapolis, MD 21401, U.S.A.

1. INTRODUCTION

Plant Certification is the process by which plant material is documented as being produced and tested by a set of standards, defined by regulation, which assure that such material is essentially free of specified diseases and pests.

Many disease and pest problems are of limited distribution and are known to occur only in specific, geographically isolated areas (e.g., peach stem pitting and apple union necrosis viruses in North America, apple proliferation in Europe). Plant certification programs are designed to protect nursery stock from infestation or infection by local pests and plant pathogens, thus reducing or preventing the possible movement of the pest/disease out of the region.

Other plant diseases and pests, although widely distributed, are known to have serious effects on the growth and potential productivity of crops (e.g., red stele of strawberries [causal agent: Phytophthora fragariae], root knot nematode on various crops [causal agent: Meloidogyne sp.], orange rust of brambles [causal agent: Gymnoconia peckiana]). Again, it is important that nursery stock be produced under conditions that will minimize the chances for infection or infestation by such pests.

The certification programs that are presently operational in Maryland and supervised by the Plant Protection Section of the Maryland Department of Agriculture form the basis for this discussion. These programs include production of Fragaria, Rubus, Malus, Pyrus, Prunus, and Vitis nursery stock.

2. CURRENT PLANT CERTIFICATION PROGRAMS

2.1. Plant propagation

Specific cultivars of most tree and small fruit nursery stock cannot be propagated by seed, due primarily to their genetic heterogeneity.

Zimmerman, R.H. et al. (eds), Tissue Culture as a Plant Production System for Horticultural Crops.
 ISBN 90-247-3378-2.

Vegetative propagation methods such as grafting or budding, tip layering, root suckers, runners, and cuttings have been used traditionally.

Vegetative propagation results in a product that is identical to the parent in every aspect, i.e., a clone. Cloning can be detrimental as well as advantageous, as the products of vegetative propagation are identical to the parent stock in all characteristics. Any abnormality in a source plant, including off type, disease infection, or pest infestation may be amplified by the number of propagules produced in each successive generation.

The use of vegetative propagation methods requires establishment and maintenance of some form of a nuclear or foundation stock. Such material must be produced and maintained under conditions that minimize off types, pest infestations, and diseases. If a particular plant has a large market demand, or is difficult to propagate, a large number of source plants may be necessary, or the plants may be produced through a series of successive generations.

2.2. Methods used in plant certification programs

The final step in the plant certification process is the issuance of a phytosanitary certificate. This certificate states that the nursery stock meets the requirements or standards established by the state or country of destination. A variety of methods are utilized by regulatory officials to verify that these standards have been met.

2.2.1. Field inspections are usually made at times when disease problems or pest infestations are readily detectable. These inspections may involve a representative sample of the planting or a row by row inspection of the plants.

Red stele of strawberry (causal agent: *Phytophthora fragariae*) can be first detected in plants during the spring of the second season of growth. However, strawberry propagating stock is planted, and the resulting nursery stock harvested, in a single growing season. To inspect fields for the presence of red stele, low areas, particularly those planted to red stele susceptible cultivars, are identified and portions of rows marked with wire flags. These areas are not dug during the harvesting of nursery stock. Plants from these areas are hand dug during the spring by an inspector, who examines the roots for symptoms of red stele. Suspected diseased plants are microscopically examined in the laboratory for evidence of disease. If red stele is detected, the field can no longer be used for strawberry nursery stock production. Plants can be inspected at the same time for other disorders that also typically show symptoms in the spring such as June yellows.

Inspection for root knot nematode infections are conducted in strawberry production fields in the fall, prior to digging. Teams of inspectors systematically sample daughter plants in each row across a field. If root knot is detected, the limits of the infested area are determined, the area is flagged, and nursery stock is not dug from the infested area.

Foliar symptoms of virus infection, if ever produced, will often appear in new flushes of growth. Field inspections are timed to maximize the chances of observing these symptoms. Rubus nursery stock is typically inspected during the late spring and early fall for this reason. During the spring inspection, symptoms of orange rust can also be observed, and plants rogued out before spores are produced.

Nursery stock can be a vehicle for moving pests of other crops such as soybean cyst nematode (Heterodera glycines), and corn cyst nematode (Heterodera zeae). If non-host nursery stock is produced in fields infested with these nematodes, there is a chance that cysts in soil clinging to the roots could be moved with the stock. Where such pests are known to occur, soil samples are taken from nursery stock production fields prior to planting and harvest to assure freedom from these pests.

2.2.2. Greenhouse and laboratory testing. Commercial cultivars of many fruit trees and small fruits will often not show obvious, easily detected, symptoms of virus infection, particularly under nursery stock growing conditions. However, some cultivars or species of plants show striking symptoms of infection by these agents. Such plants, known as indicators, are used to test nursery stock, particularly source plants, for the presence of infectious agents. This testing can be done under field conditions, or, in many cases in the greenhouse or laboratory.

Laboratory techniques such as ELISA are also used to screen source plants and nursery stock for the presence of infectious agents. These techniques have been discussed by others in this session.

Some infectious agents can be transmitted from test plants to appropriate indicators by means of mechanical inoculations. Such tests involve the collection and homogenization of leaf, stem, flower, or root tissue in an appropriate buffered solution, followed by gentle rubbing of the leaf surfaces of indicator plants with a sterile gauze pad soaked in the homogenate. Often the leaf surfaces are lightly dusted with a mild abrasive to promote the transmission of any viruses in the inoculum. Results can usually be obtained from such tests in 2-3 weeks. This method is used to screen plants for certain viruses in several of the Maryland certification programs including *Rubus*, *Vitis*, and fruit tress.

Many viruses cannot be successfully transmitted using mechanical inoculation techniques. Various grafting techniques are employed to test for the presence of a variety of virus and virus-like disorders. In general, the procedure involves the insertion of a piece of test plant material into an appropriate indicator, and observing the indicator during a period of time for symptom development.

One such technique, called petiole grafting, is used to insert a portion of a leaf from the test plant into the indicator. The leaf from the test plant is removed, and a significant portion of the leaf surface trimmed. A razor blade is used to shape the petiole into a narrow wedge, and the resulting piece of tissue is inserted into the cleft of a petiole on the indicator plant. This cleft is formed by splitting the petiole with a razor, after the leaf blade has been removed or trimmed. The graft is sealed with latex bandaging and the test plant observed for symptom development for up to 13 weeks. Such techniques are used to screen for selected viruses in the *Fragaria* and *Rubus* certification programs.

Fruit trees can be tested in the greenhouse for certain virus diseases by chip budding both indicator and test buds onto a suitable rootstock. Indicators are budded above the test buds. After the buds have taken, the rootstock is topped, and the indicators allowed to grow. Tests are observed for a period of 4-10 weeks for the development of virus symptoms. Many of these tests were originally conducted under field conditions. Techniques for greenhouse testing of fruit trees were developed and refined by Dr. Paul Fridlund, plant pathologist in charge of the IR-2 Fruit Tree Repository in Prosser, Washington. In general, field testing of fruit trees is considerably more expensive than testing in the greenhouse, on a per test basis.

2.2.3. *Field testing*. Certain virus diseases of fruit trees will produce symptoms only on the fruit or limbs of older indicator plants. It is therefore necessary to conduct tests in the field with test plants held until one or more crops of fruit are produced. To perform these tests, buds from both the indicator cultivar and the budwood tree to be tested are grafted onto the same rootstock, with indicator buds located above the test buds. The indicator buds are allowed to grow and produce fruit, whereas the test buds are not. The two indicators used to test for fruit and limb deforming, infectious agents are 'Golden Delicious' and 'Lord Lambourne'.

Another test conducted in the field utilizes 'Shirofugen' flowering cherry as the indicator. This cultivar of ornamental cherry is sensitive to certain viruses including Prunus necrotic ringspot and prune dwarf viruses. When virus infected material is grafted onto this tree, a necrotic reaction occurs under the bark in the area of the graft within 6-8 weeks of grafting.

2.2.4. The Fragaria nursery stock certification program can be used to illustrate how inspections, testing, and production standards are integrated into a successful program.

The Fragaria nuclear stock is maintained in screenhouses located on the University of Maryland Vegetable Research Farm in Salisbury, Maryland. All nursery stock produced in the Maryland program and sold as "essentially virus-free" (i.e., registered stock) is derived from this single source. Growing the plants in a screenhouse protects them from possible virus vectors such as aphids. New plant material introduced into the screenhouse is obtained from the USDA strawberry program at Beltsville, MD, and has been tested and found free of detectable virus and virus-like agents.

Cultivars are separated from each other by pieces of aluminum siding partially embedded in the soil, and mother plants destined to produce the planting stock for the following year are isolated from other plants of the same cultivar by circular, fiberglass collars. Each of the collars is identified by a unique code and number. Several daughter plants from each mother plant in a collar are collected by the Maryland Department of Agriculture in the fall, and tested in the greenhouse on one or more indicators, using the petiole grafting technique discussed earlier. If these test plants are free of detectable virus and virus-like agents, the remainder of the daughter plants become the planting stock for the following year.

Plants propagated in the screenhouses are sold to participating nurseries to serve as their foundation planting. Typically, the foundation planting is in an isolated field of approximately 4 acres. Plants produced in the foundation planting form the propagation stock for production fields.

All fields in the program, whether foundation or production, are located in isolation from any fruiting strawberry fields. Fields are treated in the fall prior to spring planting with a soil fumigant such as Telone II or Vorlex for nematode control. After planting, the fields are treated on a regular schedule with insecticides, miticides, fungicides, and nematicides.

Prior to digging, each field is inspected for important insect and disease problems, and is sampled to detect root knot nematode infestations. At this time, the areas to be sampled for red stele are designated, but this sampling occurs early in the following spring.

Nursery stock is also inspected routinely in the packing house during the grading process, and prior to shipment for those plants requiring an additional inspection at shipping time.

3. HOW TISSUE CULTURE CAN AND DOES AFFECT PLANT CERTIFICATION PROGRAMS

Plant certification programs are expensive for both growers and regulatory agencies. Plant tissue culture presents opportunities to reduce costs and to improve the quality of nursery stock in several of the certification programs in Maryland.

The *Rubus* nuclear planting, like *Fragaria*, is maintained in screenhouses. However, each mother plant occupies a larger area and is considerably less productive than its *Fragaria* counterpart. It has thus been possible to maintain only a very small nuclear planting with few propagules produced from each mother plant. This situation has necessitated sub-propagating cultivars for several successive generations in the field before sufficient propagating stock could be produced.

Several thousand plants may be produced from a single mother plant in one year by using plant tissue culture techniques. There is virtually no chance for the introduction of plant pathogens to plant materials handled and maintained in tissue culture. This results in nursery stock that is much "closer" to the indexed mother plant, which greatly reduces the chances for virus infection. Explants made from "virus-tested" plants will be maintained in this condition through successive generations in tissue culture, a condition which cannot be met for successive generations of field grown plants.

Certain cultivars of strawberries, particularly everbearers, produce few runners under nursery stock production conditions when using field or screenhouse produced planting stock. However, when tissue culture derived planting stock is used, runner production is greatly increased and the higher yields of nursery stock more than off-set the higher cost of the tissue culture planting stock. The use of tissue culture produced planting stock rather than stock produced by conventional vegetative propagation can reduce the amount of plant material that must be maintained in a nuclear planting. Any time that the amount of plant material kept in a nuclear planting can be reduced, there can be appreciable savings in maintenance, testing, and inspection costs.

Another important contribution of plant tissue culture is its utilization in combination with other techniques such as thermotherapy to eliminate viruses from desirable cultivars. This routine is well established with Fragaria.

At present, the extensive use of plant tissue culture in nursery stock production is limited primarily by economics. When the return from a crop can off-set the cost of the planting stock, tissue culture planting stock will be more widely used.

4. SOURCES OF INFORMATION ON PLANT CERTIFICATION PROGRAMS

Your state department of agriculture can provide information on existing plant certification programs, or can address any inquiries concerning the establishment of such programs in your area.

The references listed below are sources of additional information concerning some of the testing methods discussed.

REFERENCES

1. Frazier NW (ed): 1970. Virus diseases of small fruits and grapevines. Berkeley: University of California.

2. Fridlund PR: 1980. Glasshouse indexing for fruit tree viruses. Acta Phytopathologica Academiae Scientiarum Hungaricae 15:153-158.

3. Dunez J (Chairman): 1983. Indexing of virus and virus-like diseases of fruit trees. Working Group on Virus Diseases, Intl. Soc. Hort. Sci., 22 p.

4. Maas JL (ed): 1984. Compendium of strawberry diseases. The American Phytopathological Society, St. Paul.

5. Posnette AF (ed): 1963. Virus diseases of apples and pears. Headley Bros., Ltd., Ashford, Kent.

6. United States Department of Agriculture: 1976. Virus diseases and noninfectious disorders of stone fruits in North America. Handbook #437.

PLANT QUARANTINE AND INTERNATIONAL SHIPMENT OF TISSUE CULTURE PLANTS

ROBERT P. KAHN

Plant Protection and Quarantine, Animal and Plant Health Inspection Service, U.S. Department of Agriculture, Federal Center Building, Hyattsville, MD 20782, U.S.A.

ABSTRACT

The plant quarantine aspects of the international exchange of plant tissue cultures for scientific or commercial purposes are discussed. As a background for quarantine considerations, the author discusses quarantine principles, natural and manmade pathways for the entrance of exotic pests, pest risk analysis, and safeguards. Quarantine regulations are established by governments to lower the chances (= risk) that an imported article may be carrying a pest or pathogen of quarantine significance (= hazard). The entry status of imports ranges from liberal to conservative as determined by hazard/risk considerations as influenced by biological data and benefit/risk factors. Safeguards, particularly when modes of action are independent, can lower the risk and, therefore, biologically justify a more liberal entry status. The role of tissue cultures in providing safeguards that lower risk and facilitate international movement of plant materials is discussed.

1. INTRODUCTION

1.1. Objectives

The objective of this paper is to discuss the plant quarantine aspects of the international movement of plant tissue cultures. The consensus of scientists and commercial plant propagators, as well as quarantine officials is that tissue cultures represent plant material of significantly improved health status. If everything else were equal, tissue cultures should enjoy a more liberal entry status than the same species imported in other forms. In fact, most genera and their species do enjoy membership in this elite category. However, hazards and pest risk considerations make it impossible for "everything else to be equal" for many other plant genera

Zimmerman, R.H. et al. (eds), Tissue Culture as a Plant Production System for Horticultural Crops.
 ISBN 90-247-3378-2.

and species. For these genera, tissue cultures are not necessarily superior to their counterparts outside the glass or plastic wall of *in vitro* culture.

As a background to understanding the quarantine aspects of tissue culture imports and exports, the author discusses 1) quarantine regulations and principles in general, 2) natural and man-made pathways for the entrance of exotic pests, 3) pest risk analysis and safeguards, and 4) tissue cultures as safeguards.

1.2. Scope

This presentation discusses the plant quarantine aspects of tissue culture imports and exports and is not intended to provide specific information about the entry status of tissue cultures under the rules and regulations of the United States or other countries. Information about the entry status of importations into the United States may be obtained from the Permit Unit, Plant Protection and Quarantine (PPQ), Animal and Plant Health Inspection Service (APHIS), U.S. Department of Agriculture, Federal Building, Hyattsville, Maryland. Information about the rules of other countries may be obtained by writing to the PPQ Export Certification Unit, which maintains digests of foreign quarantine regulations, at the address given above or from the plant quarantine service of the importing country. For exports, information furnished should be used only as guidelines. As discussed later in this report, it is the permit issued by the importing country that determines whether plants, including tissue culture, are enterable. Information about exports and imports may also be obtained through any PPQ, APHIS office listed under "U.S. Government" in the white pages of telephone directories in many cities in the United States.

The word "tissue culture" is used herein in a generic sense to cover all forms of aseptic propagation whether started from cells, protoplasts, embryos, meristem, meristem tip cuttings or various tissues, buds or other organs, as well as final products whether tissues, cells, organs, or rooted plants, or other forms. The discussion covers both imports and exports as well as movement for commercial or scientific purposes.

The term "pest" is used in its broad or regulatory sense as stated in U.S. statutes (12) and regulations, as follows: A plant pest is "any living stage of any insects, mites, nematodes, slugs, snails, protozoa, or other invertebrate animals; bacteria, fungi, other related plants or reproductive parts thereof, viruses, or any other organisms allied with any of the foregoing, or any infectious substances which can directly or indirectly injure or cause plant disease." The definition of "pests" in other countries includes these same groups of organisms, but in many cases also includes birds, plant-feeding bats, weeds, and rodents.

2. PLANT QUARANTINE

2.1. Concepts

Plant quarantines, rules, actions, or procedures are established by a government or a group of governments to prevent the entry of pests of quarantine importance in or on articles imported along man-made pathways. Among the concepts used by governments in promulgating rules or taking regulatory actions are the following:

1) Quarantines should be biologically based and not used as trade barriers or for other political and economic reasons. Nevertheless, it is alleged that the quarantine fabric contains some nonbiological factors (5).

2) Actions taken should be the least drastic action commensurate with the known or perceived level of risk associated with the article to be imported. The least drastic action is the least restrictive action that will effectively exclude the pest (6).

3) Benefit/costs and benefit/risk considerations should be included in taking any regulatory position. Under benefit/risk considerations, risks are often taken for importations for scientific purposes (imported under specified safeguards) that are not taken for commercial importations.

2.2. Legal basis of quarantine

The legal basis for quarantine action in any country is either the legislation passed by National and sometimes State or Provincial Governments as acts, orders, rules, or regulations, or enabling legislation that authorizes the Minister of Agriculture to issue rules, orders, or directives. In the United States, the legal bases are the Plant Quarantine Act, the Federal Plant Pest Acts (FPPA), and other related Acts. The regulations promulgated under these acts have been published (12).

The legal umbrella under which international plant quarantine activities are covered is the International Plant Protection Convention (IPPC) of 1951, also known as the "Rome Convention." The IPPC provided the international model for the phytosanitary certificate which accompanies certain articles in international movement. The IPPC is a treaty and the United States is a signatory nation. The phytosanitary certificate is an instrument of that treaty. Groups of governments such as the European Economic Community (EEC) are bound by quarantine rules promulgated by the EEC Parliament.

2.3. Pathways for the entry of plant pest

Exotic pests can be moved along natural or man-made pathways (Table 1) (2, 4, 6, 7). Whether a pest moves along a natural pathway depends primarily on the life cycle of the pest. Some species produce airborne or vector-borne propagules and are readily spread over long distances. Others may move by the cumulative effect of a series of short distance movements over a long period of time until a natural barrier (including the absence of a susceptible host) is reached. However, many of the viruses, virus-like pathogens, bacteria without mobile vectors, most nematodes, many fungi without airborne spores, and some insects such as scales are seldom spread far naturally, although they may be moved over long distances by man along man-made pathways. Whether a pest moves along man-made pathways depends on activities of the traveling, scientific, and commercial publics; the hitchhiking ability of some pests; and the effectiveness of quarantine safeguards. When nature does have an effective way of moving pests long distances, quarantine restrictions become less meaningful and less successful. Consequently, quarantine rules and restrictions are usually directed towards pests that are moved primarily by man rather than by nature.

The hazard to U.S. agriculture is that imported articles can serve as a pathway for the entry of pests. Safeguards can lower the risk or chances that a pest will be moved. Tissue cultures are one category of articles moved by man. What is the risk that they might be transporting pests of quarantine significance?

2.4. Pest risk analysis

The biological basis for promulgating regulations, rules, and regulatory actions is the perceived or known pest hazard and risk associated with the entry of an imported article including plants, seeds, cuttings, tubers, tissue cultures, etc., (3, 4, 5). The "hazard" is the danger attributed to an activity, i.e., importing plants that might harbor exotic pests, while

TABLE 1. List of pest groups, articles of quarantine interest, and natural and man-made pathways along which pests may move.

Pest groups	Articles of quarantine interest that might carry pests on man-made pathways
Insects	Packing materials
Mites	Bagging
Snails and slugs	Soil, sand, gravel
Fungi	*Plants and plant parts
Bacteria	Growing media
Nematodes	Cultures of microorganisms and other pests
Parasitic plants	Commodities
Noxious weeds	Agricultural raw materials
Protozoa	Containers
Algae	Crates, pallets, dunnage
Viroids	Used vehicles
Viruses	Used farm equipment
Spiroplasmas	
Mycoplasma-like organisms	

Natural pathways	Man-made pathways
Wind and storms	*Agricultural cargo
Air currents	Nonagricultural cargo
Ocean currents	*Mail
Surface drainage	*Baggage
Seed dispersal	Common carriers including ships, airplanes, boxcars vehicles
Root grafting	Ships' stores
Migration	*Smuggling
Vectors	
Self-locomotion	

* Denotes direct pathways for plants and plant parts including plant tissue cultures

the "risk" is the chance that the hazard will be expressed. Pest risk analysis is a thought process whereby the entry status of imported articles is determined based on the calculated or perceived risk (= chances) of inadvertently introducing hazardous pests and pathogens on these items as moved by man.

"Entry status" refers to the entire range of decisions, policies, rules, or regulations that govern whether or not an article is enterable, and if enterable, under what safeguards. Attitudes toward entry status may range from "conservative" to "liberal," which in this context does not imply "lax." The most liberal attitude is that plant propagative material may enter freely without agricultural regulatory restrictions. The most conservative attitude is that the material is excluded without any exceptions even for scientific purposes.

It is possible to diagram the interaction of entry status and pest risk (Figure 1). When entry status is "matched" with estimated pest risk, a biologically sound pest risk analysis is made. Thus, when pest risk is low, the entry status should be liberal. When pest risk is high, the entry status should be conservative. If these biologically sound "matchings" are plotted, a line such as the one in Figure 1 can be developed.

2.5. Independent safeguards

A safeguard is an action taken to reduce the chances that an article, in general, or a plant, in particular, might be carrying or harboring a plant pest of quarantine significance. An independent safeguard is one that has a different mode of action than another safeguard. Thus, inspection, treatment, and tissue cultures are independent safeguards. Inspection at origin and inspection upon entry are also safeguards, but their modes of action are the same. Therefore, they are not independent safeguards. The greater the number of independent safeguards, the lower the chance is of a pest entering on an imported but safeguarded article (Figure 2).

2.6. Characteristics of plant quarantine regulations

Quarantine regulations and procedures in the United States and in most other countries of the world--particularly those that adhere to, or are guided by, the IPPC--have the following safeguarding characteristics (2):

1) Specify prohibitions, exclusions, or other regulatory actions.
2) Grant exceptions to prohibitions for scientific purposes to allow entry under safeguards.
3) Require import permits.
4) Require phytosanitary certificates.
5) Stipulate inspection upon arrival.

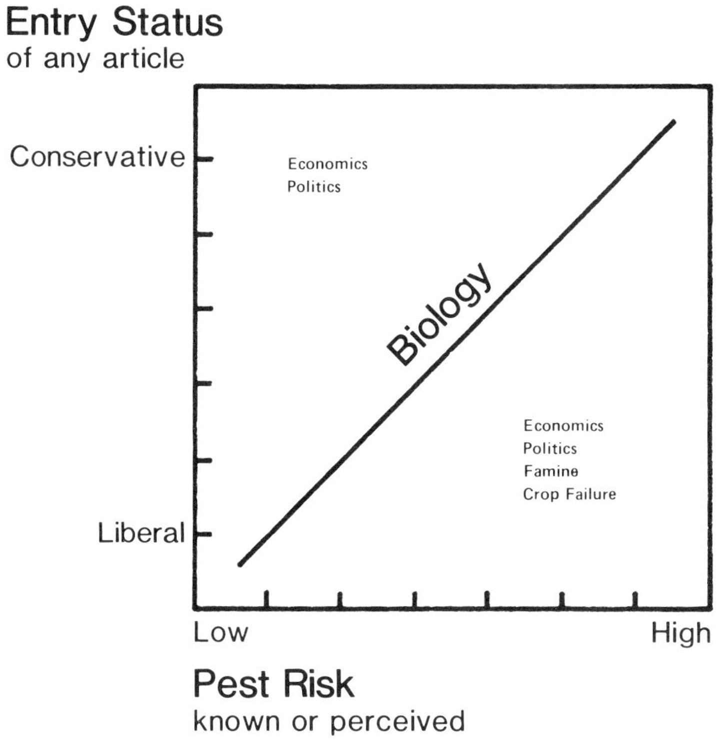

FIGURE 1. A diagrammatic representation of the relationship between pest risk (PR) and entry status (ES) of imported articles including plant tissue cultures. When PR and ES are matched; e.g., high PR with conservative ES or low PR with liberal ES, points developed by this matching fall on the biology curve as shown. When PR and ES are mismatched; e.g., high PR with liberal ES or low PR with conservative ES, points developed by this mismatching do not fall on the curve. Such mismatching may be required when famine occurs. Mismatching, allegedly, may have been used for economic policy but such positions are not biologically based and are contrary to plant quarantine principles.

6) Prescribe treatment upon arrival to eliminate a risk.
7) Prescribe quarantine, postentry quarantine, isolation, or other safeguards.

2.7. Regulatory actions

Among the regulatory actions that can be taken against imported articles, in general, or plants or plant parts, in particular, are the following

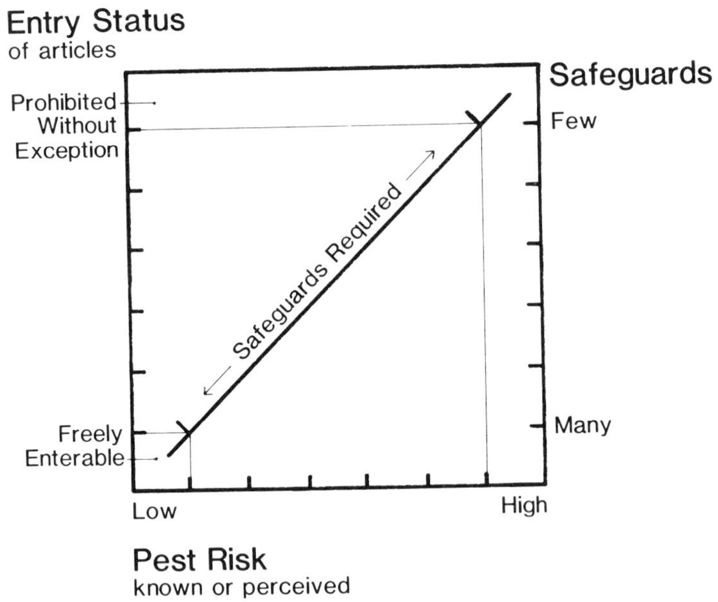

FIGURE 2. A diagrammatic representation of the relationship between entry status, pest risk, and safeguarding. As the number of safeguards increases, pest risk decreases. Consequently, the entry status may become more liberal and articles may be more freely enterable.

arranged in the order of the most drastic action first:

1) Prohibit or exclude the plant material.
2) Allow entry of prohibited plant material, but after inspection (and treatment, if necessary) require the material pass through a quarantine station or other special containment (usually allow for scientific materials).
3) Allow entry for planting in isolation under postentry quarantine on the property of the approved importer.
4) Require a mandatory treatment at origin or upon entry whether or not a pest is found.
5) Require inspection as a condition of entry and if a pest of quarantine importance is found:
   (a) Destroy the consignments;
   (b) Deny entry and allow the consignment to be returned or to be reexported;

(c) Treat the entire consignment, if an effective treatment is known;

(d) Treat only the plants or plant parts that are infected or infested; or

(e) Excise infected or infested portions of a plant or remove infected or infested plants and release the remainder.

Of these regulatory actions, 1, 2, 3, 5a, 5b, and 5e often apply to tissue cultures and the role of these actions is discussed in section 4.

3. THE THREAT OF EXOTIC PESTS

Quarantine regulations are usually promulgated on the basis of how each country views the threat of exotic pests. Some countries take regulatory actions or otherwise restrict importations on the basis of a wide spectrum of exotic pests including species that already occur in the country. Others less conservative are likely to take action on imports only against pests that do not occur in their country.

In the United States, rules and procedures are designed to ensure or to increase the chances that imported articles are free of exotic pests of economic and quarantine significance. An exotic pest may be considered to be of quarantine significance if:

1) It is known to cause economic damage, or based on its life cycle is capable of causing such damage;
2) It is not known to occur in the United States; or, if it does occur, it is not widely distributed; or is under a Federal domestic quarantine program for eradication, containment, or suppression; or represents a particularly virulent exotic strain of a domestic pest; and
3) It cannot be inspected for by procedures routinely available at a Plant Inspection Station at a U.S. port of entry; or, if the pest can be inspected for, there is no known effective eradicant quarantine treatment to apply should a significant pest be found as a result of inspection.

Plant pests have been intercepted on imported plant material other than tissue cultures. The incidence of the detection of pests that require a regulatory action at U.S. ports of entry has been reported periodically (1, 11). For example, during the period between fiscal years 1980 and 1984 (inclusive), 10,563 pests of quarantine importance were detected on importations of plants or plant parts. The pests intercepted included 183 nematodes, 81 mites, 1,724 disease agents, and 8,627 insects. During the period between 1981-1983, there were 8,307 interceptions of quarantine significance, i.e., a regulatory action was required on plants and cuttings received at U.S. ports of entry (Table 2) (7).

Pests have also been detected in the prohibited plant material in quarantine at the USDA Plant Quarantine Facility at Glenn Dale, Maryland, and other locations. These importations have been tested for obscure pests such as viruses. For example, between 1957 and 1967, 62 percent of the 1,277 introductions of *Citrus*, *Prunus*, *Malus*, *Pyrus*, *Solanum*, and *Vitis* were infected with viruses or virus-like pathogens (8). Between 1968 and 1978, 53 percent of 518 introductions of fruit trees (*Citrus*, *Cydonia*, *Olea*, *Prunus*, *Pyrus*, and *Malus*) were infected with one or more viruses or virus-like pathogens (9).

4. EXCLUSION AS A REGULATORY CONTROL MEASURE

Exclusion or prohibition of articles including plant materials is the most drastic action that can be taken in a regulatory sense. The impact of exclusion on plant materials, in general, is discussed in this section and on tissue cultures, in particular, in section 5. The author reviewed (6) the quarantine regulations of 125 countries to determine the methods by which plants were prohibited, the most frequently prohibited plant genera, and the most frequently named pests in these regulations.

4.1. Methods of exclusion or prohibition

The author's review (6) of the quarantine regulations of 125 countries showed that there were at least 11 methods by that plants were excluded or prohibited. The most frequently used method was to require a permit for the entry of all plants and then not issue permits for certain species or genera from some or all countries. Shipments that arrive without permits are denied entry. Among the other methods of direct or indirect prohibition are to require that the phytosanitary certificate contain an added declaration that a named pest is not known to occur in a country or designated area and deny entry if the certificate cannot be furnished; to deny entry if an otherwise enterable plant is accompanied by another species that is prohibited or is accompanied by prohibited packing materials (e.g., hay, straw); to deny entry, if upon presentation for

TABLE 2. Pests of quarantine significance[1] intercepted on plants at U.S. ports of entry (1981-1983).

Pest group	Man-made pathways				Total
	Baggage	Cargo	Mail	Non-entry[2]	
	(number of interceptions)				
Fungi	143	474	4	0	621
Insects	3,092	3,899	212	2	7,205
Mites	28	12	0	0	40
Mollusks	122	138	8	0	268
Nematodes	149	6	18	0	173
Total	3,534	4,529	242	2	8,307

[1] Includes pest interceptions for which regulatory actions are required in response to the interception.

[2] Interceptions on plants used as decoration on airplanes or ships.

entry, the plants are infected or infested with a pest of quarantine importance for which there is no effective eradicant treatment.

4.2. Plant genera prohibited in quarantine regulations

In the quarantine regulations of 125 countries, 246 different plant genera or crops were named on lists of prohibited articles (6). The average number of genera prohibited was 15 per country but the range was 0 to 132.

The 38 most frequently prohibited genera (including all their species) and crops were: Citrus (citrus); Cocos (coconut); Fragaria (strawberry); Musa (banana); Malus, apple; Pyrus, pear; Prunus (cherry, plum, peach, etc.); Ribes (current); Rubus (raspberry); Vitis (grapevine); Ipomoea (sweet potato); Solanum (Irish potato); Acer (maple); Castanea (chestnut); conifers; Crataegus (hawthorn); Juglans (walnut); Populus (poplar); Quercus (oak); Salix (willow); Sorbus (ash); Ulmus (elm); Coffea (coffee); Gossypium (cotton); Helianthus (sunflower); Hevea (rubber); Nicotiana (tobacco); Elaeis, (oil palm); Rosa (rose); Theobroma (cacao); Saccharum (sugar cane); and Camellia sinensis (L.) Ktze. (tea).

In the absence of any information to the contrary in regulations, it is likely that if a given plant species or genus was prohibited, tissue cultures of that species or genera would be prohibited. However, exceptions may be made for tissue cultures imported for scientific purposes and even for commercial purposes. As stated earlier, publications about regulations may only be used as guidelines because they represent the status quo at the time of publication. The definitive document that informs the exporter that articles are enterable (subject to pest finds during inspection upon entry) is the permit issued by the importing country.

4.3. Species named in quarantine regulations

The number of different pest species named in the quarantine regulations of the 125 countries was 1,588, which were cited 5,303 times. The total number of species and the number of citations of pests were respectively as follows: Insects and mites, 614 listed, 2,475 citations; nematodes, 46 and 225; fungi, 537 and 1,444; and bacteria, 95 and 514. For viruses, 292 different names were listed in 645 citations. The 10 most frequently named fungi, nematodes, bacteria, insects, and viruses are listed in the review (6).

5. TISSUE CULTURE AS A SAFEGUARD

5.1. Introduction

In section 2.4, the interaction of hazard, risk, safeguards and entry status were discussed. Safeguards lower the risk (= chances) that an article would be carrying hazardous pests. Several safeguards acting in concert can lower the pest risk so as to improve the entry status of imported plant materials. The aseptic features of tissue cultures often serve as a safeguard (3, 4). For many pests, the tissue culture itself by virtue of its aseptic methodology does offer an improvement in health status, particularly in combination with other regulatory actions, but there are other limitations. For the hazard of obligate parasites such as viruses, the tissue culture is only as healthy as its mother plant from which it was derived. However, for many other pests, the tissue culture is far superior to the mother plants as discussed in the sections that follow.

5.2. Port of entry inspection of tissue cultures

A commercial tissue culture consignment accompanied by the importer's permit is subject to inspection at a port of entry. The plant quarantine officer examines a sample of the consignment or the entire consignment to determine whether, in his judgment, entry presents a risk to U.S. agriculture as a result of the presence of a hazardous pest. The officer may decide to open a sample to permit a closer examination, but this is not often done. Inspectors look for signs of pests, symptoms on the plant tissue, or for contamination of the medium. Such contamination may be due to a breakdown in the aseptic procedures used to produce the tissue culture or perhaps due to saprophytes carried by an undetected pest in the consignments. In some regulations, there is a requirement that media be transparent (e.g., no charcoal in the media) to facilitate observation of roots.

Tissue cultures may be rejected if contaminated. In a 100 percent inspection, usually only vessels containing contaminations are discarded. In sample inspection, policy or guidelines would determine the entry status.

The groups of pests that may be detected by observations or contamination are shown in Table 3. Since these groups contain a wide spectrum of families, genera, and species, it is only possible to generalize knowing that there may be exceptions not shown in Table 3. Four types of detection available to an inspector are shown. These are (a) signs, (b) symptoms, (c) containment of the medium by fungal or bacterial pests, and (d) containment of the medium by saprophytes carried by pests. Of course, containment of the medium may occur due to a breakdown in aseptic procedures, but an inspector may not be able to differentiate such contaminants from 3 or 4. For plants or tissue cultures, symptoms are not satisfactory as a sole safeguard because the absence of symptoms does not mean that the plant material is necessarily free of pests. Some pests have long incubation periods (i.e., time between infection and appearance of symptoms), while other pests may be latent in some varieties of a plant species or in some species in a genus. Often plant genera are prohibited because they are known to be hosts of pests that cannot be detected by inspection at ports of entry.

Consequently, observations of tissue cultures to detect obscure pests such as certain bacteria and obligate parasites including viruses or rust fungi are an unsatisfactory safeguard. For obligate parasites, the tissue culture is often not much better than the mother plant from which it was derived insofar as plant health is considered.

However, the tissue culture is superior to the mother plant from which it was derived insofar as health status is concerned when pests other than

TABLE 3. Detection of pests of quarantine interest by inspection of samples of imported tissue cultures.

Pest groups[1] of quarantine interest	Methods of detection[2]			
	Observation of plants		Contamination of media	
	Signs[3]	Symptoms[4]	by pest[5]	by non-pests[6]
Insects	+	-	-	+
Mites	+	-	-	+
Nematodes	-	-	-	+
Bacteria				
fastidious	-	#	-	-
non-fastidious	-	#	+	-
Fungi				
Obligate parasites	#	#	-	-
Non-obligate parasites				
fastidious	#	#	-	-
non-fastidious	#	#	+	-
Viruses and viroids	-	-	-	-
Mycoplasm-like organisms	-	-	-	-
Spiroplasms	-	-	-	-
Protozoa	-	-	-	-

[1] Does not include the following pest groups: algae, weeds, parasitic plants, snails and slugs.

[2] If the pest were present in the tissue culture and based on inspection--
+ = the pest would be detected
- = the pest would not be detected
= some pests in the group might be detected but only with difficulty

[3] Observation of the pest directly.

[4] Absence of symptoms does not necessarily mean absence of the pest because of long incubation periods or latency.

[5] The pest organism may grow on the tissue culture medium.

[6] The pest organism might carry a non-plant pest organisms that might contaminate the medium. Presence of a non-pest contaminant might infer the presence of a pest or a failure in aseptic procedure.

obligate or fastidious pathogens are considered. Since the tissue culture may start from a meristem tip or a small piece of a plant organ, it should be free of insects and mites and, even if these pests were present, they would be detected visually or they would contaminate the media with organisms they may be carrying. The tissue culture should also be free of nematodes, eggs, or wind-blown spores that may be present on the usual plant importation. Some pests would also be detected in tissue culture as shown in Table 3. Most of the 8,307 pests tabulated in Table 2 could have been excluded by aseptic procedures (i.e., all but most of the pathogens).

5.3. The phytosanitary certificate and tissue cultures as safeguards

The phytosanitary certificate is a document that certifies that the consignment of plant material that it accompanies was inspected within 2 weeks prior to shipment and (a) was found to apparently be free of important pests and (b) is shipped in conformance to the quarantine requirements of the importing country. It is addressed to the quarantine service of the importing country and is signed by an authorized officer of the quarantine service of the exporting country. The phytosanitary certification is based on observation for signs and symptoms and some form of a statement that the consignment is believed to be free of harmful pests. The weight given to this certificate as a safeguard by the importing country depends on a number of variables the discussion of which are beyond the scope of this paper.

When the phytosanitary certificate and tissue culture are used as safeguards, the health status is improved over plants or plant parts entering with a phytosanitary certificate. As stated in the previous section, inspection is not a satisfactory safeguard for obscure quarantine pests, but the tissue culture *per se* offers added protection against many pests.

5.4. The phytosanitary certificate with an added declaration and tissue cultures as safeguards

The health status of plant consignments and the safeguarding features of phytosanitary certificates may be improved when the importing countries require "added declarations." "Added declarations" are typed statements added to the printed phytosanitary certificate in accordance with the quarantine requirements of the importing country. The importing country may require as evidence by statements on the permit that the exporting country certifies that a certain pest is not known to occur, or certain treatments have been administered at origin, or that the plants in the consignment descended from tested mother plants.

The combination of the phytosanitary certificate, added declaration, and tissue culture can significantly improve the health status of plant material particularly if the added declaration rule requires that the mother plants from which the tissue cultures were descended were indexed for viruses or other agents.

5.5. Special certification at origin and tissue culture as safeguards

Certification at origin is a useful safeguard for the importation of plant material, but the procedure has had limited application. A clear understanding must be developed between the exporting and importing plant quarantine service because the certification discussed here extends beyond the routine phytosanitary certificate. Certification refers to a government sponsored or monitored program in the exporting country to detect and eliminate certain pests named in the program. To understand the meaning of "certified" one must know the range of pests covered by the certification. Certified material may be tested to be free of one or two pests or all important pests or even "pest free" depending on the testing program.

Special certification at origin may be useful for importation or exportation of tissue culture of plant genera that are considered to be in a higher hazard and risk category. If the importing country is willing to accept the certification of the exporting country that the mother plants were indexed by methods accepted by the importing country, it may be possible to import material that might have otherwise been subject to prohibition or other regulatory restriction. To gain acceptance, usually there are exchanges of correspondence, reviews of procedures, and other actions prior to the issuing of a permit which may call for an added declaration as to the certification. For this venture to succeed, the importing country must have confidence in the ability of the exporting quarantine service to monitor the production and certification procedure.

5.6. Heat therapy, meristem tip culture, and tissue culture as safeguards

Heat therapy can significantly improve the healthy status of mother plants that give rise to tissue cultures (2, 3, 4). The heat therapy process is directed towards developing plantlets or other propagations that are free of certain viruses and mycoplasma-like organisms. The plant to be "cured" is grown in a heat chamber, often at a continuous temperature of about 38-40°C or under a diurnal fluctuation with about 16 hours at 35-40°C and 8 hours at 32° for 30, 60, or 90 days. Meristem tips are dissected from these plants and aseptically later transferred to agar. Sometimes "micro-grafts" are used. Plantlets are developed and transferred to soil. Each plant must be indexed to locate any that are free of the agent since not every meristem that is cultured will be free of the agent. Heat therapy without subsequent testing is not often considered as an adequate safeguard.

Heat therapy plus tissue culture may improve the health status of tissue cultures. If an added declaration as to heat therapy plus indexing can be accepted by the importing country, it may be possible to expedite the movement of both high and low risk materials. The fact that the mother plant has been subject to heat therapy does not constitute carte blanche for entry. Not all plants can withstand heat treatment. Not all viruses can be eliminated from the meristem by heat treatment. Not all meristem-tips obtained from plants infected with an agent that can be eliminated by heat treatment will be free of the agent after such heat treatment.

5.7. Tissue cultures imported for scientific purposes

Most quarantine regulations allow the entry of high risk plant materials imported for scientific purposes provided safeguards can be instituted by the importer that will lower the pest risk. Usually only small amounts of plant materials of these high risk or prohibited genera are admissible. Indexing may also be required after entry. Isolation on the property of the importer may also be an acceptable safeguard.

REFERENCES

1. Foster JA: 1982. Plant quarantine problems in preventing the entry into the United States of vector-borne pathogens, p. 151-185. In: KF Harris, K Maramorosch (eds). Pathogens, vectors and plant diseases: approaches to control. Academic Press, New York.

2. Kahn RP: 1977. Plant quarantine: principles, methodology, and suggested approaches, p. 289-307. In: WB Hewitt, L Chiarappa, (eds). Plant health and quarantine in the international transfer of genetic resources. CRC Press, Inc., Cleveland.

3. Kahn RP: 1978. International exchange of genetic stocks, p. 233-245. In: KW Hughes, R Henke, M Constantin (eds). Propagation of higher plants through tissue culture: A bridge between research and application. U.S. Department of Energy, Washington, D.C.

4. Kahn RP: 1979. Tissue culture applications for plant quarantine, p. 185-201. In: K Maramorosch, H Hirumi (eds). Practical tissue culture applications. Academic Press, New York.

5. Kahn RP: 1979. A concept of pest risk analysis. Bull. Eur. Med. Plant Protection Org. 9:119-130.

6. Kahn RP: 1982. The host as a vector: Exclusion as a control, p. 123-149. In: KF Harris, K Maramorosch (eds). Pathogens, vectors, and plant diseases: Approaches to Control. Academic Press, New York.

7. Kahn RP, CE Miller: 1985. International transportation of cut flowers and ornamental plants: Threats to the United States. Acta Horticulturae, No. 164, pp. 131-138.

8. Kahn RP, RL Monroe, WB Hewitt, AC Goheen, JM Wallace, CN Roistracher, EM Nauer, WL Acherman, HF Winters, GA Seaton, WA Pifer: 1967. Incidence of virus detection in vegetatively propagated plant introductions under quarantine in the United States, 1957-1967. Plant Dis. Reptr. 51: 715-719.

9. Kahn RP, HE Waterworth, AG Gillaspie, Jr, JA Foster, AC Goheen, RL Monroe, WL Povich, RG Mock, CF Luhn, EC Calavan, CN Roistracher: 1980. Detection of virus or virus-like agents in vegetatively propagated plant importations under quarantine in the United States, 1968-1978. Plant Dis. Reptr. 63: 775-779.

10. U.S. Department of Agriculture, Animal and Plant Health Inspection Service: 1933-1985. Export Manuals, Vols 1 and 2 (looseleaf) Washington, D.C.

11. U.S. Department of Agriculture, Animal and Plant Health Inspection Service: 1973-1984. List of Intercepted Pests (various dates between 1973-1984), Washington, D.C.

12. U.S. Department of Agriculture: 1984. Code of Federal Regulations, Title 7, Agriculture, Chapter III, Animal and Plant Health Inspection Service, Washington, D.C.

SUMMARY OF PANEL DISCUSSION ON PLANT QUARANTINES AND INTERNATIONAL SHIPMENT OF TISSUE CULTURED PLANTS

R. P. KAHN (moderator), J. VAN AARTRIJK, L. K. C. CLAY, R. A. DE FOSSARD, AND G. HENNEN

U. S. Department of Agriculture, Hyattsville, MD 20853, U.S.A.; Bulb Research Centre, Lisse, The Netherlands; Les Clay & Son, Inc., Langley, BC, Canada V3A 7B9; Tamborine Mountain Plants, Eagle Heights, Queensland 4621, Australia; and Oglesby Plant Laboratories, Altha, FL 32421, U.S.A., respectively.

No special quarantine problems exist for the international shipment of tissue cultured plants *in vitro*. If a particular plant genus is enterable into a country as a bare root plant, then it is also enterable as a tissue cultured plant. On the other hand, no special dispensation is given to a plant merely because it is shipped *in vitro*. Variation occurs among countries as to whether they will inspect an entire consignment of tissue cultured plants or only a sample.

Perhaps the main reason why tissue cultured plants do not and should not receive special consideration is the question of virus status of plants *in vitro*. Presence of virus may not be detectable while the plants are in culture or even for several years afterwards, depending upon the crop and the virus. Thus several years of post-entry quarantine would be required to ascertain the virus status of plants. The alternative of virus-indexing before shipment, or before placing the plants in culture, exists, but obtaining appropriate and acceptable certification of such indexing is often difficult.

Specific information on quarantine restrictions must be obtained from the appropriate quarantine authority for any country into which one desires to ship cultures or plants. Generally speaking, cultures must be

Zimmerman, R.H. et al. (eds), Tissue Culture as a Plant Production System for Horticultural Crops.
 ISBN 90-247-3378-2.

shipped in rigid, clear plastic or glass containers with tight fitting covers. The medium must also be clear, e.g. contain no activated charcoal. No measures to suppress microbial growth, e.g. deletion of sucrose from, or inclusion of antibiotic in, the medium, is permitted.

Problems sometimes arise between quarantine officers and commercial exporters/importers, because quarantine officers tend to be conservative with regard to entry status whereas the commercial firms tend to be liberal. In addition, quarantine officers will often consider previous history when inspecting shipments. Thus if shipments from a particular firm have been found to be contaminated or infested often in the past, then future shipments will be subjected to more rigorous scrutiny by quarantine inspectors.

Plant quarantine inspection plays no role in determining the trueness to type of cultivars or varieties. Quarantine regulations require only that plants must be labeled correctly by genus and, where required, by species.

DETERMINING AND MINIMIZING PRODUCTION COSTS

ALVAN DONNAN, JR.

Plant-Tech, Inc., Apopka, FL 32703, U.S.A.

1. INTRODUCTION

Little information has been published on determining costs in operating a plant tissue culture laboratory, and more specifically on individual costs for a given crop. Anderson et al. (1) provided useful information on costs of tissue-cultured plants of one crop several years ago. Since then few documents have been published. This is understandable since commercial laboratories are often unwilling to discuss information on their specific costs for a particular crop. Partly this may be because their records are incomplete and they cannot determine the exact costs. I would encourage the research community to consider costs and cost accounting when conducting research on new crops. Although many of the research results are interesting from an academic standpoint, they are not of much benefit to the commercial laboratories if the technology cannot be applied economically.

I have been involved in the commercial application of tissue culture techniques since 1974 and this has provided me with some insight into the economic operation of a laboratory and the main factors to be considered when evaluating a potential new crop or technique. The information presented here is not new, but hopefully it is organized in such a manner as to focus on some general concepts that can be applied to specific cases.

2. COSTS

2.1. Selecting crops to culture

In order to operate a tissue culture laboratory economically, one has to know the costs not only of propagating a crop in tissue culture, but also

Zimmerman, R.H. et al. (eds), Tissue Culture as a Plant Production System for Horticultural Crops.
 ISBN 90-247-3378-2.

the costs of propagating the same crop by conventional means. These costs may vary depending on the specific geographical location. For example, certain plants that could be produced most economically by conventional techniques in central or south Florida might be better produced in tissue culture in a less hospitable climate such as in northern U.S. or Canada. Similarly, when we were recently considering whether to propagate a certain flowering plant in tissue culture, we looked at recent trends in the seed industry. For this crop, seed costs have declined by 50% in the last 3 years and will probably continue to go down. In addition, seed quality has improved and new cultivars that will come true from seed are becoming available. Hence, the decision was made that tissue culture would not be competitive economically in this particular case. The point is that you must have sound reasons for deciding to propagate a specific crop in tissue culture.

Another reason for propagating plants in tissue culture is to provide plants in short supply. For example, citrus in Florida was in high demand several years ago following severe freezes. The demand was met by increasing production of seedling rootstocks. Since then however, a serious outbreak of citrus canker disease has resulted in the destruction of thousands of trees and a great demand for clean planting stock. Had commercial laboratories pursued development of micropropagation techniques for citrus several years ago, they might now be in the position of being able to supply plants in a time of shortage.

A third reason for using tissue culture propagation is to overcome a disease problem in a specific crop. Use of tissue culture together with appropriate disease organism identification and indexing techniques has permitted the production of plants free of specific pathogens. The final production method for these crops may not be in vitro once the specific pathogens have been eliminated. The decision on a production method will depend upon the costs of tissue culture and conventional propagation and the chance of reinfection of conventionally propagated plants.

2.2 Determining costs

How do we determine the cost of a plant being produced in tissue culture? One of the simplest ways is to take the entire cost of operating the tissue culture laboratory for a given amount of time and divide this cost by the the total number of units sold. A longer time, e.g. 3 months,

6 months or 1 year, is better to get a more reliable estimate of costs. It is essential that all costs, such as salary of manager or owner, and the plants produced but not sold, be included.

A second method is to take the entire operating cost and divide by the number of transfer hood positions times the number of hours per position for the time in question. This gives the cost of operating a hood position for 1 hour. Then examine the crops being handled by the hood operators. If the value of the explants being handled by the operator in 1 hour does not equal or exceed the cost of operating the hood position, then you are losing money on that crop. For example, suppose that it costs $25 per hour to operate a hood position. Then assume that an operator can handle 4 stage III culture vessels per hour, each with 20 plantlets valued at $0.30, for a total of $24 per hour. Obviously, there is a loss of $1 per operator hour in such a situation.

Another factor one needs to know when looking at production systems is the cost per square foot (ft^2) of shelf space. In studies I have made during the past several years, costs have ranged from $\$0.75/ft^2$/week to $\$3.40/ft^2$/week. A rough approximation for most laboratories would be $\$1.75/ft^2$/week. At the higher costs, the lab operator must produce either high value plants or a large volume in order to make money. Doing a study of this type makes one very aware of the importance of keeping shelf space full at all times. Similarly, it becomes clear that being able to add an additional shelf to each rack will reduce the cost per ft^2. When establishing a new laboratory, costs per ft^2 of shelf space will be high until enough cultures have been produced to fill all shelves in all growth rooms.

2.3. Categories of costs

Costs are generally divided into indirect, or fixed, and direct costs. Indirect costs include such items as utilities, depreciation, taxes, cost of borrowing money, sales and general administrative expenses, general supplies, dues, subscriptions, and meeting attendance. Direct costs include labor, media, and production supplies.

Labor is by far the largest cost involved in producing tissue culture plants at the present time, accounting for 60% up to 85% in extreme cases. One must consider the type of labor needed, the labor pool available, the training required, and the convenience of the laboratory to the labor pool. Wages for most workers in U.S. tissue culture laboratories begin around the minimum wage of $3.35 per hour to a maximum of $4.00 per hour. In contrast, a laboratory in Costa Rica can draw upon a labor force where the minimum wage is $0.25 per hour.

The cost of medium is only a fraction of the cost of labor, probably in the range of 5 to 10%, although sometimes undue emphasis is put on the cost of particular ingredients of the medium. It is more important to optimize the medium to produce healthy cultures with the maximum number of usable shoots. For example in one laboratory, one more microcutting per culture vessel per subculture dropped the cost per microcutting by $0.018. In contrast, the cost of media ingredients will influence cost only in the range of $0.001 to $0.009, with the effect usually in the lower part of that range.

Other expenses include utilities, about 5%, supplies other than media accounting for 10 to 15%, and miscellaneous expenses about 15%. These percentages will vary with the laboratory and its location since items such as taxes and utilities will vary greatly from one location to another.

2.4. Managing and motivating workers

Many laboratories have considered some sort of incentive or piece work plan to help motivate workers and to enable the supervisor to manage without intensive supervision. These laboratories have obtained information from other industries using similar plans as well as observing inovative management techniques in other tissue culture laboratories. Oglesby Plant Labs has been a pioneer in this area in Florida.

There are three general ways that an incentive plan can be set up. In the first, all employees, i.e. hood operators (cutters), media preparation person, dishwasher, etc., are placed in a pool and everyone is paid based upon the productivity of the whole laboratory. This works well if the owner/manager of the laboratory is the person taking care of the records. This method tends to penalize the fast workers, the most productive people, and tends to reward the slower employees. However, there is also a greater effort among the workers to stimulate one another to maximize productivity.

The second system, which works quite well, is that of paying an incentive bonus for individual propagules that are harvested or cut or separated. First, there is a need to set a minimum standard below which an incentive bonus is not paid. This system needs to be set up on a long-term basis, e.g. 20 days (= 4 work weeks), to reduce the potential for absenteeism among workers who have met the preset minimum standard and prefer to take time off rather than earn a bonus. Variations on the plan include paying workers a straight salary plus a bonus for every unit produced above a minimum. The main disadvantage of this system is the need for close supervision to make certain that cutters maintain the quality of their work when striving to attain the bonus. Poor quality work includes increased contamination rates and damaged mother cultures that die or are unproductive in succeeding subcultures. Some systems accept a minimum level of contamination, pay no bonus for an intermediate level, and impose a pay penalty for high levels, with the levels established beforehand. Another problem with this system is equitable assignment of cutters to crops so that the same person does not always get an "easy" crop and another always a "difficult" one.

The third system, and the one that I use, is to pay individuals for the culture vessels they handle. We picked a standard in the crop cycles, e.g. 50 jars per hour for African violet, and assigned a point total per vessel, e.g. 6 points per jar. Each point was made equivalent to a certain value in cents. If for the next crop the cutters could handle only 25 jars per hour, then each jar would be worth twice as many points, i.e. 12. It makes no difference then what crop an operator gets since an easy crop is worth fewer points per jar and a difficult one more points. At the beginning, it was essential to obtain accurate records; once that has been achieved, the system is easy to manage. With this system, the cutters tend to leave the smaller shoots on the mother cultures ensuring a good crop in the next subculture. Using this method, productivity increased 37% with about a 10% increase in payroll. Record keeping manually has been difficult, but it will become easier since records are now being transferred to a computer system.

Use of an incentive system will depend upon the individual laboratory situation. Incentives can work very well when properly managed in terms of maintaining standards for quality of workmanship and for contamination. They are very useful for eliminating absenteeism and tardiness. Money is the motivator in these systems and it can be a very effective one.

2.5. Mechanization

Since labor is the largest cost, mechanization can help to a certain degree. As an industry, we need to look at media preparation and filling of culture vessels. Other possibilities including robotics were discussed earlier by Jan Rowe (2). However, we can over mechanize with the result that overhead costs become so high that they are not supportable. Movable shelves have been considered for improving handling of culture vessels, but their cost is nearly triple that of stationary ones. Use of carts in place of movable shelves may be nearly as efficient at a fraction of the cost.

A concept currently gaining favor is to use the trays for vessels as the shelves in the growth room as well as the basic unit for handling culture vessels. In our laboratory, we use a tray available in the nursery trade that holds 30 to 33 baby food jars, depending upon the arrangement of the jars. Containers are washed, stored, filled with medium, autoclaved, transported, and placed in the growth room in these trays, being removed only for the actual transfer of the cultures. This technique increases efficiency of handling the jars, uses space more efficiently and reduces cost of shelving construction.

3. PRODUCTION PLANNING

Since the operation of a plant tissue culture laboratory is so expensive, production planning is absolutely essential. To generate a production plan, one must know the market for which plants will be produced. Then it is necessary to project how many plants of each crop will be needed and to calculate how much space and labor are required, and the cost of the materials needed. First attempts at planning may result in underestimating space and labor needed for the amount of product desired. Thus, it is essential to have accurate records on which to base the plan. Planning is not as critical when production is only 5,000 to 10,000 units per week. However, there are now laboratories in Florida that are producing 100,000 to 130,000 units per week. The planning involved in these large labs is very intense and is absolutely essential to keep them operating profitably.

One way to estimate space and labor requirements is to determine the number of jars needed each week to produce a certain crop, e.g. there are 1000 jars of Syngonium 'White Butterfly' to be done at the rate of 20 jars per hour. One must then decide if space and labor exists to devote the required 50 hours to this crop each week. In addition, media preparation must be coordinated with the transfer schedule so that the cutters always have the fresh medium when needed.

Start up costs for a new crop are high and must be considered carefully when broadening a product line or doing contract work. One laboratory operator estimates the costs currently as $75 per hour and that may be low. Thus the total cost of producing 100 plants may differ very little from the costs of producing 10 to 50 times as many plants.

A sales plan is necessary to put together a meaningful production plan. The sales plan must take into account many factors including the seasonality of the market. Fall is the worst season for selling foliage plants. Inadequate planning means that the entire supply line becomes filled with plants and cultures backing up into the laboratory. Then there are two options, discard the excess material or sell it for less money, perhaps at no profit.

REFERENCES

1. Anderson WC, GW Meagher, AG Nelson: 1977. Cost of propagating broccoli plants through tissue culture. HortScience 12:543-544.

2. Rowe WJ: 1986. New technologies in plant tissue culture, pp. 35-51. In: RH Zimmerman, RJ Griesbach, FA Hammerschlag, RH Lawson (eds). Tissue culture as a plant production system for horticultural crops. Martinus Nijhoff, Dordrecht.

DETERMINING MARKETS AND MARKET POTENTIAL OF HORTICULTURAL CROPS

J.B. JONES

HERS, 4247 Brookview Drive, Dallas, TX 75220 U.S.A.

1. INTRODUCTION

With more than 250 tissue culture laboratories in the United States today, we have recognized the shift from "production capability" to "products offered" in many horticultural crops. The next typical pattern of change from "products offered" to "market needs" is occurring now. This presentation addresses those major products offered and today's market needs.

2. INDUSTRY STRUCTURAL ANALYSIS

Agricultural production in the United States is a fragmented industry populated by a large number of small- and medium-sized companies, many of them privately held. The emergence of plant tissue culture technology in the 1970's caused fundamental changes in the way plants are propagated and distributed. There continues to be a restructuring in the business of growing and distributing horticultural plants. This restructured plant growing industry can be segmented to include laboratories as follows:

1. Laboratory Propagator/Shipper with no Greenhouse.
   Sells Stage II and/or Stage III cultures in vessels,
   Stage II microcuttings and/or Stage III plants.
2. Laboratory Propagator/Grower with Greenhouse.
   Uses all lab production internally.
   Sells mature plants, 2-1/4's, liners, cell packs, plug trays, flats.
3. Laboratory Propagator/Grower/Shipper with Greenhouse.
   Sells primarily Stage IV acclimated plants, but could sell all the same plant stage types as a laboratory propagator/shipper. Can grow plants through maturity.
4. Propagator/Grower with No Laboratory
   Buys from #1 and #3.
   Sells mature plants.

Zimmerman, R.H. et al. (eds), Tissue Culture as a Plant Production System for Horticultural Crops.
 ISBN 90-247-3378-2.

5. Propagator/Grower/Shipper with No Laboratory
   Buys from #1 or #3.
   Sells Stage IV acclimated plants, 2-1/4's, liners, cell packs, plug trays, flats, as well as mature plants.
6. General Grower Consumer
   Buys Stage IV acclimated plants from #1, #3, #5.
   Resells nursery and greenhouse products wholesale.

The majority of the commercial horticulture micropropagation facilities in the U.S. are in California, Florida, Texas, and New York (Figure 1), the top four production states in the nursery business (Table 1).

The 1984 Nursery Business Wholesale (NBW) 100 report lists the 100 largest wholesale nurseries in the United States and the two largest wholesale growers and the amount of the sales in each state. Members of the NBW 100 account for $630.3 million in sales from a production of 198,735 acres (80,427 hectares). This is 25% of all wholesale nursery sales. The top 10 reported $231.9 million or 9% of all sales. Additional breakdowns of floricultural crops and producers have been published (1,6).

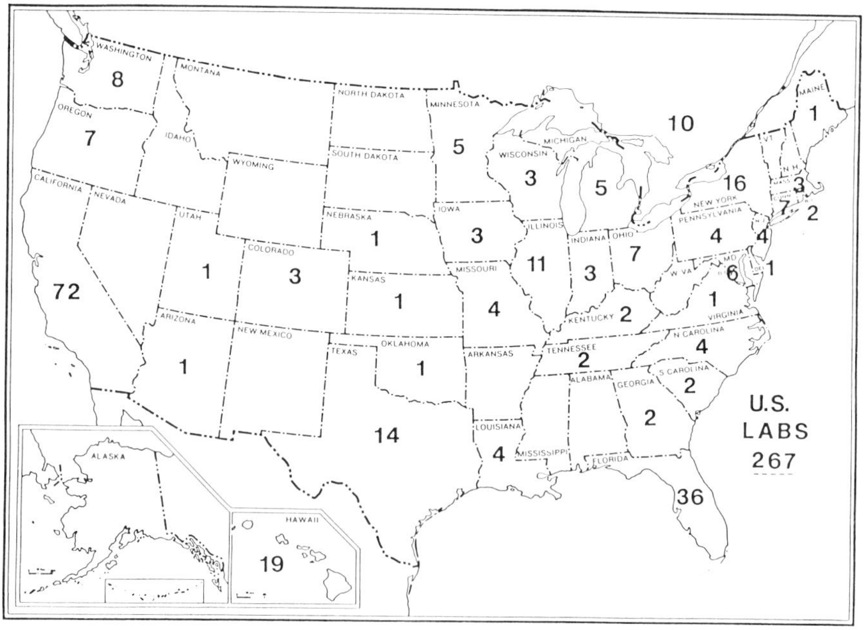

1985 Directory of Commercial T.C. Labs.

FIGURE 1. Location of commercial tissue laboratories in the U.S.

TABLE 1. Leading states in U.S. nursery production.

Rank	State	Value of state production ($1,000,000)	Total area (acres)	Total area (hectares)	Total nurseries
1.	California	395	15,000	6,070	1,900
2.	Florida	375	16,000	6,475	2,300
3.	Texas	175	7,000	2,833	700
4.	New York	127	8,800	3,561	1,500

In contrast, the top 12 laboratory producers (Table 2) account for 83% of all projected sales for 1985 (Table 3). Assuming that two-thirds of the total plants produced in these units (Table 3) are sold at an average wholesale price of $0.45 and one-third at $0.21, the present wholesale market value is $22.51 to 24.35 million for these laboratory products.

3. STRATEGIC GROUPS

The major U.S. laboratory producers are listed in Table 2. The reported production volume is highly variable because output depends on sales of microcuttings vs. Stage II vessels vs. Stage III vessels vs. Stage III bare rooted plants vs. liners of various sizes requiring varying periods of time for growing on lighted shelves or greenhouse benches.

Total units produced do not represent total production capacity. Weyerhaeuser, Oglesby and Twyford each project a capability of 20-25 million units per year with their newly expanded facilities. Therefore, the total units produced could easily double as these three firms increase production. Planned increases by George J. Ball, Hartman's Plants, Inc. (CIBA-GEIGY), Native Plants Inc., Armstrong Nurseries, Plant Reproduction Intl. and Briggs support increased growth in the total number of tissue culture plants produced to 125 million units. The difference of 10 million units between Tables 2 and 3 is due to a number of medium (0.5 million units) and small (0.25 million units and fewer) laboratories not accounted for in Table 2.

4. PRODUCTS-MARKETS

The products listed in Table 3 represent the major groups of plants being micropropagated for sale. This list does not include the number of units

TABLE 2. Major U.S. laboratory producers.

Company	State	1985 Production (million units)
Weyerhaeuser Company	FL	7 - 8
Hartman's Plants, Inc.	FL	6.2
Phyton Technologies, Inc.	TN	6
Oglesby Plant Laboratories, Inc.	FL	5
Santa Rosa Tropicals	CA	5
Twyford Plant Laboratories	CA	4 - 5
HyClone, Inc.	TX	5
Briggs Nursery, Inc.	WA	2 - 4
Ceres 2000, Inc.	FL	2.6
Plant-Tech, Inc.	FL	2.5
B & B Labs	WA	2
Native Plants, Inc.	UT	1
Knight Hollow Nursery	WI	0.5
Hartmann Lab.	MI	0.5
Oregon Bulb Farm	OR	0.5
Microplant Nurseries, Inc.	OR	0.5
Armstrong Nurseries	CA	0.075
Plant Reproductions Intl.	TX	NA
Total		50.38 - 54.38

produced for in-house recovery of specific pathogen-free nuclear stock (e.g., chrysanthemum, geranium, Begonia hiemalis, and carnation), or for breeding and new plant development. Additionally, orchids and vegetables are not included.

According to The Census of Horticultural Specialities (1), the total value of all industry segments (cut flowers, potted flowering plants, bedding plants, foliage plants) more than tripled from 1970 to 1979. It is estimated that this figure will more than triple again by 1988, placing the wholesale value of this part of the horticultural industry at approximately $4.6 billion.

Both potted flowering plants and foliage plants are likely to maintain or slightly increase their 1977 market shares, while bedding plants are expected to gain the most. Most bedding plants are produced from seed so that growth in sales of tissue culture-derived plants will be greater in both potted flowering and foliage plants. The potential for hybrid seed improvements using tissue culture techniques is great. Since flowering pot plants paralleled the total industry growth rate between 1970 and 1979, a

TABLE 3. Major U.S. laboratory products.

Crop	1985 Production (millions of units)	
Foliage plants		32.5
Syngonium	15	
Spathiphyllum	6+	
Fern	5	
Dieffenbachia	3	
Philodendron	2+	
Ficus	1	
Misc. Foliage	0.5	
Flowering plants		7.7
Gerbera (pot, landscape, cut flw)	4	
African Violets	1.5	
Roses (pot, landscape, cut flw)	0.175	
Lilies and other bulb crops	2	
Woody ornamentals and fruit crops		
Azalea	2.5	
Rhododendron	8	
Fruit crops, misc. ornamentals	10-15	20.5 - 25.5
Total		60.7 - 65.7

continuation for the 1979-to-1988 period is expected. Sales of flowering pot plants could do better than that since per capita consumption was only 0.95 in 1979, or less than one flowering pot plant per person per year. In April, 1985, the Wall Street Journal reported an average sales of $20.00 per capita, about three pot plants per year. Increased production of tissue cultured plants used for cut flower production (carnations, chrysanthemums, roses, gladioli) will be limited because of declining domestic cut flower producers.

The number of consumers in the market to buy floral items in supermarkets also has grown over the period from 1976 to 1985. Floral Index, Inc. reports that the 12.5 percent of U.S. households counted during May 1976 who buy floral items increased to 15.5 percent for May 1985. In terms of real population numbers this gain represented a 52 percent increase in customer buying during May 1985 versus May 1976. Similar large increases in sales were made to commercial interior landscape customers.

5. GLOBAL INDUSTRY

Plant production using tissue culture technology is a world-wide business. Major laboratories outside the U.S. are located in the Netherlands, Israel, Belgium, United Kingdom, Italy, France, Canada, Australia and Japan. In a review of the 33 Dutch micropropagation businesses, Pierik (4) listed a total in vitro production of 30 million units for potted plants, bulbs and plants for cut flower production. He identified roses and landscape ornamentals as the future growth segments for the industry.

The growth and development of horticultural businesses in newly industrialized countries and third world nations will further fragment tissue culture business activities. Interest in micropropagation of tropical food and oil crops has increased and some of these crops are now being micropropagated commercially.

6. MARKET STRATEGY

Four market strategies are to build, to hold, to harvest, and to divest. The years 1984 and 1985 have been a time for building capacity. The most notable new laboratories are those of Weyerhaeuser, Oglesby, HyClone, Twyford, Plant Reproductions Intl. and Geo. J. Ball. New laboratory products, foliage liners in particular, are moving from an introduction stage to a growth stage of the product life cycle. The saturation of the Syngonium 'White Butterfly' market in 1985 illustrates how quickly production can reach current grower demand with growers buying at the lowest possible prices.

Building capacity to meet market demands (a 1983 market research survey indicated 15% of market demand being met) means that major laboratory producers must plan on gaining market share. The following five key strategies are most important for gaining market share:

1. Price --- lower prices below competitive levels to take business away from competitors among price-conscious customers.
2. New Products --- introduce product modifications or significant innovations that meet customer needs better and displace existing products or expand the total market by meeting and stimulating new needs.
3. Service --- offer more rapid delivery than competitors to service-conscious customers; improve the type and timeliness of information that customers need from the service organization,

information such as items in stock, delivery promise dates, invoice and shipment data, etc.

4. Strength and Quality of Marketing --- field a larger, better-trained, higher-quality sales force targeted at customers who are not getting adequate quality or quantity of attention from competitors; build a larger or more effective distribution network.
5. Advertising and Sales Promotion --- increase advertising and sales promotion of superior product, service, or price benefits to underpenetrated or untapped customers, advertise new or improved benefits to all customers.

7. OTHER CONSIDERATIONS

The structure of the industry has evolved from the independent laboratory operation that ships Stage II microcuttings and Stage III plantlets to one that grows on plantlets as Stage IV or liner products. The reason is that laboratories must push product out the door to make room for the constantly increasing numbers of plantlets in the lab. Otherwise laboratory efficiency declines and costs rise to a noncompetitive product price. The multiplication process is difficult to slow down without changing the nutrient formulation in the vessel. Temperature modifications to slow down growth are difficult to achieve in large culture rooms. The laboratory of the future requires several culture rooms for temperature controlled production as well as means to chemically induce more control for inventory purposes without altering quality and consistently performing outputs.

The forward integration of a laboratory to include a greenhouse operation provides the flexibility to meet changing buyer habits without altering the course of events in the laboratory. The greenhouse also gives the operator a place to refine the difficult acclimation, re-establishment procedures needed to get plants back to the real world. It also broadens the product line. It gives the operator an opportunity to test new products. The greenhouse is a low cost ($0.15 - 0.29/sq. ft./week) production area adding higher values to plant products.

8. CONCLUDING REMARKS

Production of large quantities of laboratory products with frequent sales is the growth area for the micropropagation business. Long-run average cost curves will become important operating management tools. High volume and frequent shippers will dominate future growth in the industry. Small volume and infrequent shippers will fall into specialty niches.

Continued forward vertical integration of the laboratory-greenhouse operation into production of four-inch pot size plants will meet the emerging trend of end product consumer sales through mass market floral departments.

The future dominant classification of laboratory business should be the laboratory propagator/grower/shipper (see Section 2, number 3 above). This business will ship Stage II culture vessels and microcuttings worldwide to specialized propagator/grower/shippers and propagator/growers as well as domestically growing on product for liner sales.

In contrast to their present limited use, plant brokers will not only be the "finders" of new plants for cloning in the laboratories, but they will also distribute these "new plants." A global network of vertically integrated companies should develop in 1985-186.

9. SPECIAL ACKNOWLEDGEMENTS

The author wishes to thank the numerous industry associations, industry leaders and laboratory producers for sharing marketing information for this presentation.

REFERENCES

1. American Association of Nurserymen. Census of horticultural specialties, research summary of the Horticultural Research Institute.

2. Fogg CD: 1974. Planning gains in market share. J. Marketing. July, pp. 30-38.

3. Jones JB: 1985. Raising plants by tissue culture. In: P Rotheim (ed), Biotechnology in plant and animal agriculture. Business Communications Co., Inc., Stamford, Connecticut, in press.

4. Jones JB: 1985. Directory of U.S. Laboratories.

5. Pierik RLM: 1985. Vermeerdering in kweekbuizen stijgt tot 30 miljoen. Vakblad voor de Bloemisterij 26: 44-45.

6. U.S. Dept. of Agriculture. 1984. Floriculture crops 1984 Summary intentions for 1985. Crop Reporting Board.

PROPAGATION OF FRUIT, NUT, AND VEGETABLE CROPS - OVERVIEW

RICHARD H. ZIMMERMAN

U. S. Department of Agriculture, Agricultural Research Service,
Fruit Laboratory, Beltsville, MD 20705, U.S.A.

1. INTRODUCTION

In April, 1980, a conference on nursery production of fruit plants through tissue culture was held at Beltsville to stimulate interest among U.S. nurserymen in the potential of micropropagation for temperate tree and small fruit crops (48). At that time, use of tissue culture for propagation of fruit crops was well under way and expanding rapidly in Europe but was used very little in North America. In the succeeding five and one-half years, micropropagation of these crops has increased significantly throughout the world, both in terms of crops being propagated and in total number of plants produced. Micropropagation methods are now employed for a number of vegetable crops as well, while production of ornamental plants using this technology continues to expand and remains the largest segment of this young industry.

The reasons for using tissue culture propagation for these crops can be grouped as follows: (a) production of pathogen-free plants, whether for direct planting or for further increase after tissue culture; (b) rapid increase of select genotypes for use in breeding programs; (c) large-scale production of new cultivars for more rapid introduction into commercial trade; (d) propagation of plants for which conventional vegetative methods are ineffective or unprofitable; and (e) production of selected parents for F_1 hybrid seed production.

This paper will survey briefly the subject matter treated in more detail in the following five chapters, then discuss those characteristics that are common to most, if not all, plants propagated in tissue culture.

Zimmerman, R.H. et al. (eds), Tissue Culture as a Plant Production System for Horticultural Crops.
© 1986. Martinus Nijhoff Publishers, Dordrecht. Printed in The Netherlands. ISBN 90-247-3378-2.

Finally, the role and importance of evaluating tissue cultured (TC) plants in the field will be explored, emphasizing tree crops, since field data are required before tissue culture propagation can be recommended, but results to date have been limited.

2. CURRENT STATUS OF TISSUE CULTURE PROPAGATION

2.1 Small fruits and grapes

Strawberry was the first fruit crop to be micropropagated on a large scale (5, 6, 7) and considerable field experience with TC plants has been accumulated in the past 10 years. Meristem-tip culture has been used extensively to produce virus-indexed plants that are then micropropagated. Currently TC plants are used primarily as mother plants for nursery runner propagation, usually in conjunction with a scheme for production of virus-indexed plants. TC plants are usually not used for fruit production because the cost of such plants is too high and fruit size on plants directly from tissue culture tends to be smaller than normal the first year (42).

Production of brambles (raspberry, blackberry) through the use of tissue culture has increased rapidly in the past few years, particularly in the United States. One reason has been that the price differential between TC plants and conventionally propagated ones is far less than for strawberry. In addition, TC plants have performed so well in the field for those growers that have tried them that demand is increasing rapidly.

Blueberry plants produced from tissue culture are also becoming more important, especially in the U.S., but perhaps not to the same extent as the brambles. The same reasons apply here as for the brambles. Applications of tissue culture to this crop were reviewed recently (39).

Use of tissue culture for production of grapes has lagged behind other crops in this category for several reasons. Perhaps the main one is concern about the phenotypic stability and field performance of plants produced by micropropagation, particularly those of wine producing grapes. It seems likely that TC plants will be used first for table grape cultivars and possibly rootstocks. Some new cultivars have already been increased by tissue culture prior to introduction. Uses of tissue culture for grape have been reviewed (19).

Other crops such as currants and gooseberries can be micropropagated but the extent of commercial application is still limited. Kiwifruit are also micropropagated commercially, but there have apparently been some problems with the field performance of kiwi plants. Until further testing is done, it will not be clear how useful the procedure is for this crop.

2.2 Temperate tree fruits and nuts

The greatest use of TC plants for tree fruits to date has been the production of clonal rootstocks, particularly for peach, in Europe, especially Italy (24). Production of rootstocks has been less successful in North America, mainly because the cost of TC plants has been too high. Self-rooted cultivars hold great promise for the future if early field evaluation results are confirmed by longer term evaluations. At the present, very few self-rooted TC trees are produced commercially but there are indications that this situation could change rapidly within the next several years. Tissue culture applications have been reviewed recently for apple (38, 50), pear (36), cherry (16), peach (13), plum (11), stone fruits (37) and olive (35). The production of Paradox walnut rootstocks through tissue culture has been the chief commercial application for nut crops to date (10), since nut crops have been generally very difficult to establish and grow in vitro (17, 45). Micropropagation techniques have been developed for filbert (3) but commercial production is still limited.

2.3 Tropical and subtropical crops

Numerous plants in this broad category have been propagated successfully in vitro, but actual commercial production is probably still limited to only a few of them. Bananas from tissue culture have been used to establish large scale field plantations in several countries (15, 28) and it is likely that this use will expand in the future. Plants of other crops such as pineapple and some spices, e.g. cardamon, are being produced in vitro. A significant obstacle to further use of TC plants will be the cost of individual plants. In those situations in which it can be shown that TC plants are more productive, the economic justification will exist for increased usage of TC plants. Recent reviews have been published on tissue culture methods for citrus (4, 41), banana (8, 18), pineapple (32), mango (23), papaya (21, 22), coconut palm (30), date palm (44), cacao (12), coffee (40) and oil palm (31).

2.4 Vegetables

Most vegetable crops are seed propagated so that somatic embryogenesis is the only method with potential for supplanting conventional seedlings. Although somatic embryogenesis systems have been developed for some crop plants (e.g. celery, 29), many details for workable commercial systems remain to be resolved (25). Potatoes are micropropagated on a limited commercial scale now, but production of minitubers *in vitro* appears to be superior for many purposes to rooted shoots that then need to be acclimatized before planting out. Superior strains of asparagus are also being micropropagated successfully. Another use of micropropagation has been the production of parent lines needed for F_1 hybrid seed production in crops such as cucumber. Recent reviews have been published of tissue culture methods for asparagus (33), beans (1), celery (29), cole crops (47), potato (27), tomato (20), sweet potato (14) and yams (2).

3.RESPONSES IN COMMON

3.1 Rejuvenation

When most, if not all, crops are established *in vitro*, one obtains a reversion to the juvenile phase, at least with respect to leaf type. This can be seen in strawberry where a monophyllous leaf develops rather than a trifoliolate one. A similar phenomenon is observed with thornless blackberry, which produces monophyllous leaves in culture. When the plants are rooted, acclimatized and begin to grow, trifoliolate leaves are produced, followed somewhat later by pentafoliolate leaves characteristic of mature plants. With blueberry, established cultures produce shoots with thin stems and small leaves resembling the seedling growth habit. In apple, changes in leaf serration, size and shape are observed that are more typical of the seedling stage than of the mature phase from which the explants were obtained.

3.2 Branching and plant vigor

Concomitant with rejuvenation comes increased branching and greater vigor in the plants produced via tissue culture. The greater branching is probably the result of the cytokinin treatment *in vitro*. In the case of strawberry, total runner-plant production from TC mother plants of 3 cultivars was 60% to 130% greater than from standard (runner produced) mother plants (9). The increase in number was not at the expense of size either since runner plants in all size classes increased, with those in the largest size class (crown diameter >8 mm) increased 40% to 150% (9). Similar results were found when TC and standard strawberries were planted in the field and allowed to grow so as to fill a square 60 cm on a side (42). The number of crowns per m^2 was more than 50% greater for TC plants; similar increases were noted in stand (% area filled) and vigor. Differences in vigor were maintained during the second year in the field.

Thornless blackberry plants propagated in TC were also more vigorous than plants propagated by cuttings (43). By the end of the first year in the field, TC plants had 35% more laterals and produced 45% more growth than the cutting propagated plants. Leaflet length and width were 6 to 8% greater, resulting in greater leaf area per node. At the end of the second year, all plants were pruned to a maximum of 3 canes per plant, each about 2 m long. Not all control plants had 3 canes (mean = 2.2) and many of the canes were shorter than 2 m. Differences within the 6 cultivars tested were not always significant for the growth parameters measured, but the conventionally propagated plants were never more vigorous than TC plants for any measurement.

Apple trees propagated by TC also illustrate the vigor that this technique provides to the resulting plants. Tissue cultured trees rooted in January or February, 1984, then planted in the field in early June, had reached a height of 1.3 m for 'Gala' and 1.1 m for 'Triple Red Delicious (52). Trees of both cultivars budded on M.7a rootstock averaged 1.7 m in height at this time, but had been budded on 1-year-old rootstocks in 1982 and had then grown 1 year in the nursery (1983) before being planted in the orchard in the spring of 1984. TC trees of 'Gala' averaged 5 laterals and 3.7 m total shoot growth compared to 13 laterals and 7.3 m total shoot growth for trees on M.7a; comparable figures for 'Triple Red Delicious' were 1 and 1.5 m for TC trees and 12 and 6.3 m for trees budded on M.7a. Trunk cross-sectional area of the TC trees was 1/2 ('Gala') or 1/4 ('Triple Red Delicious') that of trees on M.7a. By the end of the second year in the orchard, the differences between the TC and budded trees had narrowed considerably. Differences in growth are not surprising, however, when one considers that the budded trees had a 3-year-old root system and more than 1 year's shoot growth at the time of orchard planting. In contrast, 5 months, at most, had elapsed from the time at which roots had been initiated on TC trees and they were only 40 to 60 cm tall when planted in the orchard.

3.3 Early flowering and fruiting

Despite the facts that TC plants are much more vigorous than ones propagated by conventional methods and that rejuvenation occurs, at least in leaf characteristics, early flowering and fruiting still occurs with TC plants. Strawberries have produced flowers and fruits in less than 3 months after unrooted TC shoots were placed for rooting under mist in the greenhouse (Zimmerman, unpublished data).

In the strawberry field performance trial described earlier (42), yield and number of fruit per m^2 were increased in the 60 x 60 cm plots, but size of individual fruits was reduced (Table 1). Part of the plots were retained and harvested the following year. The increases in yield by plants of TC origin were maintained but the size of individual fruits was not reduced the second year.

TABLE 1. Fruit yield and size of tissue cultured and standard runner propagated strawberries - 1979

Cultivar	Type	Yield per m^2 (kg)	No. of fruit per m^2	Fruit weight (g)
Earliglow	TC	7.8	1080	7.2
	ST	5.6	572	9.9
Redchief	TC	6.7	1047	6.5
	ST	5.6	669	8.5
Guardian	TC	7.0	947	7.5
	ST	6.1	586	10.4

All differences between TC and ST within a cultivar are significant.

Peaches and nectarines propagated by TC in the fall of 1980, and planted to the field in 1981, began fruiting in 1982 producing yields of 10-11 kg of fruit per tree (26). Although flowering has occurred as soon as the second growing season for 'Jonathan' apple (Table 2), most of the apple cultivars have started to flower at the start of the third or fourth growing season (49). The flowering response seems to be characteristic of

TABLE 2. Flowering of own-rooted tissue cultured apple cultivars field planted in spring 3 to 6 months after rooting

Cultivar	Year planted	No. of trees	Trees flowering (%) 1981	1982	1983	1984
Golden Delicious	1979	41	41	100	100	100
Spartan	1979	54	0	13	65	96
Stayman	1979	38	3	95	100	100
Nugget	1980	26	0	96	100	100
Rome Beauty	1980	52	0	98	94	100
Spuree Rome	1980	41	2	100	100	100
Gala	1981	61	-	0	64	95
Jonathan	1981	14	-	93	93	100

the cultivar; those that are precocious when budded on a rootstock are also precocious following TC propagation. Once flowering has begun, trees continue to bloom annually for all cultivars observed to date.

Fruit yields vary with cultivar and increase as tree age increases (Table 3). No tendency to biennial bearing has been noted yet. Since the first really heavy crop was picked in 1985, an indication of whether biennial bearing will be a problem should be evident in 1986. As would be expected, size of individual fruits was inversely related to the number of fruit per tree.

Standard and spur-type trees seem to respond somewhat differently following TC propagation, at least with 'Rome Beauty' (standard) and

TABLE 3. Fruit production of tissue cultured self-rooted apples-1985

Cultivar	Year planted	No. of trees	Yield per tree (kg)	Mean fruit weight (g)
Golden Delicious	1979	9	52.2	142
Ozark Gold	1979	6	31.6	149
Nugget	1980	26	38.0	152
Delicious	1980	4	16.2	210
Gala	1981	18	31.0	129
Jonathan	1981	14	2.9	170

'Spuree Rome' (spur-type). Trees of both cultivars were planted in 1980 and began flowering in 1982. Flowering intensity in both cultivars has been equal each year. Some fruit were harvested from 'Spuree Rome' in 1982 and in every year thereafter with a cumulative yield of more than 20 kg per tree through 1985. Trees of 'Rome Beauty' set no fruit until 1983 and cumulative yield through 1985 has been slightly more than 10 kg per tree. Trees of 'Rome Beauty' are taller, have greater spread, and have a larger trunk cross-sectional area. Yields for both cultivars have been lower than anticipated based on the amount of bloom, probably because they flower so late in the season that pollination has been inadequate.

4. FIELD EVALUATION

Evaluation of the performance of plants in the field is an essential part of developing any propagation system based on tissue culture. Such data are needed before the technology will be fully accepted by the producers who will be the ultimate users of the plants. This information is particularly important for tree crops because of the investment required to establish a planting and because of the length of time needed before the trees begin bearing and the years for which crop production is anticipated.

Problems with the acceptance of TC-propagated plants by growers have arisen in the past and these problems are attributable in large part to inadequate controlled testing before distribution of plants and/or inadequate quality control during the production of the TC plants.

4.1 Rootstocks

Rootstocks for peach, apple, cherry and plum have been TC-propagated on a commercial scale for several years (24) and annual production in Italy alone is now several million trees per year. Of the rootstocks propagated, those for peaches are produced in largest quantities. There have been no apparent problems with the use of TC-propagated rootstocks, but little testing was done before these trees were introduced into commercial usage. The first comparison of TC and conventionally propagated peach rootstocks has only recently been made (34). In September, 1980, 'Flavorcrest' peach was budded on GF 677 rootstock and 'Springcrest' on GF 43 for this study. 'Flavorcrest' had significantly larger trunk cross-sectional area on TC-propagated GF 677 each year from

1982 through 1985. With 'Springcrest', trees budded on cutting-propagated rootstocks had larger trunk cross-sectional area although the difference was significant only in 1982, 1983, and 1985. Prunings removed from the trees from 1981 through 1985 and fruit yield for 1983-1985 did not differ between trees on rootstocks propagated by the two methods. These results certainly are encouraging, but longer term studies with rootstocks for both peach and other fruit trees are still needed.

4.2 Cultivars

The comparison of TC-propagated scion cultivars as self-rooted trees with trees budded or grafted on rootstocks is more complex than merely comparing rootstocks propagated by two methods. The problem arises because the root system on a budded tree is 3 years old and the top nearly 2 years old by the time the tree is planted in the orchard. In contrast, TC trees can be established successfully in the field only a few months after being rooted or they may be grown in a nursery for 1 year before planting in the ochard. In any case, both root system and top are younger, hence less extensive, than for budded trees. If TC trees are maintained in the nursery longer than one season, they become too large to be transplanted conveniently because of their vigor. Furthermore, TC trees not grown in a nursery will develop differently in the field depending upon their size and growth status at the time of planting. These problems can be illustrated by the following studies with apple undertaken in several countries having quite different climatic conditions.

4.2.1. East Malling, England. Self-rooted TC trees of 4 cultivars, one of which was 'Golden Delicious', were propagated in 1978, and were planted in the orchard in April, 1979, while dormant when the plants were only 9 to 48 cm tall (48 cm for 'Golden Delicious') (46). Comparison trees, budded in 1977 on M.9 and MM 106 rootstocks propagated in 1976, were also planted in the orchard in April, 1979. As would be expected, the TC-propagated trees produced the least amount of vegetative growth the first several years; the TC trees did not produce more shoot growth than trees budded on M.9 until 1981 for 'Greensleeves' and 1982 for 'Cox', 'Golden Delicious' and 'James Grieve'. TC trees never produced as much total shoot growth as trees budded on MM 106 through 1982, the last year for which data were reported. TC trees produced very few shoots in the first two seasons following planting.

Flowering of the TC trees was delayed 1 to 2 years in comparison to trees budded on M.9 and MM 106 based on floral buds per tree and per m of shoot length. Number of fruits harvested also showed that the TC trees in this study yielded less than budded trees. Although the general pattern was the same for each cultivar, interaction between the cultivar and rootstock clearly existed.

At Beltsvile, we have also observed that small TC trees (10 to 50 cm tall) allowed to go dormant while still in pots, then planted in the nursery, did not perform well in the first, and sometimes the second, year after planting out. Less extension growth was produced and branches produced were shorter and weaker. Trees of this type after 2 years in the nursery were still smaller than trees grown one summer in the nursery following June planting of actively growing 4- to 5-month-old plants. The early growth of these trees obviously has a significant effect on the way they perform in the field.

4.2.2. Bologna, Italy. Self-rooted TC trees of the Perleberg 3 strain (spur-type) of 'Golden Delicious' were propagated in 1979, grew in the nursery in 1980, and were planted in the orchard in spring, 1981 (34). Scions were grafted on MM 106 and MM 111 rootstocks in the nursery in 1980 and the trees were moved to the orchard in spring, 1981. As measured by trunk cross-sectional area, the TC trees were significantly smaller than trees grafted on either rootstock in 1981 and 1982, did not differ in 1983, and were significantly larger in 1984 and 1985. The weight of pruned wood removed each year was significantly greater from the TC trees than from trees grafted on either rootstock. The spur-type character changed somewhat following TC propagation with TC trees having only 2/3 as many spurs as the grafted trees. Flowering of TC trees began in 1983, one year later than for the grafted trees, and did not equal flowering on rafted trees until 1984. Fruiting of TC trees also began in 1983. Although it was very light that first year, it increased rapidly so that cumulative yields for 1982-1985 were about 54 kg per tree for both TC trees and scions grafted on MM 106, but only 44 kg per tree for those grafted on MM 111. Fruit size was not reduced on TC trees and no differences in fruit quality were apparent.

4.2.3. Beltsville, MD, and Kearneysville, WV, U.S.A. Self-rooted TC trees were propagated in 1980, planted in a nursery in June, 1980, and grown there until fall, 1981 (51). Comparison trees were budded on

TABLE 4. Total length of primary shoots and total shoot length in 1982 of 'Stayman' apple planted at Beltsville and Kearneysville in spring, 1982.

Rootstock	Total-primary shoots (m)		Total shoot length (m)	
	Beltsville	Kearneysville	Beltsville	Kearneysville
Own-TC	3.9 a[1]	5.9 a	4.0 a	6.6 a
Seedling	1.6 b	4.4 b	1.6 b	4.4 b
MM 106	1.3 b	3.2 b	1.3 b	3.3 b
M.26	1.2 b	4.3 b	1.3 b	4.3 b

[1]Column means followed by different letters differ significantly, p=0.05.

seedling, MM 106, or M.26 rootstocks in 1980 and were dug in the fall of 1981. At the time of establishment in the field in the spring of 1982, the TC trees were somewhat larger and had considerably more branches than the budded trees. It was then decided to head back all trees to 1 m and to remove all lateral branches in an effort to equalize the trees. Results with the cultivar 'Stayman', one of four in the planting, are representative.

In the first year, total shoot growth was greatest on the TC trees at both sites, but all trees grew more at Kearneysville, particularly the budded ones (Table 4). The differences have persisted for several years as shown by the trunk cross-sectional area (Table 5). Data for tree height and spread in 1984 showed a similar relationship (Table 6). With more time in the field, the difference between TC and budded trees is increasing at Beltsville and diminishing at Kearneysville. We now think that these growth differences may result from the fact that the Beltsville site has been planted to orchard for many years and has a heavy nematode

TABLE 5. Stem cross-sectional area (cm^2) of 'Stayman' apples planted at Beltsville and Kearneysville in spring, 1982.

Rootstock	Beltsville			Kearneysville		
	1982	1983	1984	1982	1983	1984
Own-TC	3.1 a	8.2 a	20.1 a	5.1 a	13.1 a	28.0 a
Seedling	1.9 b	4.7 b	12.9 b	2.9 b	10.1 b	19.7 a
MM 106	1.4 b	2.8 c	6.1 c	2.5 b	8.1 bc	20.7 a
M.26	1.5 b	3.1 bc	5.7 c	2.1 b	6.8 c	15.6 a

[1]Column means followed by different letters differ significantly, p=0.05.

TABLE 6. Height and spread in row in 1984 of 'Stayman' apple planted at Beltsville and Kearneysville in spring, 1982.

	Height (m)		Spread in row (m)	
Rootstock	Beltsville	Kearneysville	Beltsville	Kearneysville
Own-TC	2.8 a[1]	3.4 a	1.8 a	1.8 a
Seedling	2.6 a	3.2 a	1.4 b	1.7 a
MM 106	2.0 b	2.6 b	1.0 c	1.5 a
M.26	1.9 b	3.1 a	1.0 c	1.9 a

[1]Column means followed by different letters differ significantly, p=0.05.

infestation through most, if not all, of the planting. This appears to affect the budded trees, particularly those on MM 106 and M.26 rootstocks, more than the TC trees. In contrast, the Kearneysville planting, on a site not previously in orchard, has very few nematodes. The reason for this apparent difference in response to nematode infestation between budded and TC trees is currently being explored.

Flowering and fruiting of these TC trees were delayed at least one year compared to the budded trees and compared to sibling TC trees that were not pruned back at time of planting. Flowering and fruiting have been greater at Kearneysville than at Beltsville for all trees, probably because of the greater size of the trees at Kearneysville.

4.2.4. Comparison of early cultural practices. From the three preceding studies it is clear that the growing conditions to which trees are subjected prior to orchard planting is an important factor in subsequent growth and field performance. To evaluate this, three groups of 'Jonathan' TC trees were compared. The first group was propagated in winter, 1982, and planted to a nursery in June, when approximately 50 cm tall, where they grew until the following spring. The second group was propagated in spring, 1982, grown in the greenhouse throughout the summer until approximately 2 m tall, then placed in cold storage until spring. The third group were propagated in summer, 1982, grown in the greenhouse until the following spring at which time they were 2 m or more in height. Trees from all 3 groups were planted in the orchard in spring, 1983. At the time of planting, they were pruned to either 35 or 90 cm above the ground and any laterals were removed.

In the first year, the length of the new terminal shoot averaged 93 cm for trees pruned at 35 cm and 70 cm for trees pruned at 90 cm. Trees from the nursery averaged 96 cm for the new terminal, those from the greenhouse plus cold averaged 71 cm, and those from the greenhouse averaged 78 cm. Total lateral shoot length was 6.7 m, 5.8 m, and 2.8 m, respectively, for the same three groups, with little difference between trees pruned at 35 cm and those pruned at 90 cm at planting time. Trunk cross-sectional area for the 3 groups was 2.0, 1.9, and 1.2 cm^2, respectively, with those trees cut back only to 90 cm tending to have somewhat larger areas. After 1 more season's growth, differences in height (2.1 to 2.3 m) and spread (1.3 to 1.5 m) were minimal but some differences in trunk cross-sectional area persisted, although these were proportionally much smaller than in 1983. First flowering occurred in some trees in 1985 but these trees were scattered among all treatments. Flowering of all trees is anticipated in 1986.

4.2.5. Other considerations. Finally, when evaluating TC self-rooted trees for their orchard potential, one must look at other factors such as response to environmental conditions (e.g. vulnerability to winter injury), susceptibility to diseases and insects, and anchorage of the root systems. With regard to the last characteristic, we have observed that TC trees that produce very vigorous top growth are susceptible to leaning if there are high winds following or accompanying heavy rains. It is apparent that a better balance between top and roots needs to be established in these trees; growing them in a nursery for 1 year before transplanting to the orchard may be necessary to achieve this balance. The influence of TC propagation on tree nutrition needs to be evaluated. The absence of a graft union might well change the uptake of various mineral elements with consequent effects on fertilizer practices. The influences of TC propagation on season of flowering, time of fruit ripening and fruit quality need examination. Clearly, studies to investigate all these factors will take a number of years, so it will probably be some time before use of self-rooted TC trees becomes wide-spread.

5. CONCLUSIONS

Propagation of fruit crops by tissue culture has become an established industry that continues to grow each year, both in number of plants and number of crops being produced. Use of the technique has depended upon the demonstration of its advantages to the ultimate user of the plants produced. Accordingly, applications have come first with those plants in which a complete cycle from propagation to harvest of a crop and replanting with new plants is shortest, e.g. strawberry. Adoption of the technique for propagation of tree crops, at least on a massive scale, awaits the results of the necessary field evaluations that are now under way for a number of crops, e.g. apple, banana and peach.

REFERENCES

1. Allavena A: 1984. Beans (Phaseolus), pp. 137-168. In: WR Sharp, DA Evans, PV Ammirato, Y Yamada (eds). Handbook of plant cell culture, Vol. 2. Crop species. Macmillan Publishing Co., New York.

2. Ammirato PV: 1984. Yams, pp. 327-354. In: PV Ammirato, DA Evans, WR Sharp, Y Yamada (eds). Handbook of plant cell culture, Vol. 3. Crop species. Macmillan Publishing Co., New York.

3. Anderson WC: 1983. Micropropagation of filberts, Corylus avellana. Comb. Proc. Intl. Plant Prop. Soc. 33:132-137.

4. Barlass M, KGM Skene: 1986. 6. Citrus (Citrus species), pp. 207-229. In: YPS Bajaj (ed). Biotechnology in agriculture and forestry 1. Trees I. Springer-Verlag, Berlin.

5. Boxus P: 1974. The production of strawberry plants by in vitro micropropagation. J. Hort. Sci. 49:209-210.

6. Boxus P, C Damiano, E Brasseur: 1984. Strawberry, pp. 453-486. In: PV Ammirato, DA Evans, WR Sharp, Y Yamada (eds). Handbook of plant cell culture, Vol. 3. Crop species. Macmillan Publishing Co., New York.

7. Boxus P, M Quoirin, JM Laine: 1977. Large scale propagation of strawberry plants from tissue culture, pp. 130-143. In: J Reinert, YPS Bajaj (eds). Applied and fundamental aspects of plant cell, tissue and organ culture. Springer-Verlag, Berlin.

8. Cronauer SS, AD Krikorian: 1986. 8. Banana (Musa species), pp. 233-252. In: YPS Bajaj (ed). Biotechnology in agriculture and forestry 1. Trees I. Springer-Verlag, Berlin.

9. Damiano C: 1980. Strawberry micropropagation, pp. 11-22. In: Proc. conf.on nursery production of fruit plants through tissue culture - applications and feasibility. U.S. Dept. of Agriculture, Sci. & Educ. Admin., Agr. Res. Results, ARR-NE-11.

10. Driver JA: 1985. Direct field rooting and acclimatization of tissue culture cuttings. In Vitro 21(3, Pt.2):57A. (Abstract).

11. Druart P, R Gruselle: 1986. 1. Plum (Prunus domestica), pp. 130-154. In: YPS Bajaj (ed). Biotechnology in agriculture and forestry 1. Trees I. Springer-Verlag, Berlin.

12. Dublin P: 1984. Cacao, pp. 541-563. In: PV Ammirato, DA Evans, WR Sharp, Y Yamada (eds). Handbook of plant cell culture, Vol. 3. Crop species. Macmillan Publishing Co., New York.

13. Hammerschlag F: 1986. 3. Peach (Prunus persica L. Batsch), pp. 170-183. In: YPS Bajaj (ed). Biotechnology in agriculture and forestry 1. Trees I. Springer-Verlag, Berlin.

14. Henderson JHM, BR Phills, BT Whatley: 1984. Sweet potato, pp. 302-326. In: WR Sharp, DA Evans, PV Ammirato, Y Yamada (eds). Handbook of plant cell culture, Vol. 2. Crop species. Macmillan Publishing Co., New York.

15. Hwang SC, CL Chen, JC Lin, HL Lin: 1984. Cultivation of banana using plantlets from meristem culture. HortScience 19:231-233.

16. Ivanicka J, A Pretova: 1986. 2. Cherry (Prunus avium L.), pp. 154-169. In: YPS Bajaj (ed). Biotechnology in agriculture and forestry 1. Trees I. Springer-Verlag, Berlin.

17. Kester DE, L Liu, CAL Fenton, DJ Durzan: 1986. 12. Almond (Prunus dulcis (Miller) D. A. Webb), pp. 414-430. In: YPS Bajaj (ed). Biotechnology in agriculture and forestry 1. Trees I. Springer-Verlag, Berlin.

18. Krikorian AD, SS Cronauer: 1984. Banana, pp. 327-348. In: WR Sharp, DA Evans, PV Ammirato, Y Yamada (eds). Handbook of plant cell culture, Vol. 2. Crop species. Macmillan Publishing Co., New York.

19. Krul WR, GH Mowbray: 1984. Grapes, pp. 396-434. In: WR Sharp, DA Evans, PV Ammirato, Y Yamada (eds). Handbook of plant cell culture, Vol. 2. Crop species. Macmillan Publishing Co., New York.

20. Kut SA, JE Bravo, DA Evans: 1984. Tomato, pp. 247-290. In: PV Ammirato, DA Evans, WR Sharp, Y Yamada (eds). Handbook of plant cell culture, Vol. 3. Crop species. Macmillan Publishing Co., New York.

21. Litz RE: 1984. Papaya, pp. 349-368. In: WR Sharp, DA Evans, PV Ammirato, Y Yamada (eds). Handbook of plant cell culture, Vol. 2. Crop species. Macmillan Publishing Co., New York.

22. Litz RE: 1986. 7. Papaya (Carica papaya L.), pp. 220-232. In: YPS Bajaj (ed). Biotechnology in agriculture and forestry 1. Trees I. Springer-Verlag, Berlin.

23. Litz RE: 1986. 10. Mango (Mangifera indica L.), pp. 267-273. In: YPS Bajaj (ed). Biotechnology in agriculture and forestry 1. Trees I. Springer-Verlag, Berlin.

24. Loreti F, S Morini: 1982. Mass propagation of fruit trees in Italy by tissue culture: present status and perspectives. Comb. Proc. Intl. Plant Prop. Soc. 32:283-291.

25. Lutz JD, JR Wong, J Rowe, DM Tricoli, RH Lawrence Jr: 1985. Somatic embryogenesis for mass cloning of crop plants, pp. 105-116. In: RR Henke, KW Hughes, MJ Constantin, A Hollaender (eds). Tissue culture in forestry and agriculture. Plenum Press, New York.

26. Martin C, M Carre, R Vernoy: 1983. La multiplication vegetative in vitro des vegetaux ligneux cultives: cas de arbres fruitiers et discussion generale. Rev. Agron. 3:303-306.

27. Miller SA, L Lipschutz: 1984. Potato, pp. 291-326. In: PV Ammirato, DA Evans, WR Sharp, Y Yamada (eds). Handbook of plant cell culture, Vol. 3. Crop species. Macmillan Publishing Co., New York.

28. Oglesby RP, JL Griffis Jr: 1986. Commercial in vitro propagation and plantation crops, pp. 251-255. In: RH Zimmerman, RJ Griesbach, FA Hammerschlag, RH Lawson (eds). Tissue culture as a plant production system for horticultural crops. Martinus Nijhoff, Dordrecht.

29. Orton TJ: 1984. Celery, pp. 243-268. In: WR Sharp, DA Evans, PV Ammirato, Y Yamada (eds). Handbook of plant cell culture, Vol. 2. Crop species. Macmillan Publishing Co., New York.

30. Pannetier C, J Buffard-Morel: 1986. 13. Coconut palm (Cocos nucifera L.), pp. 430-450. In: YPS Bajaj (ed). Biotechnology in agriculture and forestry 1. Trees I. Springer-Verlag, Berlin.

31. Paranjothy K: 1984. Oil palm, pp. 591-605. In: PV Ammirato, DA Evans, WR Sharp, Y Yamada (eds). Handbook of plant cell culture, Vol. 3. Crop species. Macmillan Publishing Co., New York.

32. Rangan TS: 1984. Pineapple, pp. 373-382. In: PV Ammirato, DA Evans, WR Sharp, Y Yamada (eds). Handbook of plant cell culture, Vol. 3. Crop species. Macmillan Publishing Co., New York.

33. Reuther G: 1984. Asparagus, pp. 211-242. In: WR Sharp, DA Evans, PV Ammirato, Y Yamada (eds). Handbook of plant cell culture, Vol. 2. Crop species. Macmillan Publishing Co., New York.

34. Rosati P, D Gaggioli: 1986. Field performance of micropropagated peach rootstocks and scion cultivars of sour cherry and apple. Acta Hort. (in press).

35. Rugini E: 1986. 9. Olive (Olea europaea L.), pp. 253-267. In: YPS Bajaj (ed). Biotechnology in agriculture and forestry 1. Trees I. Springer-Verlag, Berlin.

36. Singha S: 1986. 5. Pear (Pyrus communis), pp. 198-206. In: YPS Bajaj (ed). Biotechnology in agriculture and forestry 1. Trees I. Springer-Verlag, Berlin.

37. Skirvin RM: 1984. Stone fruits, pp. 402-452. In: PV Ammirato, DA Evans, WR Sharp, Y Yamada (eds). Handbook of plant cell culture, Vol. 3. Crop species. Macmillan Publishing Co., New York.

38. Skirvin RM, M Kouider, H. Joung, SS Korban: 1986. 4. Apple (Malus xdomestica Borkh.), pp. 183-198. In: YPS Bajaj (ed). Biotechnology in agriculture and forestry 1. Trees I. Springer-Verlag, Berlin.

39. Smagula JM, PM Lyrene: 1984. Blueberry, pp. 383-401. In: PV Ammirato, DA Evans, WR Sharp, Y Yamada (eds). Handbook of plant cell culture, Vol. 3. Crop species. Macmillan Publishing Co., New York.

40. Sondahl MR, T Nakamura, HP Medina-Filho, A Carvalho, LC Fazuoli, WM Costa: 1984. Coffee, pp. 564-590. In: PV Ammirato, DA Evans, WR Sharp, Y Yamada (eds). Handbook of plant cell culture, Vol. 3. Crop species. Macmillan Publishing Co., New York.

41. Spiegel-Roy P, A Vardi: 1984. Citrus, pp. 355-372. In: PV Ammirato, DA Evans, WR Sharp, Y Yamada (eds). Handbook of plant cell culture, Vol. 3. Crop species. Macmillan Publishing Co., New York.

42. Swartz HJ, GJ Galletta, RH Zimmerman: 1981. Field performance and phenotypic stability of tissue culture-propagated strawberries. J. Amer. Soc. Hort. Sci. 106:667-673.

43. Swartz HJ, GJ Galletta, RH Zimmerman: 1983. Field performance and phenotypic stability of tissue culture-propagated thornless blackberries. J. Amer. Soc. Hort. Sci. 108:285-290.

44. Tisserat B: 1984. Date palm, pp. 505-545. In: WR Sharp, DA Evans, PV Ammirato, Y Yamada (eds). Handbook of plant cell culture, Vol. 2. Crop species. Macmillan Publishing Co., New York.

45. Vieitez AM, ML Vieitez, E Vieitez: 1986. 11. Chestnut (Castanea spp.), pp. 393-414. In: YPS Bajaj (ed). Biotechnology in agriculture and forestry 1. Trees I. Springer-Verlag, Berlin.

46. Webster AD, VD Oehl, JE Jackson, OP Jones: 1985. The orchard establishment, growth and precocity of four micropropagated apple scion cultivars. J. Hort. Sci. 60:169-180.

47. Zee SY, BB Johnson: 1984. Cole crops, pp. 227-246. In: PV Ammirato, DA Evans, WR Sharp, Y Yamada (eds). Handbook of plant cell culture, Vol. 3. Crop species. Macmillan Publishing Co., New York.

48. Zimmerman RH (ed): 1980. Proc. conf. on nursery production of fruit crops through tissue culture - applications and feasibility. U. S. Dept. of Agriculture, Sci. and Educ. Admin., Agr. Res. Results, ARR-NE-11. 119 pp.

49. Zimmerman RH: 1983. Factors affecting in vitro propagation of apple cultivars. Acta Hort. 131:171-178.

50. Zimmerman RH: 1984. Apples, pp. 369-395. In: WR Sharp, DA Evans, PV Ammirato, Y Yamada (eds). Handbook of plant cell culture, Vol. 2. Crop species. Macmillan Publishing Co., New York.

51. Zimmerman RH, SS Miller: 1985. Field growth of tissue-cultured compared to budded apple trees. HortScience 20:584 (abstr.).

52. Zimmerman RH, BW Yae, I Fordham: 1986. Comparison of rooting methods for apple cultivars in vitro. Acta Hort. (in press).

SMALL FRUIT AND GRAPE TISSUE CULTURE FROM 1980 TO 1985: COMMERCIALIZATION OF THE TECHNIQUE

HARRY JAN SWARTZ AND JON T. LINDSTROM

Department of Horticulture, University of Maryland, College Park, MD 20742, U.S.A.

1. INTRODUCTION

Tissue culture has become an accepted, profitable and established technique for the propagation of strawberries, blueberries, and brambles. This paper reviews the status of small fruit plant production via tissue culture (TC) and discusses several recent refinements in small fruit micropropagation. Data were obtained from a questionnaire distributed in September 1985 to 34 scientists, tissue culture labs and nurseries. Additional information from the literature, lab/nursery visits and our experience is also presented. Six areas of interest are addressed: 1) the use of TC production in the nursery, 2) *in vitro* establishment, 3) proliferation in culture, 4) greenhouse establishment, 5) phenotypic stability and field performance and, 6) the occurrence of off-types.

2. THE USE OF TISSUE CULTURE PRODUCTION IN THE NURSERY

2.1. Strawberries

There are an estimated 700,000,000 plants produced from runners in North America, of which over 70% are produced in California. In California, tissue culture is beginning to be used for production of foundation plants for nurseries (Table 1).

In other areas of the United States, the use of tissue culture varies by nursery and is dependent on the economics of the situation. Strawberry plants produced from tissue culture are 4-7 times more expensive than runner stock ($0.04-0.08 per plant), and 2.5 times more expensive than screenhouse foundation stock ($0.125 per plant). Therefore, the economic incentive to use TC propagation is due to (a) the ability to rapidly propagate valued stock by TC (9), and (b) the temporarily improved performance of TC material in the nursery bed (8, 11, 16, 43, 50, 58, 65).

Zimmerman, R.H. et al. (eds), Tissue Culture as a Plant Production System for Horticultural Crops.
 ISBN 90-247-3378-2.

TABLE 1. Production figures for tissue culture and standard propagation techniques in 1984 (estimated from questionnaires).

Crop or region	Yearly production from tissue culture	standard	Ratio of standard plants to those produced by TC
Strawberry:			
California	2,000	500,000,000	1:250,000
USA-elsewhere	350,000	200,000,000	1:570
Canada	25,000	20,000,000	1:800
Italy	30,000	210,000,000	1:7,000
France and Belgium	1,200,000	200,000,000	1:170
Raspberry and Blackberry:			
US West Coast	400,000	2,500,000	1:6
Midwest	0	1,100,000	-
Southeast	0	250,000	-
Northeast	775,000	4,740,000	1:6
Canada	12,000	1,200,000	1:100
Italy	20,000	50,000	1:3
Blueberry: (highbush and rabbiteye only)			
US West Coast	49,000	-	-
Midwest	224,000	2,335,000	1:10
Southeast	150,000	2,500,000	1:17
Northeast	0	2,800,000	-
Grape:			
US	30,000	15,000,000[1]	1:500

[1] Average of the last three years (82-84); this figure was much lower in 1984.

Of the 350,000 strawberry plants produced in tissue culture in the United States, the majority are produced in a few labs that use hundred thousand quantities of TC propagated plants (as their French counterparts do) to replace screenhouse production of foundation stock. A few plants are sold directly to fruit growers. These nurseries take advantage of the improved nursery bed performance of TC propagated strawberries.

The remaining US production is sold or used in 10- to 30-thousand lots. These nurseries use TC because of its rapid proliferation--i.e., it "gets a nursery back into production"--of a valued cultivar. Many nurseries also use TC-produced plants of shy runnering cultivars, especially the everbearers. Tissue culture-produced everbearers not only runner more, but they also temporarily produce fewer flower trusses, which inhibit runnering and are expensive to remove in the nursery bed (58).

2.2. Raspberries and blackberries

Bramble plants cannot be as rapidly field propagated as strawberries unless single node cuttings are used (71). For plant certification or registration, field production should be no longer than 3 years from protective isolation or the screenhouse (26). For a one million unit per year nursery, sufficient field production within this period frequently requires over 5000 foundation plants per cultivar or at least 40,000 to 50,000 plants per year per nursery. This requires at least a 1 acre area screenhouse costing between $100,000 and $200,000. In addition, TC production of bramble plants is relatively problem-free. Wholesale costs compare favorably with field production; tissue culture plant prices average $0.45 whereas field produced plants average $0.35 each. Performance of tissue culture plants is equal to standard plants under good management and superior under environmental stresses (47, 66). Off-type plant production is not a problem (66).

Thus, TC production of bramble plants is widespread (Table 1). The markets for certified TC plants are not saturated nationwide, however; two nurseries will sell, at a premium, TC propagated plants or TC plants "field hardened" for one growing season. Like strawberry, tissue culture has proven extremely useful for increasing unique plants, i.e., virus-indexed stock or new cultivars.

2.3. Blueberries

Blueberries propagate readily from cuttings and tissue culture (1, 5, 14, 31, 40, 41, 57, 70). The cost of roughly similar sized plants from each propagation method is equivalent (about $1.20 for a 2-year plant); however, much of the cost is not in propagation, but in raising the stock to size. TC-produced "liners" are available for around $0.75. TC plants root faster and at a higher percentage, thus are ideal material to use for cuttings (41). Two respondents predicted tissue culture will commonly be used to produce one-year plants that will then be used to make cuttings. Several nurseries use TC-plants for this purpose (Table 1). In addition, TC will continue to be used to produce new cultivars. Blueberry viruses can spread rapidly, but unlike viruses of brambles, infected blueberries react with characteristic symptoms. Thus the chance of latent virus occurrence is reduced and certification is not yet so costly (26).

2.4. Grapes

Tissue culture has not gained wide acceptance in grape nursery stock production for several reasons: 1) the use of scions and rootstocks (i.e., graftage), 2) the ease of propagation of hardwood cuttings, 3) problems with explant establishment *ex vitro* (37), and 4) changes in TC plant performance (25, 29, 35, 56, 62). Thus, of the 15,000,000 grape plants produced in the United States (90% from California), only 30,000 are produced in tissue culture (Table 1). All of these are from the Eastern United States and are mainly new table grape releases. Scientists are using TC to produce plants from embryos because this may "rejuvenate" cultivars (29, 35, 56). After considerable experience with rejuvenated clones, perhaps continuous embryogenic processes will be used more commonly to inexpensively and mechanically produce plants (34, 35).

2.5. Other crops

The methods for TC production of currants and gooseberries have recently become available (68, 69). Again, a portion of the nursery production of these plants will be through tissue culture since viruses are found in many cultivars and can effectively spread in nursery plantings (26). In addition, TC-derived plants may be easier to propagate by standard techniques. Unfortunately, growth habit changes might prevent TC-propagated gooseberry plants from being readily used in fruit production (69).

There are several tens of thousands of *Actinidia* plants, both *A. chinensis* (= kiwi) and *A. arguta* (= native), being TC propagated. Both species readily propagate from cuttings.

3. ESTABLISHMENT IN VITRO

Small fruit plant establishment in culture can occasionally be unsuccessful; more frequently, it is successful but a bit inefficient for some cultivars. All commodities are normally established on the same basal media used for proliferation. Raspberries, *Vaccinium* species, strawberries, and *Ribes* seem to be more sensitive to higher nutrient levels and/or osmotic strength (2, 3, 9, 14, 17, 22, 33, 40, 57, 59, 66, 68, 69, 70). Blackberries and grapes can be established on higher concentrations of major salts for example, Murashige and Skoog (MS) major and minor salts (6, 10, 12, 13, 24, 28, 34, 35, 38, 45, 51, 52, 54, 55, 64, 65).

Benzyladenine (BA) (4.4-8.8 μM, 1-2 mg/1) and indolebutryric acid (IBA) or naphthaleneacetic acid (NAA) (0.05-1.3μM, 0.01-0.25 mg/1) are used in the above-mentioned establishment media, except for strawberries (9, 65). In strawberries, where explant size (weight) is frequently smaller than for other crops, the addition of cytokinins often leads to very rapid proliferation (33, 59). Though numbers of explants are increased quickly, this method is not used because of concerns with production of off-type plants (42). Greenhouse-grown plants from rapidly proliferating stage I strawberry explants did not have any altered disease reaction when compared to non-proliferating TC-produced plants (59). The addition of 80 mg/1 adenine sulfate improved the establishment percentage from 3% to 23% of Fragaria vesca 'Albo-marginata', but only when cytokinins were not added (42). Adenine sulfate did not cause proliferation in vitro, but is an organic amendment to other establishment media (30).

The condition of the source plant for in vitro establishment is very critical. In strawberry and Rubus, virus-infected material seems to be more difficult to establish, especially as small explants. Asexual vegetative primordia are the explant of choice for small fruit crops. The choice of apical primordia versus lateral bud is species, cultivar and lab dependent. For example, 'Royalty' purple raspberry is often cultured from lateral buds while other raspberries are started from shoot tips. Black raspberries and blackberries are commonly started from lateral vegetative buds. As flowers are initiated in the summer or fall, lateral buds are best taken in the spring from current season primocanes. Other considerations dictate the choice of spring as the best time for explanting. First, plants have had their chilling requirement satisfied and moderately vigorous growth establishes with greater frequency. Second, vitrification in some crops occurs more during establishment in the summer. Finally, contamination tends to worsen as the season progresses.

Strawberry establishment is relatively successful throughout the year, although contamination is frequently a problem in late summer to winter. Perhaps the constant runnering and rooting processes, as contrasted with the annual root-growth patterns in other small fruit crops, is responsible for this establishment insensitivity to the environment.

Source plants should be greenhouse grown or, as used commonly in grapes (13, 28, 34), cuttings can be forced in water in the greenhouse after sufficient 4°C chilling (4 weeks for strawberry, 6-8 weeks for other crops). The explants can be taken from any subsequent vegetative lateral branch growth or new Rubus primocanes.

Contamination in culture is frequently the cause of major episodic losses in establishment. Sodium hypochlorite remains the most common surface sterilant (0.5% for 20 min). Incyte and calcium hypochlorite (44) have been tried, but are not preferred over sodium hypochlorite. Dr. B. Reisch, (NY Agr. Expt. Sta.) uses rifampicin (53) as a prophylactic antibiotic in grape cultures.

Meristem-tip cultures and heat treatments in vitro have been used to eliminate small fruit plant viruses (7, 30, 48). Unfortunately, all viruses are not sensitive to both treatments (26). Virazole, an anti-virus compound, has been used successfully to delay virus symptoms, but it did not eliminate the virus without the use of embryogenesis or meristem-tip culture (48, 64). Further research is needed in this area.

On occasion, establishing plants will turn brown, produce tannins or vitrify, i.e., become translucent. Necrosis often follows. Several solutions to these problems were obtained from the literature and on returned questionnaires (69). Suggestions include reducing nitrate levels or increasing calcium salts. Strawberry meristem-tips are particularly sensitive to high light intensity (> 30 $\mu mol/m^2/sec$) and will readily produce tannins if not kept at a moderate light level. Affected explants can be moved on the same medium or recultured if necessary. An ascorbate or 1-day liquid pre-culture soak (10) has also been used to reduce tannin or phenolic compound problems. A small amount of NAA (0.1 μm) helps prevent browning (69), while reducing the BA level (< 2.2μM, 0.5 mg/1) often reduces the incidence or severity of vitrification. In one lab, 0.6mM (100 mg/1) phloroglucinol is being used prophylactically to prevent vitrification of strawberry explants; it will not help once vitrification has occurred. Some labs report Sigma and Difco Bacto-agar results in less vitrification than other agars. More commonly, explants are recultured frequently, within 14 days if liquid media are used. Although these remedies can help (but are not guaranteed), several labs report the condition of the explant and tissues explanted has much to do with the chance of success of culture initiation. Vigorous growth is preferred.

4. PROLIFERATION

Although no effort was made to survey the industry, it became obvious that several media are being employed successfully for small fruit proliferation *in vitro*. Major and minor salts of common woody plant-adapted media are usually satisfactory. Strawberries are commonly proliferated in MS or Boxus media with or without 0.21M (16.5 g/l) ammonium nitrate added (9, 42, 43, 54, 56). Raspberries are frequently proliferated in Anderson's medium (3, 4, 17, 55), while blackberries are successfully cultured in either MS or Anderson's media (3, 6, 45, 49, 66). *Vaccinium* species are cultured in Woody Plant Medium (WPM) or various rhododendron media (2, 14, 22, 31, 39, 40, 57). Wolfe et al. (70) compared 6 media for proliferation of blueberry: WPM, modified MS (inc. KH_2PO_4), Zimmerman Z-3 and 1/2 strength MS media produced greater shoot growth and rooting than other media. Grapes are proliferated on either Nitsch, MS or Chee's media (higher nitrates, lower Mn and no KI or $CaCl_2$) (7, 12, 13, 28, 34, 38, 49, 51, 54, 61, 62). One lab uses one part MS mixed with one part Chee medium. *Ribes* grows better on MS with nitrate levels reduced 90% (49, 68, 69).

Proliferation on the above media is generally accomplished with the addition of generally 2.2-8.8μM (0.5-2 mg/l) BA and around 0.5μM (0.1 mg/l) IBA (or NAA) (12, 28, 43). The exception is *Vaccinium* species where 25-98μM (5-20 mg/l) 2-isopentenyl adenine (n^6-γ,γ-dimethylallylamino purine) is used in place of BA (14, 31, 40, 57, 70). Media for *Vaccinium* cultures are also adjusted to a lower pH (4.5-5.2).

Most proliferation occurs under 16 hr/day fluorescent lighting. Chee and Pool used 10 hr/day light, but only a few of the grape cultivars grew better (12, 13). The proliferation doubling time of gooseberries was reduced with increased irradiation; however, there was no additional effect above 18 W/m^2 (69).

Proliferation of small fruit plants involves basal branching, although a portion of the shoots obtained at moderate to high cytokinin levels are probably adventitiously derived. Various labs have used chimeral small fruit plants to study the origin of *in vitro* shoots or displacement of histogenic layers. In strawberry, addition of 1.3-13.3μM (0.3-3 mg/l) BA caused proliferation and a high rate (> 86%) of histogen reorganization (42). Serial sections of these explants indicated a similar rate of adventitious bud formation. In grape, displacement of the epidermal layer occurred in 52% of the tissue culture propagated 'Meunier' plants (62). In Rubus, a 'Thornless Evergreen' blackberry (L1-thornless) chimera regularly produces three phenotypes: one thornless and normal, one (1%) thorny and one (50%) thornless and dwarf (46). Rosati et al. (55) observed 0.07% genetically thornless plants when chimeral 'Thornless Loganberry' (L1-thornless) was proliferated. Thorny sectors or patches were found on both standard and in vitro-propagated plants. Besides generating completely thornless or altered plants by reorganizing histogens, this work demonstrates the effect of relatively high levels of cytokinins on small fruit explants. Some adventitious shoot formation and histogen displacement can be expected during proliferation.

In all of the above cases, standard plant phenotypes are relatively stable in the field, except when shoots arise adventitiously, e.g., from Rubus roots (55). If thornless, hairless or chlorotic layers can be reorganized and be expressed as a new phenotype ex vitro, it should not be surprising that slight variation in subclone or plant performance could be related to rearrangement of chimeras with only a few genes or a single gene mutation. Thus, germplasm preservation in vitro should not include proliferation and cytokinin level should be reduced as much as possible. Adenine sulfate (30, 42, 64) is suggested where cytokinins improve, or are necessary for, survival of cultures. The addition of 3.8μM (1 mg/l) abscisic acid (perhaps after establishment) will also help prevent proliferation (60).

5. ROOTING AND EX VITRO ESTABLISHMENT

Small fruit crops generally root well in vitro, or out of Stage II culture in high humidity chambers or in a fogged room (36). Intermittent mist is frequently less ideal for establishment. However, losses up to 60-70% have been reported for each crop, usually during the summer or if Stage II shoots are subdivided or are too small (36). Grapes are notoriously difficult to establish in large quantities, yet researchers have no trouble obtaining > 90% success in grape establishment in high humidity chambers inside growth chambers (37). Temperature, light, water stress and sanitation differences between the two environments, as well as the reduction of individual attention, are undoubtedly related to these differences.

Light levels are reduced from ambient during the early stages of *ex vitro* establishment. Donnelly et al. (18, 19, 20) studied the effect of various levels of light *in vitro* and immediately after, during establishment. When compared to those raised at lower light levels, raspberry leaves grown in 80-120 $\mu mol/m^2/sec$ light developed the highest photosynthetic rate, chlorophyll content and dry weight accumulation. Leaves developed *in vitro* had low rates of photosynthesis (1-2 mg/dm/hr), whereas those initiated in culture, but developed in the acclimation chamber, had progressively higher photosynthetic rates and normal anatomy. In strawberry (23) and *Rubus* (19, 20), this progression towards normal morphology included an increase in number or development of the palisade cell layers and a decrease in intercellular air space.

In several field experiments, saturating light intensity for small fruit crops was 25% full sun. Thus, no benefit can be gained by keeping acclimating plants > 350 $\mu mol/m^2/sec$. In addition, respiration increases under higher levels of light (18). High light intensities can not only cause trouble with photo-oxidation and excess heat loads, but, when combined with high temperatures, can lead to respiratory exhaustion and necrosis before the development of transitional leaves. High temperatures can also preferentially stimulate callus formation to the detriment of rooting.

Given the low rates of photosynthesis and CO_2 compensation point *in vitro* and in early Stage IV raspberries, it is not surprising that carbon dioxide enrichment (1200 ppm) in grape establishment does not stimulate early grape plant growth *ex vitro* (37). After 30 days, however, carbon dioxide enrichment resulted in a 6x increase in root dry weight accumulation and a 4x increase in plant dry weight.

Labs usually attempt to keep temperatures during establishment less than 80°F. In the summer, shading is a necessity. Frequently, white or translucent polyethylene, shade cloth or microfoam (27) are used for this purpose. Temperatures less than 60°F have caused rot, particularly *Botrytis*. Captan, benomyl and Ornalin have been used as fungicides (36). In Maryland, the use of Ornalin, a sterol-inhibiting fungicide, has resulted in petiole shortening, reduced growth and mid-summer flowering of TC-produced strawberries.

If rooting is accomplished in culture, low levels of auxins are used and cytokinins are generally eliminated (2, 3, 9, 10, 17, 31, 38, 40, 49, 63,

69). If MS medium is used, salt levels (primarily nitrates) are frequently lowered to 1/2 strength (30, 48, 52, 62, 64). High auxin levels should be avoided as callus formation can disrupt the root-shoot connections. In *Prunus*, lower auxin levels resulted in more viable roots (21). Charcoal (3) or 50 mg/l ascorbate has been used to improve rooting or root growth, but neither is used routinely.

6. PHENOTYPIC STABILITY AND FIELD BEHAVIOR

Tissue culture propagated small fruit plants frequently exhibit traits reminiscent of juvenile seedlings. Yet, they are not truly juvenile as was demonstrated in everbearing strawberries (58). The effects of "rejuvenating" propagation stock by reducing the size of the plant by radical or severe pruning have been demonstrated (5). Tissue culture propagation seems to be an extension of this process.

Individual crops manifest their juvenile-like characteristics differently (Table 2). Some form of increased lateral bud behavior (branching) is common to all small fruit derived from tissue culture, with the possible exception of *Rubus*. Red raspberries from tissue culture made more (adventitious) root suckers, but each cane does not branch more. Branching is neither beneficial nor common in red raspberries. The increased branching of tissue culture propagated thornless blackberries may have been due to their increased growth (66). Other juvenile characters include thorniness, which was preferentially increased in basal sections of TC-propagated gooseberry plants (69), and variation in disease susceptibility (59). TC-propagated strawberries of a susceptible cultivar were temporarily as susceptible as seeding material to red stele and *Verticillium* wilt. Runner-propagated stock was more resistant (59). In grape and blueberry, juvenile and tissue culture stock root earlier and at a greater percentage (41). Finally *Vaccinium* species propagated by tissue culture have a juvenile habit, with increased branching or rhizomatous growth (31) and a narrower stem, shorter internodes and smaller leaves (1, 41).

Some tissue culture-related changes in plant behavior (increased vigor, branching, ease of rooting, delayed or reduced flowering) increase the efficiency of plant propagation, so tissue culture plants are more valuable to the nursery trade. The same traits can lead to reduced performance in the fruit grower's field however. TC-produced plants of strawberries

TABLE 2. The temporary effects of tissue culture propagation on the phenotypic stability and field performance of small fruits.

Strawberry: (general duration of effects: 4 months to 1 year)

1. Approximately 50% more runners from tissue culture propagated plants of over 30 cultivars tested; however the amount of response is cultivar dependent. The response is temporary (3-4 months) but may be reactivated (US) or perhaps deactivated (France) with cold treatment on some cultivars. The response is observed only on the mother plant and its first runners and is not sensitive to *in vitro* cytokinin levels (8, 11, 16, 43, 50, 56, 58, 65).
2. Increased lateral bud activity (more crowns and flower trusses) and a greater productivity in matted rows with tissue culture propagated plants. Increased competition and yield are correlated with reduced fruit size. Everbearing cultivars directly out of culture will fruit, but at a reduced rate for several months (8, 11, 50, 56, 58, 65).
3. Phenotype changes have been observed, plants are smaller, petiole length is different (mostly longer) and there can be a change in bloom date (50, 56, 65).
4. Tissue culture propagated plants of susceptible cultivars are temporarily more susceptible to red stele and *Verticillium* wilt. Juvenile seedlings also as susceptible as tissue culture plants. Plants given many subcultures *in vitro* were once found to be more susceptible to mildew (50, 59).
5. Stage IV plants are sensitive to agrichemicals. Ornalin, a sterol-inhibiting fungicide related to paclobutrazol, stimulated flower bud formation in June-bearing strawberry. Plictran and Buprimate (Nimrod) caused necrosis of Stage IV leaves.

Raspberry, Blackberry: (general duration of effects: 1 year).

1. Greater vigor under field stress observed as a reduced coefficient of variation (amount of variability) for growth, yield and other variables (66).
2. Increased sensitivity to daylength and herbicides, especially Surflan (47).
3. Increased suckering of red raspberries, increased branching of thornless blackberries (possibly related to increased vigor) (66).

Blueberry, Cranberry, Lingonberry: (general duration of effects: 1 year)

1. Earlier and a larger percent rooting of tissue culture propagated plant hardwood cuttings (41).
2. Increased basal branching or rhizomatous growth (31, 41).
3. Smaller leaves, shorter internodes and a possible 1-year delay in fruiting. These traits are similar to juvenile seedlings (1, 41).

Grape: (general duration of effects: 1 year or longer)

1. Increased rooting from cuttings from tissue culture propagated plants.
2. Increased vigor, lateral bud break and cane pigmentation (25, 56).
3. Flower abortion of basal nodes associated with a change in leaf sinus morphology. Cuttings taken from the apical (typic) portion of cane will develop into completely normal plants (25).
4. A change in fruit phenotype on some cultivars (29, 56).

(wider mother plant spacing), blueberries (different pruning), certain grape cultivars (a change to cane pruning) and gooseberries (development of a non-"leg" training system) all would require changes in field management.

Tissue culture plants could actually reduce costs and management in lingonberry (rhizomatous habit) and Rubus (increased uniformity) (31). Most tissue culture-propagated cultivars of strawberries and thornless blackberries are more productive; however, this increased yield and its greater inter- or intra-plant competition results in a slight reduction in (primary) fruit size (8, 11, 16, 50, 65, 66). An increase in truss or fruit fasciation has been reported in strawberry and raspberry. Several plant phenotype and yield performance traits measured in these studies were almost always less variable in TC-propagated plants (65, 66). In grape, a change in plant phenotype from pubescent to hairless is accompanied by a tendency toward flower abortion (29). This results in reduced yields if the affected basal portion of the cane is all that remains after pruning (spur pruning). If plants are cane pruned, yields are not reduced. Hardwood cuttings taken from the apical (normal) phenotype, produce plants typical of that cultivar.

Grape plants from tissue culture can have a general change in phenotype (29, 56). In one case, the tissue-cultured plants more closely resembled the original cultivar description than the mother plant (56). This has caused concern regarding the enologic character of TC-produced plants, especially those produced by embryogenesis. It also provides an opportunity to improve the cultivar. Embryogenic techniques are being employed to improve 'Chardonnay' in France, because of its poor performance.

Care needs to be taken in the area of pathogen or pesticide interaction with tissue culture-propagated plants. At this point, the increase in susceptibility of TC-propagated strawberry to root rots (59) and mildew (50) or of TC-propagated raspberry and strawberry to pesticides (47), should not limit the use of tissue culture. Most TC material is being planted in isolation and there are alternative herbicides or pesticides. An increase in TC raspberry plant rosetting in Stage IV might be eliminated by increased light intensity or photoperiod or other treatments (32).

The cause of this transient variation in plant performance is unknown. Many researchers and labs consider it to be an effect of the cytokinin added in vitro. Similar effects can be obtained with exogenous sprays of cytokinins or gibberellins (15). In strawberry, added cytokinin in stage

II does not result in increased runnering the field (43). In our lab, we are trying to determine the effect of abscisic acid (ABA) in this response. During strawberry in vitro establishment, the explant is excised from its sources of ABA. Since ABA is a cytokinin antagonist (60), a depletion of ABA could have a growth-promoting effect. ABA reduces in vitro branching.

Dr. Owen P. Jones is presently investigating the effect of daylength on in vitro and ex vitro plant performance. Finally, CO_2 levels have had morphogenic effects on grape in vitro (25). After 1 hr in the light, the ambient CO_2 level fell to 100 ppm in parafilm-sealed cultures of grape. Grape plant morphology in these tubes was juvenile. If 200 ppm CO_2 were added, or if the parafilm was not used (ambient CO_2), in vitro grape morphology was more typical. The effects of these treatments have not yet been determined on ex vitro plants.

7. OFF-TYPE PLANT OCCURRENCES

The occasional occurrence of long-lasting, non-chimeral, off-types from TC-propagated plants is not limiting to tissue culture production (Table 3). Almost all are relatively rare and easily rogued on sight in Stages II, III, and IV. Problems occur only when the symptoms appear after the plant is in the field, e.g., June yellows or sterility (56, 65).

The off-types that have been noted in TC plants are readily observable in bramble, strawberry and blueberry seedling populations and are perhaps expressed in juvenile plants or part of the repressed or recessive genome of these crops. Several labs suggest reduction of time in culture (no. of subcultures) will reduce off-type occurrence. As certain cultivars seem to be more unstable in terms of a particular off-type, e.g., white streak in 'Earliglow' strawberry, the use of tissue culture for these cultivars should be avoided. Some off-types do not cause economic losses. Strawberry rows with either a few white streak or sectoral chlorotic plants tend to be as productive as normal tissue culture or runner produced plants (65). In addition, individual plants of many off-types do not survive transplanting.

Care should be exercised when proliferating chimeral plants as 50-80% off-type production can be observed (43, 46, 62). Histogenically pure sports do not seem to revert. A 'Tophat' blueberry sport (different vigor) and its original form both propagated true-to-type. Plants produced

TABLE 3. Episodes of off type plant occurrence.

Crop	Off Type	Frequency (%)	Possible Solutions
Strawberry:			
	Chlorotic[1] sectors	(0.05%), (0.1%), (0.05%), (0.5%), (4%), (7%), (0.03%), (0.8%)	limit subcultures
	White streak	(8.4%), (1%)	use safe cultivars; 'Earliglow' is unstable
	Dwarfs	(1.2%)	
	Multi apexing	(2-5%)	lower the BA *in vitro*
	Small fruit size or sterility	(0.2%)	limit subcultures
Raspberry and Blackberry:			
	Chlorotic sectors	(0.5%), (0.05%), (0.0.5%)	rogue
	Rudimentary leaves	(0.01%)	*in vitro* and in Stage IV
	Mosaic leaves 'Darrow'	(15%)	rogue
	Breaking periclinal chimeras	(47%) (0.1%), (50%)	use low rates of cytokinins
Blueberry:			
	Albinos	(< 0.1%)	rogue
	Dwarfs	(< 0.1%)	rogue
	Polyploids	(< 0.1%)	rogue
Grapes:			
	Breaking periclinal chimeras	(52.2%)	use low rates of cytokinins

[1] Each parenthesis represents an episode of off-type occurrence and there are two types of chlorotic leaves, June "yellows" and the sectoral chlorotic chimera.

adventitiously from the cotyledons of blackberry seeds were also over 99% true-to-type, i.e. identical within seed (24).

8. CONCLUSIONS

Tissue culture will continue to be a valuable tool to aid nursery production of small fruit plants, especially when the production is within a pest-certification program. Expansion will probably continue until the needs for foundation, nuclear or propagation stock are met. Further expansion will depend on new mechanically adapted proliferation techniques. Such techniques are being developed (34, 35). Other research areas of future interest will include: the cause of the juvenile-like behavior of TC-propagated plants (and methods to manipulate field performance), causes and solutions of off-type plant occurrence, improved media and growth regulators to reduce the lag phase to onset of proliferation (28, 67), the use of antibiotics and antiviral compounds to reduce or eliminate *in vitro* pathogens or virus infections, *ex vitro* establishment physiology, and the use of tissue culture for genetic manipulation to aid small fruit breeding programs. Tissue culture will also allow fruit breeders to select cultivars that are compatible with the cultural systems used by fruit growers. Breeders will be able to place less emphasis on the ease of propagation of a selection. Often a cultivar that propagates well becomes overcrowded in the fruit grower's field.

ACKNOWLEDGEMENTS

The authors would like to express their gratitude to those who contributed information presented in this paper:
W.C. Anderson, Washington State University; M. Beech, East Malling Research Station; M. Chu, University of Illinois; A. Dale, Horticultural Research Institute of Ontario; D. Donnelly, Macdonald College of McGill University; J. Frett, Texas A & M University; R.C. Funt, Ohio State University; H.G. Hughes, Colorado State University; J. Janick, Purdue University; C. Klein-Cox, Congdon and Weller Wholesale Nursery; W. Krul, University of Rhode Island; A. Kyte, Cedar Valley Nursery; J. Luby, University of Minnesota; P. Lyrene, University of Florida; S. McCulloch, Briggs Nursery; J.N. Moore, University of Arkansas; J.C. Navatel, CTIFL, France; N. Nickerson, Agriculture Canada; T. Nourse, Nourse Farms; E.B. Poling, North Carolina State University; D. Ramming, USDA-ARS, Western Region;

B. Reisch, New York State Agricultural Experiment Station; P. Rosati, CNR, Italy; A. Rutledge, University of Tennessee; J. Sanford, New York State Agricultural Experiment Station; J.G. Strang, University of Kentucky; T.M. Sjulin, Washington State University.

REFERENCES

1. Aalders LE, IV Hall, LP Jackson: 1972. Growth of low bush blueberry seedlings as compared with clonal cuttings. Can. J. Plant Sci. 52:655-656.

2. Anderson WC: 1975. Propagation of rhododendrons by tissue culture: Part 1. Development of a culture medium for multiplication of shoots. Comb. Proc. Intl. Plant Prop. Soc. 25:129-135.

3. Anderson WC: 1980. Tissue culture propagation of red raspberries, Pp. 27-34. In: Proc. conf. on nursery prod. of fruit plants through tissue culture--applications and feasibility. USDA-SEA, Agr. Res. Results, ARR-NE-11.

4. Anderson WC: 1980. Tissue culture propagation of red and black raspberries, R. idaeus and R. occidentalis. Acta Hort. 112:13-20.

5. Austin ME, WT Brightwell. 1984. Rejuvenating rabbiteye blueberry 'Tifblue' and 'Woodard'. J. Amer. Soc. Hort. Sci. 109:74-77.

6. Babic V, M Neskovic: 1984. Propagation of three blackberry cultivars from small apical buds in vitro. J. Hort. Sci. 59:183-185.

7. Barlass M, KGM Skene, RC Woodham, LR Krake: 1982. Regeneration of virus-free grapevines using in vitro apical culture. Ann. Appl. Biol. 101:291-295.

8. Bedard R, A Garneau: 1985. Field evaluation of tissue-cultured strawberry plants. HortScience 20:522 (Abstract).

9. Boxus P: 1974. The production of strawberry plants by in vitro micropropagation. J. Hort. Sci. 49:209-210.

10. Broome OC, RH Zimmerman: 1978. In vitro propagation of blackberry. HortScience 13:151-153.

11. Cameron JS, JF Hancock, TM Nourse: 1985. The field performance of strawberry nursery stock produced originally from runners or micropropagation. Adv. Strawberry Prod. 4:56-58.

12. Chee R, RM Pool: 1982. The effects of growth substances and photoperiod on the development of shoot apices of Vitis cultured in vitro. Scientia Hort. 16:17-27.

13. Chee R, RM Pool: 1983. In vitro vegetative propagation of Vitis: Application of previously defined culture conditions to a selection of genotypes. Vitis 22:363-374.

14. Cohen D: 1980. Application of micropropagation methods for blueberries and tamarillos. Comb. Proc. Intl. Plant Prop. Soc. 30:144-146.

15. Dennis FG, HO Bennett: 1969. Effect of GA and deflowering upon runnering and inflorescence development in an everbearing strawberry. J. Amer. Soc. Hort. Sci. 94:534-537.

16. Damiano C: 1980. Strawberry micropropagation, Pp. 11-22. In: Proc. conf. on nursery production of fruit plants through tissue culture--applications and feasibility. USDA-SEA, Agr. Res. Results, ARR-NE-11.

17. Donnelly DJ, R Stace-Smith, FC Mellor: 1980. In vitro culture of three Rubus species. Acta Hort. 112:69-75.

18. Donnelly DJ, WE Vidaver: 1984. Pigment content and gas exchange of red raspberry in vitro and ex vitro. J. Amer. Soc. Hort. Sci. 109:177-181.

19. Donnelly DJ, WE Vidaver: 1984. Leaf anatomy of red raspberry transferred from culture to soil. J. Amer. Soc. Hort. Sci. 109:172-176.

20. Donnelly DJ, WE Vidaver, KY Lee: 1985. The anatomy of tissue cultured red raspberry prior to and after transfer to soil. Plant Cell Tissue Organ Culture 4:42-50.

21. Dunstan DI: 1981. Transplantation and post-transplantation of micropropagated tree-fruit rootstocks. Comb. Proc. Intl. Plant Prop. Soc. 31:39-45.

22. Economou A, PE Read: 1984. In vitro shoot proliferation of Minnesota deciduous azaleas. HortScience 19:60-61.

23. Fabbri A, EG Sutter, SK Dunston: 1984. Morphological adaptation of strawberry plants grown in vitro to growth chamber conditions. HortScience 19:554 (Abstract).

24. Fiola JA, HJ Swartz: 1985. Somatic embryogenesis, organogenesis and proliferation in vitro from Rubus ovules. Acta Hort., in press.

25. Fournioiux JC: 1985. The influence of CO_2 concentration on the morphogenesis of the grapevine cultured in vitro. In: Proc. Moet-Hennessy Conference, Grapevine improvement and in vitro culture. Moet-Hennessy, 30 Ave. Hoche, 75008 Paris, France.

26. Frazier NW, JP Fulton, JM Thresh, RH Converse, EH Varney, WB Hewett: 1970. Virus diseases of small fruits and grapevines. Univ. of California, Press, Berkeley, Calif. 290 pp.

27. Gouin FR: 1980. Vegetative propagation under thermo blankets. Comb. Proc. Intl. Plant Prop. Soc. 30:301-305.

28. Goussard PG: 1981. Effects of cytokinins on elongation, proliferation and total mass of shoots derived from shoot apices of grapevine cultured _in vitro_. Vitis 20:228-234.

29. Grenan S: 1985. Morphogenic behavior of grapevine after _in vitro_ multiplication. Observations, analysis and consequences. _In_: Proc. Moet-Hennessy Conference, Grapevine Improvement and _in vitro_ culture. Moet-Hennessy, 30 Ave. Hoche, 75008 Paris, France.

30. Harris RE, JH Stevenson: 1979. Virus elimination and rapid propagation of grapes _in vitro_. Comb. Proc. Intl. Plant Prop. Soc. 29:95-106.

31. Hosier MA, G Flatebo, PE Read: 1985. _In vitro_ propagation of lingonberry. HortScience 20:364-365.

32. Howard BH, VH Oehl: 1981. Improved establishment of _in vitro_-propagated plum micropropagules following treatment with GA_3 or prior chilling. J. Hort. Sci. 56:1-7.

33. Kartha K, N Leung, K Pahl: 1980. Cryopreservation of strawberry meristems and mass propagation of plantlets. J. Amer. Soc. Hort. Sci. 105:481-484.

34. Krul WR, JF Worley: 1977. Formation of adventitious embryos in callus cultures of 'Seyval', a french hybrid grape. J. Amer. Soc. Hort. Sci. 102:360-363.

35. Krul WR: 1985. _In vitro_ propagation of grape. U.S. Patent No. 4,532,733.

36. Kyte L, R Kyte: 1981. Small fruit culture after the test tube. Comb. Proc. Intl. Plant Prop. Soc. 31:45-47.

37. Lakso AN, BI Reisch, J Mortensen, MH Roberts: 1985. Stimulation of growth of in vitro propagated grapevines after transfer from sterile medium by CO_2 enrichment. HortScience 20:521 (Abstract).

38. Li JR, GW Eaton: 1984. Growth and rooting of grape shoot apices _in vitro_. HortScience 19:64-66.

39. Lloyd G, B McCown: 1980. Commercially feasible micropropagation of mountain laurel, _Kalmia latifolia_, by use of shoot-tip culture. Comb. Proc. Intl. Plant Prop. Soc. 30:421-427.

40. Lyrene PM: 1980. Micropropagation of rabbiteye blueberries. HortScience 15:80-81.

41. Lyrene PM: 1981. Juvenility and production of fast-rooting cuttings from blueberry shoot cultures. J. Amer. Soc. Hort. Sci. 106:396-398.

42. Marcotrigiano M, HJ Swartz: 1984. Histogenic instability of tissue culture proliferated strawberry plants. Program and Abstracts, Northeast Section, Amer. Soc. Hort. Sci. Geneva, NY. (Abstract 40).

43. Marcotrigiano M, HJ Swartz, SE Gray, D Tokarcik, J Popenoe: 1984. The effect of benzylamino purine on the in vitro multiplication rate and subsequent field performance of tissue-culture propagated strawberry plants. Adv. Strawberry Prod. 3:23-25.

44. McCulloch SM, BA Briggs: 1982. Preparation of plants for micropropagation. Comb. Proc. Intl. Plant Prop. Soc. 32:297-304.

45. McPheeters K, RM Skirvin: 1983. Histogenic layer manipulation in chimeral 'Thornless Evergreen' trailing blackberry. Euphytica 32:351-360.

46. McPheeters K, RM Skirvin: 1984. Field observations of micropropagated 'Thornless Evergreen' blackberry. HortScience 19:544 (Abstract).

47. Meador DB: 1985. Impact of herbicides on field-planted, succulent, tissue-culture propagated, red raspberry plants. HortScience 20:526 (Abstract).

48. Monette PL: 1983. Virus eradication through in vitro techniques. Comb. Proc. Intl. Plant Prop. Soc. 33:90-100.

49. Murashige T, F Skoog: 1962. A revised medium for rapid growth and bioassays with tobacco tissue cultures. Physiol. Plant. 15:473-497.

50. Navatel JC, G Varachaud, P Roudeillac, A Bardet: 1985. Comportement agromonique de plants dont les pieds meres ont été obtenus par multiplication in vitro par rapport au matériel classique. In: Report of CIREF. CTIFL Centre de Balandran 30127 Bellegarde FRANCE.

51. Nitsch JP, C Nitsch: 1969. Haploid plants from pollen grains. Science 163:85-87.

52. Novak FJ, A Juvova: 1982/3. Clonal propagation of grapevine through in vitro axillary bud culture. Scientia Hort. 18:231-240.

53. Phillips R, SM Arnott, SE Kaplan: 1981. Antibiotics in plant tissue culture: Rifampicin effectively controls bacterial contaminants without affecting the growth of short term explant culture of Helianthus tuberosus. Plant Sci. Lett. 21:235-240.

54. Rajasekaran K, MG Mullins: 1981. Regeneration of grapevine by

aseptic methods. Comb. Proc. Intl. Plant Prop. Soc. 31:213-218.

55. Rosati P, D Gaggioli, L Giunchi: 1986. Genetic stability of micropropagated loganberry plants. J. Hort. Sci. 61:33-41.

56. Schaeffer GW, C Damiano, DH Scott, JR McGrew, WR Krul, RH Zimmerman: 1980. Transcription of panel discussion on the genetic stability of tissue culture propagated plants, Pp. 64-79. In: Proc. conf. on nursery production of fruit plants through tissue culture--applications and feasibility. USDA-SEA, Agr. Res. Results, ARR-NE-11.

57. Scorza R, WV Welker, LJ Dunn: 1984. The effects of glyphosate, auxin and cytokinin combinations on in vitro development of cranberry node explants. HortScience 19:66-68.

58. Scott DH, GJ Galletta, HJ Swartz: 1985. Tissue culture as an aid in the propagation of 'Tribute' everbearing strawberry. Adv. Strawberry Prod. 4:59-60.

59. Shoemaker NP, HJ Swartz, GJ Galletta: 1985. Cultivar dependent variation in pathogen resistance due to tissue culture-propagation of strawberries. HortScience 20:253-254.

60. Singha S, LE Powell: 1978. Response of apple buds cultured in vitro to abscisic acid. J. Amer. Soc. Hort. Sci. 103:620-622.

61. Skene KGM, M Barlass: 1980. Micropropagation of grapevine. Comb. Proc. Intl. Plant Prop. Soc. 30:564-570.

62. Skene KGM, M Barlass: 1983. Studies on the fragmented shoot apex of grapevine. J. Exp. Bot. 34:1271-1280.

63. Snir I: 1981. Micropropagation of red raspberry. 1981. Scientia Hort. 14:139-143.

64. Stevenson JH, PL Monette: 1983. Delay of onset of leaf roll symptom expression in Vitis vinifera 'Liemberger' from ribavirin-treated in vitro cultures. Can. J. Plant Sci. 63:557-560.

65. Swartz HJ, GJ Galletta, RH Zimmerman: 1981. Field performance and phenotypic stability of tissue culture-propagated strawberries. J. Amer. Soc. Hort. Sci. 106:667-673.

66. Swartz HJ, GJ Galletta, RH Zimmerman: 1983. Field performance and phenotypic stability of tissue culture-propagated thornless blackberries. J. Amer. Soc. Hort. Sci. 108:285-290.

67. van Nieuwkerk JP, RH Zimmerman, I Fordham: 1985. Response of apple cultivars in vitro to thidiazuron. HortScience 20:523 (Abstract).

68. Wainwright H, AW Flegmann: 1984. Influence of light on the micropropagation of black currant. J. Hort. Sci. 59:387-393.

69. Wainwright H, AW Flegmann: 1985. The micropropagation of gooseberry: In vitro proliferation and in vivo establishment. J. Hort. Sci. 60:485-491.

70. Wolfe DE, P Eck, CK Chin: 1983. Evaluation of seven media for micropropagation of highbush blueberry. HortScience 18:703-705.

71. Zimmerman RH, GJ Galletta, OC Broome: 1980. Propagation of thornless blackberries by one-node cuttings. J. Amer. Soc. Hort. Sci. 105:405-407.

TEMPERATE FRUITS AND NUTS

F. A. HAMMERSCHLAG

U.S. Department of Agriculture, Agricultural Research Service, Tissue Culture and Molecular Biology Laboratory, Beltsville, MD 20705, U.S.A.

1. INTRODUCTION

Successful tissue culture propagation of temperate fruit and nut crops has progressed rapidly since the 1980 Beltsville Conference on tissue culture of fruit plants (94). Almost every year since then a review article has surfaced on recent advances in tissue culture of fruit crops (24, 27, 70, 71, 96, 98). Few reviews have appeared on tissue culture of nut crops (18, 48). These reviews indicate that for a number of our fruit crops, we are past the "how to" stage of in vitro propagation and should concentrate on cost reduction and field performance of tissue cultured plants. Even without the availability of data on field performance, numerous commercial companies have sprung up throughout the world (24, 47, 82, 93, 98) that produce fruit tree rootstocks and to some extent scion cultivars in huge numbers. Still, the feasibility of using tissue culture techniques to propagate temperate fruit and nut crops continues to be a major concern for many research and commercial facilities, because with the exception of apple and peach, minimal numbers of cultivars have been cultured and because numerous problems continue to impede successful in vitro propagation of some fruit and nut crops.

This chapter includes a review of the current research on micropropagation of several major temperate fruit and nut crops but focuses on Malus and Prunus species because a considerable amount of research has been reported on these compared with other fruit and nut crops. Problems that need to be solved to establish the method as feasible for commercial practice will also be discussed.

2. SPECIFIC CROPS

The following genera will be included in this review: Malus, Prunus, Pyrus, Carya, Castanea, Corylus, Juglans and Pistacia. A summary of the

Zimmerman, R.H. et al. (eds), Tissue Culture as a Plant Production System for Horticultural Crops.
 ISBN 90-247-3378-2.

status of these genera with respect to in vitro propagation is included in Table 1.

2.1. Malus spp.

Most of the recent research on apples has concentrated on optimizing micropropagation of scion cultivars, (8, 19, 22, 32, 39, 40, 42, 44, 45, 79, 80, 84, 88, 95, 97, 99, 100) and rootstocks (13, 33-37, 44, 55, 59, 67, 77, 89, 90) and on field evaluation of tissue cultured plants (64, 87, 101).

2.1.1. Media considerations. Since previous reviews (70, 96) have adequately covered most aspects of media composition, only the most recent research contributions are included in this review.

2.1.1.1. Antibiotics. Contamination will always be a major problem in Stage I cultures if explants are obtained from trees in the field (70, 96). Therefore, the initial aim in micropropagation should be to obtain or produce pathogen-free stock plants that can be maintained under protected culture. With this in mind, antibiotics have been incorporated into culture media to rid explants of contaminants. However, such studies indicated that the antibiotics were bacteriostatic and not bacteriocidal (6, 16). In a recent study (92) a combination of polymyxin B, rifampicin and cefotaxime eliminated all contaminants from eleven plant types including apple, plum, pear and filbert.

2.1.1.2. Phenolics. Many of the recent studies on in vitro propagation of apple have centered around using phenolic rooting cofactors to improve rooting. Of the many phenolics studied, the most controversy has arisen over the effectiveness of phloroglucinol (PG). PG has stimulated rooting of some apple rootstocks and cultivars (33, 35, 36, 37, 39, 40, 89, 100), while having no effect on others (13, 36, 100). In addition to stimulating rooting, PG has also been shown to reduce the period of time for IAA- and IBA-induced root initiation (33). It is clear from these studies that cultivars differ in responsiveness to PG, but Zimmerman (95) points out that the different responses could be due to different: 1) environmental conditions at the time the explant is taken, 2) physiological states of the explant, and/or 3) number of subcultures prior to rooting. The relative effectiveness of PG compared with other phenolics needs to be determined. PG has been tested at several concentrations (35); however, other phenolics have been tested at only one very high concentration (36).

Table 1. Status of micropropagation of temperate tree fruits and nuts.

Genus	Stage I	Stage II	Stage III	Field studies	Cyto-genetic analyses	Commercial production
Fruits						
Malus						
Apple	+	+	+	+	-	+
Crabapple	+	+	+	+	-	+
Prunus						
Apricot	+	+	+	+	-	+
Cherry	+	+	+	+	-	+
Peaches & Nectarines	+	+	+	+	+	+
Plums & Prunes	+	+	+	-	-	+
Pyrus						
Pear	+	+	+	-	-	+
Nuts						
Carya						
Pecan	+	+	+	-	-	+
Castanea						
Chestnut	+	+	+	-	-	+
Corylus						
Filbert	+	+	+	-	-	+
Juglans						
Walnut	+	+	+	-	-	+
Pistacia						
Pistachio	+	+	|	-	-	-
Prunus						
Almond	+	+	+	-	-	-

2.1.1.3. Thidiazuron (n-phenyl-N^1-1,2,3-thiadiazol-5-ylurea), a compound shown to have a high degree of cytokinin activity (11), has also been shown to stimulate apple shoot proliferation at concentrations considerably lower than benzyladenine (BA) (84). Shoot proliferation was shown to continue after explants were transferred to cytokinin-free medium. After 1-2 subcultures on cytokinin-free medium, shoots could be rooted.

2.1.1.4. Agar. Both concentration (68) and brand of agar (69) have been shown to influence growth and multiplication of crabapple shoots as well as shoot proliferation and vitrification in apples (Pasqualetto et al., personal communication). Because vitrification can impede successful transfer of micropropagated apple plants to *in vivo* conditions, it is significant that recent studies (Pasqualetto et al., personal communication) indicate that shoot proliferation can be induced without inducing vitrification on certain combinations of agar and the gelling agent Gelrite. It is evident from all these studies that a gelling agent can influence the anatomy and physiology of cultured shoots and therefore, each specific brand of agar or gelling agent should be clearly identified in any research publication.

2.1.2. Environmental factors. Several studies have examined the effect of dark (33, 88, 97, 100) and temperature (33, 97, 100) during root initiation, on rooting of apples. A positive effect of dark treatment was more pronounced as temperature increased from 25° to 30°C (97, 100) and when shoots were exposed to PG (33, 97). Optimum length of dark at the beginning of the rooting stage was shown to range from 3 to 9 days (19, 33, 88, 97, 100). James (33) reported that temperatures from 22° to 29°C during root initiation did not influence rooting, but his experiments were carried out in light.

2.1.3. Seasonal effects. It is fairly well documented that culturing of fruit crops *in vitro* can be influenced by the time of the year the explant is taken (70). Recently, Hutchinson (32) reported that the optimum time for establishing apple shoots *in vitro*, with minimum browning and contamination, was in the spring.

2.1.4. Field performance. Studies on field performance of tissue cultured apple cultivars indicate that micro-cuttings come to flower within 2 to 2-1/2 years demonstrating that rooting ability is not associated with reversion to the juvenile form (52, 64, 101). When tissue cultured apples were compared to budded trees over several years, Zimmerman et al. (101) and Rosati et al. (64) found that the former were more vigorous, were as productive as those on clonal rootstocks and possibly more stress tolerant (Zimmerman, personal communication). Webster et al. (87) however, found that vigor and cropping efficiency was cultivar specific and that some cultivars on their own roots were less vigorous and fruited much less than trees on specific rootstocks. These authors point out though that the micropropagated trees were a year younger than the scion part of the composite tree.

2.2. Prunus spp.

2.2.1. Apricot. (*Prunus armeniaca* L.) is a good candidate for *in vitro* propagation because apricots are currently budded onto apricot seedlings (30); therefore, growing apricots on their own roots appears to be a logical undertaking. Very little, however, has been reported on micropropagation of this species (72, 76). Whether the protocols developed by Skirvin et al. (72) or Snir (76) will be useful for many apricot cultivars remains to be seen. It has been fairly well documented that culture requirements can differ for cultivars within the same species (26, 45, 88, 97, 100). One encouraging note is that, although it's been reported that many woody plants require several subcultures to reach rooting competence, Snir (76) reported that the 'Canino' apricot, in spite of being difficult to propagate as cuttings, developed a high rooting competence very early in the *in vitro* propagation process. The ease of rooting *in vitro* might be traced to grafting budwood of mature 'Canino' onto seedlings of the same cultivar. This approach has been recommended by Franclet (23) and Derbergh and Maene (16) to induce juvenile characteristics in mature scion shoots.

2.2.2. Cherries. Interest in *in vitro* propagation of sweet (*Prunus avium* L.) and sour (*P. cerasus* L.) cherries stems from the need to: 1) avoid rootstock scion-incompatibility, reported to be the cause of sudden death of sweet cherry trees (14), 2) propagate difficult-to-root mature mazzard trees selected in stands (61), 3) propagate large numbers of inexpensive identical plants for high density orchards (75), and 4) rapidly propagate selected rootstocks for cherry (38). Protocols for *in vitro* propagation have been developed for several cherry scion cultivars (61, 72, 74, 75) and a limited number of rootstocks (38).

2.2.2.1. Rooting studies. Although sweet cherry is considered to be one of the most difficult-to-root species within the genus *Prunus* (30), both it and sour cherry have achieved a very high rooting efficiency *in vitro* (74, 75). Phloroglucinol was shown to have either no effect (38) or a negative effect (75) on rooting of sour cherry. Basal wounding had a positive effect on sweet cherry (74) and high levels of sucrose enhanced significantly the rate and percentage of rooting of sour cherry (75).

Riffaud and Cornu (61) reported considerable variability among sour cherries in their response to different rooting treatments. There was no need to transfer rooted shoots to a medium without growth regulators because sour cherry, unlike apple (77), had no tendency to form callus at the base of the shoot (74).

2.2.2.2. Mycorrhizae studies. A vesicular-arbuscular (VA) mycorrhizal association with cherry rootstocks was established *in vitro* (58). This was the first report of establishing VA mycorrhizae on axenically propagated woody plants. Although preliminary studies with *in vitro* propagated cherry rootstocks showed a marked improvement in growth when inoculated with a VA mycorrhizal fungus at the time of transplanting into a soil potting mixture (54), a more recent study (58) found no enhancement of growth following *in vitro* mycorrhizal inoculation.

2.2.2.3. Field performance. A single study on field performance of self-rooted, tissue-cultured sour cherries (65) indicated that these trees were less productive and one or two years later in bearing compared with standard trees.

2.2.3. Peaches. Only a limited number of peach [(*Prunus persica* (L.) Batsch] cultivars (28, 49, 72, 73, 86) and rootstocks (50, 51, 56, 60, 102) have been propagated *in vitro* and only in a few cases has greater than 50% rooting been achieved (28, 50, 51, 86).

2.2.3.1. Rooting studies. Difficulty and inconsistency in rooting peaches *in vitro* continues to be a major deterrent to utilizing micro-propagation on a commercial scale. Hammerschlag (26) reported that peach cultivars respond variably *in vitro* and because of this, current efforts are being directed to a more extensive examination of cultivar performance *in vitro* (28). Preliminary data from her lab (28) and results from that of Mosella Chancel et al. (51) indicate the importance of rooting cofactors.

2.2.3.2. Field performance and genetic stability. Hammerschlag (unpublished data) is currently comparing 4 self-rooted scion cultivars with bud-grafted cultivars on seedling rootstocks in Maryland, West Virginia, South Carolina and New Jersey. Reports from France (49) and Italy (64) indicate that both scion cultivars and rootstocks propagated *in vitro* compare favorably with bud-grafted cultivars and rootstocks propagated by cuttings, respectively. These reports compare favorably with previous studies on field performance of conventionally propagated self-rooted peaches (15). Currently commercial production of self-rooted cultivars by micropropagation is being carried out in British Columbia and in Italy (9, 98 and Dunstan, personal communication). Cytogenetic analysis of micro-propagated scion cultivars indicated that all plants were diploid 2n=2x=16 (28). Both field studies and cytogenetic analyses demonstrate the feasibility of using micropropagation for peaches; however, protocols must be worked out for rooting sizeable numbers of cultivars and costs must be reduced. It has been suggested that the most economical peach tree is the one produced from hardwood cuttings directly rooted in the orchard (21); however, success with this approach has been variable (20), and therefore it is questionable whether this approach is truly the most economical.

2.2.4. Plums. Recent research on micropropagation of plum (*Prunus domestica* L. and *P. salicina* Lindl.) scion cultivars (2, 57) and plum (*P. cerasifera* Ehrh. and *P. insititia* L.) rootstocks (25, 31, 38) include all stages of the propagation process although considerable emphasis has been on rooting.

2.2.4.1. Rooting and acclimatization studies. Factors that had a positive effect on rooting were gibberellic acid (GA_3) (65), temperature and dark treatments (25), phloroglucinol (2, 38) and chlorogenic acid (25). Although good survival of aseptically cultured plums during transplanting has been reported (25, 65), plantlets of the dwarfing rootstock Pixy grew poorly unless chilled for 2 months at 0°C before potting or sprayed with GA_3 following potting (31).

Although a limited number of cultivars have been propagated *in vitro*, results are encouraging. According to Pietropaolo and Reisch (57), *in vitro* propagation has commercial possibilities because the technique is simple, can be done with readily available tissue culture media and most importantly, can be timed to produce plantlets ready for transplanting to a field nursery in the spring.

2.3. Pyrus spp.

Reports on commercial production of both fruiting cultivars and rootstocks of pear (*Pyrus communis* L. and *P. pyrifolia* Burm. f.) through tissue culture techniques (9, 98), together with reports on high productivity of self-rooted cultivars (83), provide some evidence that this is a feasible plant production system. Unfortunately, research reports on micropropagation of this crop are minimal (7, 43, 68, 69) and therefore tissue culture protocols for a wide number of cultivars is lacking as well as information on field performance and phenotopic stability of in vitro propagated cultivars.

2.4. Nut crops.

Information on micropropagation of almond (Prunus amygdalus (66), chestnut (Castanea saliva Mill.) (63, 85), filbert (Corylus avellana L.) (1), pecan [Carya illinoensis (Wangenh.) K. Koch)] (29, 41, 46, 91), pistachio (Pistacia vera L.) (3, 4, 5), and walnut (Juglans nigra L. and J. regia L.) (12, 17, 18, 48, 62, 78, 81) are extremely limited and because of this, have been combined in this section. A common thread runs through most of these reports. For example, in most cases shoot tips were collected from seedlings and nutrient media contained lower levels of salts compared with Murashige and Skoog (MS) salts (53). Attempts to culture explants from mature plants were unsuccessful (4, 81). Driver and Kuniyuki (17) conducted the most in depth study on nutrient requirements for walnut and reported that lowering nitrogen levels gave the same results as lowering levels of all mineral salts. They (17) also reported on the use of Gelrite and the positive effect of increased levels of sucrose on rooting. Commercial production of a number of nut crops is underway (9, 48, 98) although reports on field performance are lacking.

3. CONCLUSIONS

It is clear from the above that a considerable number of fruit and nut crops are being propagated in vitro. This trend is likely to continue in the future and will most likely be aimed at producing 1) virus-free stock plants, 2) new cultivars, 3) difficult-to-propagate elite genotypes of plants, and 4) large quantities of rootstocks. In addition, micropropagation will continue to be the logical end point for any plant genetic engineering program.

Problems still remain concerning propagation of scion cultivars. A major problem is cost and a major cost is labor (10). Micropropagation is extremely labor intensive. Labor costs can be reduced however, if procedures are shortened or simplified. Simmonds (67) shortened the procedure for apples by substituting rooting in the mist for in vitro rooting. M.26 apple rootstocks were dipped in commercial rooting powder and established in a peat:sand (1:1) substrate within a mist frame. With this procedure greater than 80% plant establishment was obtained. Fiorini and Leva (22) simplified the rooting procedure and used less expensive materials when they induced root initiation of apples in a liquid basal medium. Zimmerman and Fordham (100) further simplified the procedure by inducing root initiation in an auxin-sucrose liquid medium and root induction in preformed peat plugs. The Plant Research Laboratory, Modesto, CA, has tested a new cost efficient process (patents pending) that allows successful rooting and acclimatization in the field (18).

Suttle (82) demonstrated that labor costs as well as the cost of heating and lighting could be reduced if rooted plantlets of apple, pear, plum and cherry were stored in the dark at 2°C from October to February. This treatment also had a positive effect on subsequent growth of plants in the greenhouse.

Another important problem in propagating scion cultivars in vitro is that differences exist among cultivars (26, 45, 80, 97, 100) as well as between juvenile and adult forms of the same plant (89) in their culture requirements; therefore, attempts to culture juvenile shoots to provide background information for work with mature tissues may not be helpful and also, general recommendations for procedures and media composition based on experience with one cultivar may be ill advised. Certainly any success in improving synchrony of shoot proliferation and root development and thus reducing variability among cultivars in response to treatments will be a tremendous contribution.

It is clear that continued efforts to propagate mature scion cultivars or to streamline the micropropagation process will result in considerable reduction in cost of this approach. If reduction in costs are coupled with satisfactory field performance, then this propagation approach will be adopted for more plants. To date, field studies indicate that tissue cultured apples (64, 101) and peaches (49), are exceptionally vigorous and over a several year span produces equally well as bud-grafted trees. Large scale testing of a variety of tree fruit cultivars and rootstocks is currently being conducted in the United States, Canada, England, France and Italy.

Although a micropropagated rootstock may never be economically competitive with a conventionally propagated one unless the plant material is in short supply and there is a huge demand for it, a micropropagated, self-rooted scion cultivar stands a very good chance of being economically competitive with the high-priced bud-grafted tree.

REFERENCES

1. Anderson WC: 1983. Micropropagation of filberts, *Corylus avellana*. Comb. Proc. Intl. Plant Prop. Soc. 33:132-137.

2. Baleriola-Lucas C, MG Mullins: 1985. Micropropagation of two French prune cultivars (*Prunus domestica* L.). Agronomie 5: (In press).

3. Barghchi M, PG Alderson: 1983. *In vitro* propagation of *Pistacia vera* L. from seedling tissues. J. Hort. Sci. 58:435-445.

4. Barghchi M, PG Alderson: 1983. *In vitro* propagation of *Pistacia* species. Acta Hort. 131:49-60.

5. Barghchi M, PG Alderson: 1985. *In vitro* propagation of *Pistacia vera* and the commercial cultivars Ohadi and Kalleghochi. J. Hort. Sci. 60:423-430.

6. Bastiaens L, L Maene, Y Harbaoui, C Van Sumere, KL Vande Casteele, PC Debergh: 1983. The influence of antibacterial products on plant tissue cultures. Med. Fac. Lanbouw. Rijksuniv. Gent 48:13-24.

7. Bhojwani SS, K Mullins, D Cohen: 1984. *In vitro* propagation of *Pyrus pyrifolia*. Scientia Hort. 23:247-254.

8. Brainerd KE, LH Fuchigami: 1981. Acclimatization of aseptically cultured apple plants to low relative humidity. J. Amer. Soc. Hort. Sci. 106:515-518.

9. Briggs BA, SM McCulloch: 1983. Progress in micropropagation of woody plants in the United States and Western Canada. Comb. Proc. Intl. Plant Prop. Soc. 33:239-248.

10. Brown CL, HE Sommer: 1982. Vegetative propagation of dicotyledonous trees, pp. 109-149. In: JM Bonga, DJ Durzan (eds). Tissue culture in forestry. Martinus Nijhoff/Dr. W Junk, The Hague.

11. Capelle SC, DWS Mok, SC Kirchner, MC Mok: 1983. Effects of thidiazuron on cytokinin autonomy and the metabolism of N6-(Δ^2-isopentenyl) [8-^{14}C] adenosine in callus tissues of *Phaseolus lunatus* L. Plant Physiol. 73:796-802.

12. Chalupa V: 1981. Clonal propagation of broad-leaved forest trees *in vitro*. Commun. Inst. Forest. Cech. 12:255-271.

13. Cheema GS, DP Sharma: 1983. *In vitro* propagation of apple rootstock EMLA 25. Acta Hort. 131:75-88.

14. Childers NF: 1976. Modern fruit science. Horticultural Publications, New Brunswick, N.J.

15. Couvillon GA: 1982. Leaf elemental content comparisons of own-rooted peach cultivars to the same cultivars on several peach seedling rootstocks. J. Amer. Soc. Hort. Sci. 107:555-558.

16. Debergh P, L Maene: 1985. Some aspects of stock-plant preparation for tissue culture propagation. Acta Hort. 166:21-24.

17. Driver JA, AH Kuniyuki: 1984. In vitro propagation of Paradox walnut rootstock. Hortscience 19:507-509.

18. Driver JA, GR Suttle: 1986. Nursery handling of propagules. In: JM Bonga, DJ Durzan, (eds.). Tissue culture in forestry, 2nd edition, Martinus Nijhoff/Dr. Junk, The Hague. (In press).

19. Druart P, C Kevers, P Boxus, T Gaspar: 1982. In vitro promotion of root formation by apple shoots through darkness effect on endogenous phenols and peroxidases. Z. Pflanzenphysiol. 108:429-436.

20. Erez A: 1984. Improving the rooting of peach hardwood cuttings under field conditions. HortScience 19:245-247.

21. Erez A, Z Yablowitz: 1981. Rooting of peach hardwood cuttings for the meadow orchard. Scientia Hort. 15:137-144.

22. Fiorini P, AR Leva: 1983. Propagation of apple cultivars. Acta Hort. 131:95-97.

23. Franclet A: 1980. Rajeunissement et propagation vegetative des ligneux. Ann. AFOCEL 1980:12-14.

24. George EF, PD Sherrington: 1984. Plant propagation by tissue culture. Exegetics Ltd., Eversley, Basingstoke.

25. Hammerschlag F: 1982. Factors influencing in vitro multiplication and rooting of the plum rootstock Myrobalan (Prunus cerasifera Ehrh.). J. Amer. Soc. Hort. Sci. 107:44-47.

26. Hammerschlag F: 1982. Factors affecting establishment and growth of peach shoot tips in vitro. HortScience 17:85-86.

27. Hammerschlag F: 1985. Peach (Prunus persica L. Batsch), p. 170-183. In: YPS Bajaj (ed). Biotechnology in agriculture and forestry 1. Trees 1. Springer-Verlag, Berlin.

28. Hammerschlag F, C Bauchan: 1985. Factors influencing rooting of peach scion cultivars propagated in vitro. HortScience 20:521 (Abstr.).

29. Hansen KC, JE Lazarte: 1984. In vitro propagation of pecan seedlings. HortScience 19:237-239.

30. Hartmann HT, DE Kester: 1975. Plant propagation. principles and practices. 3rd ed. Prentice-Hall, Englewood Cliffs, N.J.

31. Howard BH, VH Oehl: 1981. Improved establishment of in vitro-propagated plum micropropagules following treatment with GA_3 or prior chilling. J. Hort. Sci. 56:1-7.

32. Hutchinson JF: 1984. Factors affecting shoot proliferation and root initiation in organ cultures of the apple 'Northern Spy'. Scientia Hort. 22:347-358.

33. James DJ: 1983. Adventitious root formation '*in vitro*' in apple rootstock (*Malus pumila*). I. Factors affecting the length of the auxin sensitive phase in M.9. Physiol. Plant. 57:149-153.

34. James DJ, IJ Thurbon: 1979. Rapid *in vitro* rooting of the apple rootstock M.9. J. Hort. Sci. 54:309-311.

35. James DJ, IJ Thurbon: 1981. Shoot and root initiation *in vitro* in the apple rootstock M.9 and promotive effects of phloroglucinol. J. Hort. Sci. 56:15-20.

36. James DJ, IJ Thurbon: 1981. Phenolic compounds and other factors controlling rhizogenesis *in vitro* in the apple rootstocks M.9 and M.26. Z. Pflanzenphysiol. 105:11-20.

37. Jones OP: 1976. Effect of phloridzin and phloroglucinol on apple shoots. Nature 262:392-393.

38. Jones OP, ME Hopgood: 1979. The successful propagation in vitro of two rootstocks of *Prunus*: the plum rootstock Pixy (*P. insititia*) and the cherry rootstock F12/1 (*P. avium*). J. Hort. Sci. 54:63-66.

39. Jones OP, CA Pontikis, ME Hopgood: 1979. Propagation *in vitro* of five apple scion cultivars. J. Hort. Sci. 54:155-158.

40. Jones OP, RH Zimmerman, I Fordham, ME Hopgood: 1985. Propagation *in vitro* of some dwarf apple trees. J. Hort. Sci. 60:141-144.

41. Knox CA, RH Smith: 1981. Progress in tissue culture methods for production of Riverside stocks. Pecan Quart. 15:27-34.

42. Lane WD: 1978. Regeneration of apple plants from shoot meristem tips. Plant Sci. Lett. 13:281-285.

43. Lane WD: 1979. Regeneration of pear plants from shoot meristem-tips. Plant Sci. Lett. 16:337-342.

44. Lane WD, NE Looney, F Mage: 1982. A selective medium for growth of compact (dwarf) mutants of apple. Theor. Appl. Genet. 6:219-223.

45. Lane WD, JM McDougald: 1982. Shoot tissue cultures of apple: comparative response of 5 cultivars to cytokinin and auxin. Can. J. Plant Sci. 62:689-694.

46. Lazarte J: 1984. Propagating pecan rootstock by tissue culture. Amer. Nurseryman 159(11):79-81.

47. Loreti F, S Morini: 1982. Mass propagation of fruit trees in Italy by tissue culture: present status and perspectives. Comb. Proc. Intl. Plant Prop. Soc. 32:283-291.

48. McGranahan GH, JA Driver, W Tulecke: 1986. Tissue Culture of Juglans. In: JM Bonga, DJ Durzan, (eds). Tissue culture in forestry, 2nd edition, Martinus Nijhoff/Dr. Junk, The Hague. (In press).

49. Martin C, M Carre, R Vernoy: 1983. La multiplication vegetative in vitro des vegetaux ligneux cultives: cas des arbres fruitiers et discussion generale. Agronomie 3:303-306.

50. Miller GA, DD Coston, EG Denney, ME Romeo: 1982. In vitro propagation of 'Nemaguard' peach rootstock. HortScience 17:194.

51. Mosella Chancel L, JJ Macheix, R Jonard: 1980. Les conditions du microbouturage in vitro du pecher (Prunus persica Batsch): influences combinees des substances de croissance et de divers composés phénoliques. Physiol. Veg. 18:597-608.

52. Mullins MG: 1985. Regulation of adventitious root formation in microcuttings. Acta Hort. 166:53-61.

53. Murashige T, F Skoog: 1962. A revised medium for rapid growth and bioassays with tobacco tissue cultures. Physiol. Plant. 15:473-497.

54. Navatel JC: 1980. L'utilization des cultures in vitro pour la multiplication de quelques especes legumineres et fruitiers. C. R. Acad. Agr. France 66:681-696.

55. Ochatt SJ: 1983. In vitro meristem culture of M.4 apple (Malus pumila Mill.) I. Optimal nutrient medium. Plant Cell Tissue Organ Culture 2:39-48.

56. Ochatt SJ, OH Caso: 1983. Dormant meristem-tips as a source for producing red-leaf peach (Prunus persica Batsch.) shoots in vitro. Phyton 43:21-28.

57. Pietropaolo PA, BI Reisch: 1984. Micropropagation of 'Stanley' plum. HortScience 19:535-536.

58. Pons FV, S Gianinazzi-Pearson, S Gianinazzi, JC Navatel: 1983. Studies of VA mycorrhizae in vitro: mycorrhizal synthesis of axenically propagated wild cherry (Prunus avium L.) plants. Plant and Soil 71:217-221.

59. Pua EC, C Chong, GL Rouselle: 1983. In vitro propagation of Ottawa 3 apple rootstock. Can. J. Plant Sci. 63:183-188.

60. Reeves DW, BD Horton, GA Couvillon: 1983. Effect of media and media pH on in vitro propagation of 'Nemaguard' peach rootstock. Scientia Hort. 21:353-357.

61. Riffaud J, D Cornu: 1981. Utilization de la culture in vitro pour multiplication de merisiers adultes (Prunus avium L.) selectionnes en foret. Agronomie 1:633-640.

62. Rodriguez R: 1982. Stimulation of multiple shoot buds formation in walnut seeds. HortScience 17:592.

63. Rodriguez R: 1982. *In vitro* propagation of *Castanea sativa* Mill. through meristem-tip culture. HortScience 17:888-889.

64. Rosati P, D Gaggioli: 1986. Field performance of micropropagated peach rootstocks and scion cultivars of sour cherry and apple. Acta Hort. (In press).

65. Rosati P, G Marino, C Swierczewski: 1982. *In vitro* propagation of Japanese plum (*Prunus salicina* Lindl. cv. Calita). J. Amer. Soc. Hort. Sci. 105:126-129.

66. Rugini E, DC Verma: 1982. Micropropagation of difficult to propagate almond (*Prunus amygdalus*) cultivar. Plant Sci. Lett. 28:273-281.

67. Simmonds J: 1983. Direct rooting of micropropagated M26 apple rootstocks. Scientia Hort. 21:233-241.

68. Singha S: 1982. Influence of agar concentration on *in vitro* shoot proliferation of *Malus* sp. 'Almey' and *Pyrus communis* 'Seckel'. J. Amer. Soc. Hort. Sci. 107:657-660.

69. Singha S: 1984. Influence of two commercial agars on *in vitro* shoot proliferation of 'Almey' crabapple and 'Seckel' pear. HortScience 19:227-228.

70. Skirvin RM: 1981. Fruit crops, p. 51-139. In: B. V. Conger (ed). Cloning agricultural plants via *in vitro* techniques. Chemical Rubber Company Press, Boca Raton, Florida.

71. Skirvin RM: 1984. Stone fruits, pp. 402-452. In: PV Ammirato, DA Evans, WR Sharp, Y Yamada (eds). Handbook of plant cell culture, Vol. 3, Crop species. Macmillan, New York.

72. Skirvin RM, MC Chu, H Rukan: 1979. Tissue culture of peach, sweet and sour cherry and apricot shoot tips. Proc. Annu. Mtg. Ill. State Hort. Soc. 113:30-38.

73. Skirvin R, H Rukan-Kerns: 1982. An improved medium for *in vitro* rooting of 'Harbrite' peach. Fruit Var. J. 36:15-17.

74. Snir I: 1982. *In vitro* propagation of sweet cherry cultivars. HortScience 17:192-193.

75. Snir I: 1983. A micropropagation system for sour cherry. Scientia Hort. 19:85-90.

76. Snir I: 1984. *In vitro* propagation of 'Canino' apricot. HortScience 19:229-230.

77. Snir I, A Erez: 1980. *In vitro* propagation of Malling Merton apple rootstocks. HortScience 15:597-598.

78. Somers PW, JW Sambeek, JE Preece, G Gaffney, O Myers, Jr: 1982. In vitro micropropagation of black walnut (Juglans nigra L.). Proc. 7th N. Amer. For. Biol. Workshop pp. 224-230.

79. Sriskandarajah S, MG Mullins: 1981. Micropropagation of Granny Smith apples: factors affecting root formation in vitro. J. Hort. Sci. 56:71-76.

80. Sriskandarajah S, MG Mullins, Y Nair: 1982. Induction of adventitious rooting in vitro in difficult-to-propagate cultivars of apple. Plant Sci. Lett. 24:1-9.

81. Steffan SJ, DF Millikan: 1983. Micropropagation of mature black walnut trees. Annu. Rept. North Nut Growers Assn. 74:26-29.

82. Suttle G: 1983. Microprogation of deciduous trees. Comb. Proc. Intl. Plant Prop. Soc. 33:46-49.

83. Thibault B, L Herman: 1982. Culture of Bartlett on its own roots: comparisons with quince and French seedling rootstocks. Acta Hort. 124:21-26.

84. Van Nieuwkerk JP, RH Zimmerman, I Fordham: 1985. Response of apple cultivars in vitro to thidiazuron. HortScience 20:523 (Abstr.).

85. Vieitez AM, ML Vieitez: 1982. Castanea sativa plantlets proliferated from axillary buds cultivated in vitro. Scientia Hort. 18:343-351.

86. Vertesy J: 1980. In vitro propagation of Prunus persica and P. persica-davidiana shoot tips in order to get virus-free plants. Acta Phytopath. Acad. Sci. Hung. 15:261-264.

87. Webster AD, VH Oehl, JE Jackson, OP Jones: 1985. The orchard establishment, growth and precosity of four micropropagated apple scion cultivars. J. Hort. Sci. 60:169-180.

88. Welander M: 1985. In vitro shoot and root formation in the apple cultivar Akero. Ann. Bot. 55:249-261.

89. Welander M, I Huntrieser: 1981. The rooting ability of shoots raised in vitro from the apple rootstock A2 in juvenile and in adult growth phase. Physiol. Plant. 53:301-306.

90. Werner EM, AH Boe: 1980. In vitro propagation of Malling 7 apple rootstock. HortScience 15:509-510.

91. Wood BW: 1982. In vitro proliferation of pecan shoots. HortScience 17:890-891.

92. Young PM, AS Hutchins, ML Canfield: 1984. Use of antibiotics to control bacteria in shoot cultures of woody plants. Plant Sci. Lett. 34:203-206.

93. Zimmerman RH: 1979. The Laboratory of Micropropagation at Cesena, Italy. Comb. Proc. Intl. Plant Prop. Soc. 29:398-400.

94. Zimmerman RH (ed.): 1980. Proc. conf. on nursery production of fruit plants through tissue culture - applications and feasibility. U.S. Dept. Agr., Sci. and Educ. Admin., Agr. Res. Results ARR-NE-11.

95. Zimmerman RH: 1983. Factors affecting in vitro propagation of apple cultivars. Acta Hort. 131:171-177.

96. Zimmerman RH: 1984. Apple, p. 369-395. In: WR Sharp, DA Evans, PV Ammirato, Y Yamada (eds). Handbook of plant cell culture, Vol. 2, Crop species. Macmillan, New York.

97. Zimmerman RH: 1984. Rooting apple cultivars in vitro: interactions among light, temperature, phloroglucinol and auxin. Plant Cell Tissue Organ Culture 3:301-311.

98. Zimmerman RH: 1985. Tissue culture propagation of woody plants: Large scale production, pp. 165-177. In: R Henke, KW Hughes, MJ Constantin A Hollaender (eds). Tissue culture in forestry and agriculture. Plenum Press, New York.

99. Zimmerman RH, OC Broome: 1981. Phloroglucinol and in vitro rooting of apple cultivar cuttings. J. Amer. Soc. Hort. Sci. 106:648-652.

100. Zimmerman RH, I Fordham: 1985. Simplified method for rooting apple cultivars in vitro. J. Amer. Soc. Hort. Sci. 110:34-38.

101. Zimmerman RH, SS Miller: 1985. Field growth of tissue-cultured compared to budded apple trees. HortScience 20:584 (Abstr.).

102. Zuccherelli G: 1979. Moltiplicazione in vitro dei portainnesti clonali del pesco. Frutticoltura 41(2):15-20.

TROPICAL AND SUBTROPICAL FRUITS AND VEGETABLES[1]

RICHARD E. LITZ, ROBERT L. JARRET AND MADHEVAN P. ASOKAN

University of Florida, IFAS, Tropical Research and Education Center, 18905 SW. 280 St., Homestead, FL 33031, U.S.A.

1. INTRODUCTION

Tropical and subtropical fruit and starchy root and tuber crops are among the most important components of diets in many parts of the world (Tables 1 and 2). They are also important sources of hard currency income, either from the export of fresh or processed food or from the sale of secondary products. Recent trends in the growth of tropical fruit production suggest that consumers in many temperate countries are discovering tropical fruit (Table 2). There is currently a massive expansion of the tropical fruit industry in Central America, South America and in many Caribbean countries. Most tropical fruit, root and tuber crops have been vegetatively propagated for several hundred, and in some cases, for several thousand years. Clonal propagation methods that have been used have ranged from relatively sophisticated grafting techniques with some fruit cultivars (12) to direct planting of tuber or stem pieces with some of the starchy root and tuber plants. In this way, the unique genetic balance of plant selections has been maintained. With the passage of time, however, many of these clones have become highly vulnerable to increased disease pressure. Classic examples of this phenomenon during the twentieth century have been the devastation of 'Gros Michel' banana production in Central America by Panama disease and the more recent epidemic of black Sigatoka disease among cooking bananas (plantains) in the same region. Finally, the spread of acquired systemic viral and bacterial infections has been facilitated by clonal propagation methods. These diseases have become the most important factors that limit production of Citrus, yam, sweet potato, cassava, and various edible aroids.

---

[1]Agricultural Experiment Stations Journal Series No. 6836.

Zimmerman, R.H. et al. (eds), Tissue Culture as a Plant Production System for Horticultural Crops.
 ISBN 90-247-3378-2.

TABLE 1. Increase in world production of major fruits since 1969-1971.[1]

Crop	1969-71 Production[2]	1983 Production[2]	% increase
Grapes	55,554	65,167	20
Bananas/plantains	47,215	60,477	28
All citrus	38,214	54,519	43
Apples	26,702	36,799	38
Mangos	12,236	13,954	14
Pears	7,936	9,523	20
Pineapples	4,911	8,665	76
Peaches/nectarines	6,269	7,406	18
Plums	5,679	6,187	9
Papaya	1,181	1,982	68
Strawberry	1,210	1,861	54
Apricots	1,527	1,746	14
Avocados	960	1,586	65

[1] Source FAO Production Yearbooks.
[2] Production in 1000 MT.

TABLE 2. Production of tropical root and tuber crops (1000 mt).[1]

	Cassava	Sweet Potato	Yams	Taro
Africa	48,251	5,081	18,643	3,397
N.C. America	898	1,289	227	-
S. America	27,839	1,396	50	-
Asia	45,929	106,315	30	699
Oceania	237	629	187	257
Europe + USSR	-	-	-	-
World	123,153	114,842	19,137	4,352

[1] Data for cassava and sweet potato are for 1983; data for yams and taro are for 1984. (FAO Production Yearbooks).

In vitro propagation has already begun to affect tropical fruit, root and tuber production, primarily as a result of international and national efforts to make disease-indexed clonal material available for distribution. Programs for production of virus-free *Citrus* have been successfully implemented in the U.S.A., Spain (48), Venezuela, Argentina, and elsewhere. In addition, there has been a strong international commitment to production and distribution of disease indexed stock of some tropical root and tuber crops, e.g., cassava (7, 8) and yam (37). The implications of *in vitro* systems for rapid propagation of new cultivars and of difficult-to-propagate fruit trees has not yet been realized. However, economic exploitation of *in vitro* cloning of plantation crops, e.g., date palm, oil palm, coffee and herbaceous tropical fruit species, including banana, plantain, papaya and pineapple has already occurred. This review will address the potential for using *in vitro* systems for propagating tropical fruit, root and tuber crops.

2. TROPICAL FRUIT SPECIES

Although several tropical fruit trees have been the objects of tissue culture studies, most of these investigations have been concerned with defining regeneration pathways from embryonic or seedling tissues and organs (33). These approaches may eventually be useful in somatic cell genetic schemes for improving tropical fruit trees; however, conditions for regeneration from explants from mature, woody species are usually very difficult and more complex than conditions that are appropriate for juvenile tissues. Consequently, there have been relatively few reports of regeneration from explants from mature woody trees. The most notable successes in the application of *in vitro* systems for propagating tropical fruit species have involved herbaceous plants, among which the problems associated with maturity versus juvenility of the explant source usually can be reconciled by minor changes in the medium. Table 3 includes those tropical fruit species that have been regenerated from shoot apices, axillary buds or from maternal tissues within the ovule, i.e., the nucellus or inner integument.

2.1. Herbaceous fruit species

In general, the herbaceous fruit species have been relatively easy to regenerate *in vitro*, either by stimulating the proliferation of axillary buds from cultured shoot apices and lateral buds or by organogenesis from callus derived from cultured tissues or organs (Table 4). Because of the constant disease pressure of tropical and subtropical environments, the condition of the stock plant can be a major limiting factor in determining success in the establishment of explants. With papaya, contamination due to systemic bacteria is particularly acute during the dry season of the subtropics. Generally, this is a season of cooler temperatures, when plant growth is suboptimal. There have also been reported differences in the responses among cultivars of the same species, e.g., papaya (30) and banana (10), and according to the sex type of the stock plants (30).

TABLE 3. Tissue culture propagation of tropical fruits.

Plant family	Genus	Species	Common Name
Anacardiaceae	Mangifera	indica	mango
Annonaceae	Annona	cherimola x squamosa	atemoya
Bromeliaceae	Ananas	comosus	pineapple
Caricaceae	Carica	papaya	papaya
		x heilbornii	babaco
Moraceae	Ficus	carica	fig
	Morus	alba	mulberry
		indica	mulberry
Musaceae	Musa	AAA group	banana
		AAB group	plantain
		ABB group	Bluggoe
Myrtaceae	Eugenia	jambos	rose apple
		malaccensis	Malay apple
	Myrciaria	cauliflora	jaboticaba
Passifloraceae	Passiflora	edulis	passion fruit
Rosaceae	Eriobotrya	japonica	loquat
Rutaceae	Citrus	aurantiifolia	lime
		aurantium	sour orange
		grandis	pummelo
		limetoides	sweet lime
		limon	lemon
		medica	citron
		paradisi	grapefruit
		reticulata	mandarin
		sinensis	sweet orange
	Poncirus	trifoliata	trifoliate orange
Solanaceae	Solanum	quitoense	naranjilla

The use of shoot apices from the main plant axis or from suckers and lateral buds as primary explants has been quite successful among herbaceous fruit species, e.g., papaya (29), passion fruit (33), naranjilla (Hendrix and Litz, unpublished data), pineapple (59) and Musa (bananas and plantains) (35). With the exception of papaya, the herbaceous fruit species have not required a special establishment medium. It is also possible to culture papaya shoot tips directly on proliferation medium, although more time is required for establishment (29). Monocotyledonous fruit species have generally been propagated in liquid medium formulations (35, 39), unlike papaya, babaco and passion fruit, all of which are propagated on solid medium. It has been reported that proliferation of pineapple and Musa cultures can be sustained for two or more years, although papaya cultures become habituated after 8-13 subcultures,

TABLE 4. Regeneration pathways for tropical fruit.[1]

Crop	Shoot tip culture M	Shoot tip culture J	Micro-grafting M	Micro-grafting J	Organo-genesis M	Organo-genesis J	Somatic embryogenesis M	Somatic embryogenesis J	Reference
Babaco	+								9
Banana	+				+				10,35
Citrus spp.			+				+		36,48,52
Eugenia spp.							+		26
Fig	+								18,45
Jaboticaba							+		25
Loquat							+		28
Mango							+		27,32
Mulberry	+				+				40,51
Naranjilla	+				+				-
Papaya	+	+				+		+	28,29,60
Passiflora spp.	+				+	+			33
Pineapple	+								39,59

[1] Explants are either mature (M) or immature (J) in origin.

depending on the cultivar (30). Proliferation is achieved with most herbaceous fruit species on Murashige and Skoog medium with low concentrations (2.7-11.4 μM, 0.5-2.0 mgl^{-1}) of auxin, usually naphthaleneacetic acid (NAA) or indoleacetic acid. Rapid proliferation of *Musa*, however, has been achieved with much higher concentrations of cytokinin (44 μM, 10 mgl^{-1} BA). Shoot regeneration from callus derived either from cultured shoot apices, e.g. *Musa* (35), and for passion fruit (Litz and Knight, unpublished data) is possible; however, the clonal stability of plants regenerated in this manner has not been determined. Apical dominance can be restored and rooting can be induced from proliferating shoots on medium with 2.5-16 μM (0.5-3.0 mgl^{-1}) indolebutyric acid or NAA; however, rooting potential declines with papaya cultures, and is associated with habituation (30).

One measure of the effectiveness of *in vitro* propagation must be the clonal fidelity of the regenerated plants. Variability resulting from passage through an *in vitro* cycle has been reported from shoot tip-derived *Musa* (Table 5), pineapple (59) and papaya (24). Characters such as leaf morphology, stature, fruit size and sex type have all been altered (Table 5). Although variability is undesirable in vegetatively propagated crop species, it must also be recognized as a potential breeding tool,

TABLE 5. Clonal stability of tropical fruit and root crops from tissue culture.

Crop	Explant	Plantlet origin	Variation	Reference
Banana	shoot tip	axillary branching (and callus?)	stature	G. Johns, pers. comm.
Banana (cooking)	shoot tip	axillary branching (and callus?)	bunch and fruit size. shape, persistence of male bract	H. Irizavvy, pers. comm.
Cocoyam	apical meristem	callus	leaf shape, size, leaf and corm no.	13
Papaya	ovules	zygote callus	petiole pigmentation, internode length	24, 31
	apical meristem	axillary branching	change of sex type	24
Pineapple	axillary buds, syncarp and crown	callus	leaf spines; epicuticular wax; leaf morphology and density	59
Sweet orange	nucellus	callus	internode length, leaf morphology, growth habit	42
Tangerine	nucellus	callus	leaf morphology, fruit characteristics	49

particularly with crop species that are intractable, e.g., banana and plantain, both of which are largely sterile triploids.

2.2. Woody fruit species

Among woody tropical fruit species, there have been only a few reports of the successful regeneration of plants from culturedshoot apices and lateral buds from mature trees (Table 4). Micrografting of *Citrus* is a standard procedure for production of disease-indexed stock plants (33), but culture of shoot tips from mature *Citrus* trees has been unsuccessful. To date, the only tropical fruit tree species that have been regenerated from cultured shoot apices have been confined to the plant family Moraceae. Muriithi *et al.* (45) and Jonard *et al.* (18) recovered single plants from cultured shoot tips of fig, utilizing similar media that contained BA, gibberellic acid and either NAA or 2,4-dichlorophenoxyacetic acid (2,4-D). Proliferating cultures of mulberry have been regenerated from cultured axillary buds (40, 51) and by organogenesis from leaf callus derived from *in vitro* grown plantlets (40) on medium with BA alone. Rooting of fig and mulberry shoots occurs on media without cytokinin.

Although the culture of shoot apices and axillary buds of other tropical woody fruit tree species has not been demonstrated, recent success with the regeneration of plants from anthers by somatic embryogenesis (47) and from seedling leaf explants by organogenesis (46) of the custard apple (Annona squamosa L.) suggest that in vitro morphogenesis may be possible with this and related tree species within the plant family Annonaceae.

Regeneration pathways that have been described involve morphogenesis from maternal tissues within cultured ovules or from isolated ovular tissues (28). Both the nucellus and the inner integuments are ovular tissues of maternal origin that possess juvenile characteristics. In many polyembryonic plant species, the nucellus can produce adventitious embryos. It is possible that the nucellus of all plant species has morphogenetic potential, but that it is inhibited chemically within monoembryonic plant species (57, 58). According to Kaur et al. (22), polyembryony may be widespread among trees, particularly fruit species, that originated in the tropical rain forests. Polyembryony is known to occur in Citrus (Rutaceae), Eugenia and Myrciaria species (Myrtaceae), mangosteen (Guttiferae) and in mango (Anacardiaceae) (28). Embryogenic nucellar and integumental callus has been stimulated from cultured ovules of polyembryonic Citrus (36), mango (27), jaboticaba (25) and Eugenia species (26) (Tables 3 and 4). Usually, callus has been obtained from cultures explanted on media containing 2,4-D. Embryogenic cultures have usually been established on Murashige and Skoog medium or a variation of it. Reduction of the concentration of the major salts by at least one half appears to be beneficial (28). Sucrose concentrations in the range of 3-6% have been used. After the induction of embryogenic callus, it is normally subcultured onto medium with 2.3-9.0 µM (0.5-2.0 mg^{-1}) 2,4-D. The efficiency of somatic embryogenesis can be greatly increased by subculturing callus into liquid medium.

Either entire ovules (25, 26, 32, 34) or the isolated nucellus (19, 52) of polyembryonic species has been used as explants. Both fertilized (32, 36) and unfertilized (23) ovules have been used. The use of fertilized ovules can result in a mixture of nucellar or integument-derived and zygote-derived somatic embryos, although there is strong evidence that the zygotic embryo aborts at an early stage in many polyembryonic species or cultivars (54). There is a direct association between the degree of polyembryony and the potential for somatic embryogenesis (32, 41). The nucellus of monoembryonic cultivars that are closely related to polyembryonic species or cultivars also possesses in vitro morphogenetic potential. If the zygotic embryo is removed from monoembryonic Citrus species and cultivars (52) or from monoembryonic mango cultivars (27), it is possible to stimulate somatic embryogenesis directly from the nucellus or from nucellar callus. The efficiency of regeneration from nucellar explants of monoembryonic plants is usually considerably lower than that of polyembryonic cultivars (27).

The nucellus of temperate woody fruit species that are solely monoembryonic also has embryogenic potential in vitro (11, 44, 61). This has also been verified with the tropical fruit tree, the loquat. Embryogenic callus has been recovered from nucellar explants (28).

It is possible that the nucellus may be the explant of choice for in vitro regeneration of tropical fruit tree species, particularly as somatic embryogenesis can be induced not only from polyembryonic cultivars but also from strictly monoembryonic cultivars and species. Whether somatic embryogenesis from ovular explants can be exploited commercially will of course depend on a number of considerations. At this time, it is very difficult to obtain efficient germination of somatic embryos of woody species, with the exception of Citrus. Moreover, plants that have been derived from somatic embryos would have to pass through a long juvenile cycle that can range from 7-20 years, depending on the species. Also, there is recent evidence that Citrus plants from nucellar callus exhibit considerable variation based upon morphological and biochemical markers (Table 5). On the other hand, most tropical fruit trees have a fairly high intrinsic value (Table 6). In south Florida, the average price of planting stock of fruit trees is approximately $10.00 per tree. Moreover, with the rapid growth of the tropical fruit industry in many parts of the world (Table 1), there is a shortage of desirable clonal material. Much of this shortage is due to quarantine restrictions on the international exchange of germplasm. The nucellus and plants derived from it by somatic embryogenesis are considered to be free of microbes that might affect the maternal or donor tree. Many tropical fruit trees that are grown in regions with heavy or poorly drained soils experience serious losses due to soil-borne pathogens that cause serious root diseases. With the notable exception of Citrus, most tropical fruit scions are grafted onto monoembryonic seedling rootstocks, resulting in undesirable differences in

TABLE 6. Average price of vegetatively propagated tropical fruit trees in South Florida.[1,2,3]

Atemoya	$10.25	Guava	$ 8.50
Avocado	7.00	Longan	14.00
Canistel	10.50	Loquat	8.50
Carambola	10.00	Lychee	10.00
Cherimola	10.00	Mamey Sapote	13.00
Citrus	8.50	Mango	9.50
Guanabana	10.00	Mulberry	8.50
		Average Price = $9.97	

[1] Tree size = 1-2 feet.

[2] Price includes $2.00 charge for preparing bare rooted plants and packaging.

[3] Delivery charge is not included.

size among grafted trees and in disease resistance of the rootstocks. Plants from nucellar somatic embryos could, therefore, expedite the distribution and exchange of tropical fruit tree germplasm, and provide a source of uniform rootstocks. In the immediate future, however, further studies are needed to increase the number of fruit species that can be regenerated and to increase the efficiency of somatic embryo germination.

3. TROPICAL ROOT AND TUBER CROPS

The conventional propagation of tropical root and tuber crops has been by vegetative means, including the use of stem pieces for cassava and sweet potato, cormels or tuber pieces for many edible aroids and tuber pieces for yams. These crops have become the most important components of diets in many of the poorest countries of the world (Table 2). Furthermore, it is possible to grow these crops under adverse conditions with minimal inputs of fertilizer, irrigation and pest control practices. Because of the combination of many factors that include constant disease pressure, vegetative propagation and primitive agricultural practices, the production of tropical root and tuber crops can be seriously threatened. Therefore, _in vitro_ techniques have been developed for disease elimination or indexing, and schemes have been established at the international research centers for the continuous indexing of clonal material, germplasm preservation and international exchange of material.

TABLE 7. Tissue culture of tropical root and tuber crops.

Crop	Explant	Result	Reference
Cassava (*Manihot esculenta*)	meristem tip	virus-free plant	1,20,21,43
Cocoyam (*Xanthosoma caracu*)	shoot tip	multiple shoots	4,56
" (*Xanthosoma brasiliense*)	shoot tip	virus-free plant	55
" (*Xanthosoma sagittifolium*)	shoot tip	virus-free plant	14
Hausa potato (*Coleus parviflorus*)	leaf	multiple shoots	6
Sweet potato (*Ipomoea batatas*)	meristem tips	virus-free plant	1,43
Sweet yam (*Amorphophallus*)	corm segments	shoots	5
Taro (*Colocasia esculenta*)	shoot tip	virus-free plant	14
" " "	shoot tip	shoots	2,16
Yam (*Dioscorea alata*)	shoot tip	shoots	15
" " "	bulbil segments	shoots	3
" (*Dioscorea microstachya*	nodal segments	shoots	38

Regeneration pathways have been described that are based on shoot tip culture, adventitious shoot development from leaf and tuber explants and from stem pieces (Table 7). However, because of the relatively low value of each propagule, *in vitro* propagation has only been used to create a foundation stock of virus-indexed plants. Increased yields of tubers have been achieved by replanting with virus-indexed stock from meristem cultures. In the Caribbean region, it has been possible to increase the yield of yam by 12-16-fold (37). Cassava yield increases of as much as 1.7-fold have also been recorded (53). Despite the impact of *in vitro* cloning on production of tropical root and tuber species, there is evidence that variation can be induced in plants derived from shoot tips of cocoyam (Table 5).

Because of the prohibitively high cost of producing plants from cultures of shoot apices, there has been considerable interest in the adaptation of other regeneration pathways, e.g., somatic embryogenesis, to facilitate the direct planting of propagules (50). The use of artificial seed (encapsulated somatic embryos) or the suspension of somatic embryos in inert gels for direct planting are alternatives that have been considered. In this way, disease-free foundation stock could be maintained under laboratory conditions, and could be increased efficiently in suspension cultures. The major costs of transportation of planting material would be eliminated. At this time, however, it has been possible to induce somatic embryogenesis from clonal selections only of sweet potato (17, 34).

REFERENCES

1. Alconero R, AG Santiago, F Morales, F Rodriguez: 1975. Meristem tip culture and virus indexing of sweet potatoes. Phytopathology 65:769-773.

2. Arditti J, MS Strauss: 1979. Taro tissue culture manual. South Pacific Commission, New Caledonia.

3. Asokan MP, SK O'Hair, RE Litz: 1983. In vitro plant development from bulbil explants of two Dioscorea species. HortScience 18:702-703.

4. Asokan MP, SK O'Hair, RE Litz: 1984. Rapid multiplication of Xanthosoma caracu by in vitro shoot tip culture. HortScience 19:885-886.

5. Asokan MP, SK O'Hair, RE Litz: 1984. In vitro plant regeneration from corm callus of Amorphophallus rivieri Durieu. Scientia Hort. 24:251-256.

6. Asokan MP, SK O'Hair, RE Litz: 1984. In vitro plant regeneration from leaf disks of Hausa potato (Coleus parviflorus). HortScience 19:75-76.

7. Centro Internacional de Agricultura Tropical: 1979. Cassava tissue culture, pp. 82-87. In: Annual Report 1979. CIAT, Cali, Colombia.

8. Centro Internacional de Agricultura Tropical: 1982. Cassava tissue culture, pp. 113-123. In: Annual Report 1982. CIAT, Cali, Colombia.

9. Cohen D, PA Cooper: 1982. Micropropagation of babaco--a Carica hybrid from Ecuador, pp. 743-744. In: A. Fujiwara (ed). Plant tissue culture 1982. Japan. Assoc. Plant Tissue Culture, Tokyo.

10. Cronauer SS, AD Krikorian: 1984. Rapid multiplication of bananas and plantains by in vitro shoot tip culture. HortScience 19:234-235.

11. Eichholtz D, HA Robitaille, PM Hasegawa: 1979. Adventive embryony in apple. HortScience 14:699-700.

12. Garner RJ, SA Chaudhri: 1976. The propagation of tropical fruit trees. Commonwealth Agricultural Bureaux, Slough.

13. Gupta PP: 1985. Plant regeneration and variabilities from tuber cultures of cocoyams (Xanthosoma sagittifolium and X. violaceum). Plant Cell Rpt. 4:88-91.

14. Hartman RD: 1974. Dasheen mosaic virus and other phytopathogens eliminated from caladium, taro, and cocoyam by culture of shoot tips. Phytopathology 64:237-240.

15. Hu CY, PJ Wang: 1983. Meristem, shoot tip, and bud cultures, pp. 177-227. In: DA Evans, WR Sharp, PV Ammirato, Y Yamada (eds). Handbook of plant cell culture. Vol. 1. Techniques for propagation and breeding. Macmillan Publishing Co., New York.

16. Jackson GVH, EA Ball, J Arditti: 1977. Tissue culture of taro, Colocasia esculenta (L.) Schott. J. Hort. Sci. 52:373-382.

17. Jarret RL, S Salazar, R Fernandez Z: 1984. Somatic embryogenesis in sweet potato. HortScience 19:397-398.

18. Jonard R, P Baud, G Valderon: 1983. Development in vitro de plantes a partir d'apex preleves sur un cultivar de Figuier (Ficus carica L.). Bull. Soc. Bot. Fr. 130:301-306.

19. Juarez J, L Navarro, JL Guardiola: 1976. Obtention de plantes nucellaires de divers cultivars de clementiners au moyen de la culture de nucelle "in vitro." Fruits d'Outre Mer 31:751-761.

20. Kaiser WJ, LR Teemba: 1979. Use of tissue culture and thermotherapy to free East African cassava cultivars of African cassava mosaic and cassava brown streak diseases. Plant Dis. Rpt. 63:780-784.

21. Kartha KK, OL Gamborg: 1975. Elimination of cassava mosaic diseases by meristem culture. Phytopathology 65:826-828.

22. Kaur A, CO Ha, K Jong, VE Sands, HT Chan, E Soepadmo, PS Ashton: 1978. Apomixis may be widespread among trees of the climax rain forest. Nature 271:440-442.

23. Kochba J, P Spiegel-Roy, H Safran: 1972. Adventive plants from ovules and nucelli in citrus. Planta 106:237-245.

24. Litz RE: Papaya, pp. 349-368. 1984. In: WR Sharp, DA Evans, PV Ammirato, Y Yamada (eds). Handbook of the plant cell culture. Vol. 2. Crop species. Macmillan Publishing Co., New York.

25. Litz RE: 1984. In vitro somatic embryogenesis from callus of jaboticaba, Myrciaria cauliflora. HortScience 19:62-64.

26. Litz RE: 1984. In vitro responses of adventitious embryos of two polyembryonic Eugenia species. HortScience 19:720-722.

27. Litz RE: 1984. In vitro somatic embryogenesis from nucellar callus of monoembryonic Mangifera indica L. HortScience 19:715-717.

28. Litz RE: 1985. Somatic embryogenesis in tropical fruit trees, pp. 179-193. In: RE Henke, KW Hughes, MP Constantin, A Hollaender (eds). Tissue culture in forestry and agriculture. Plenum Publishing Corp., New York.

29. Litz RE, RA Conover: 1978. In vitro propagation of papaya. HortScience 13:241-242.

30. Litz RE, RA Conover: 1981. Effect of sex type, season and other factors on in vitro establishment and culture of Carica papaya L. explants. J. Amer. Soc. Hort. Sci. 106:792-794.

31. Litz RE, RA Conover: 1983. High frequency somatic embryogenesis from Carica suspension cultures. Ann. Bot. 51:683-686.

32. Litz RE, RL Knight, S Gazit: 1982. Somatic embryos from cultured ovules of Mangifera indica L. Plant Cell Rpt. 1:264-266.

33. Litz RE, GA Moore, C Srinivasan: 1985. In vitro systems for propagation and improvement of tropical fruits and palms. Hort. Rev. 7:157-200.

34. Liu JR, DJ Cantliffe: 1984. Somatic embryogenesis and plant regeneration in tissue cultures of sweet potato (Ipomoea batatas Poir.). Plant Cell Rpt. 3:112-113.

35. Ma S, C Shii: 1972. In vitro formation of adventitious buds in banana shoot apex following decapitation (in Chinese). J. Hort. Sci. (China) 18:135-142.

36. Maheshwari P, NS Rangaswamy: 1958. Polyembryony and in vitro culture of embryos of Citrus and Mangifera. Ind. J. Hort. 15:275-282.

37. Mantell SH, SQ Haque, AP Whitehall: 1979. A rapid propagation system for yams. Yam Virus Project Bulletin No. 1. Caribbean Agricultural Research and Development Institute, Trinidad.

38. Mapes MO, U Urata: 1970. Aseptic stem culture of a Dioscorea clone, pp. 604-606. In: Tropical root and tuber crops tomorrow. Vol. 1. Proceedings of the second international symposium on tropical root and tuber crops.

39. Mathews VH, TS Rangan: 1979. Multiple plantlets in lateral bud and leaf explant in vitro cultures of pineapple. Scientia Hort. 11:319-328.

40. Mhatre M, VA Bapat, PS Rao: 1985. Regeneration of plants from the culture of leaves and axillary buds in mulberry (Morus indica L.). Plant Cell Rpt. 4:78-80.

41. Mitra GC, HC Chaturvedi: 1972. Embryoids and complete plants from unpollinated ovaries and from ovules of in vivo-grown emasculated flower buds of Citrus spp. Bull. Torrey Bot. Club 99:184-189.

42. Moore GA: Somatic embryogenesis in Citrus. 1985. Simposio de Interciencia. Biotecnologia en las Americas. II. Aplicaciones en Agricultura Tropical. Pp. 7-8 (abstract).

43. Mori K: 1971. Production of virus-free plants by means of meristem culture. Japan. Agric. Res. Quart. 6:1-7.

44. Mullins MG, C Srinivasan: 1976. Somatic embryos and plantlets from an ancient clone of the grapevine (cv. Cabernet-Sauvignon) by apomixis in vitro. J. Exp. Bot. 27:1022-1030.

45. Muriithi LM, TS Rangan, BH Waite: 1982. In vitro propagation of fig through shoot tip culture. HortScience 17:86-87.

46. Nair S, PK Gupta, AF Mascarenhas: 1983. Haploid plants from in vitro anther culture of Annona squamosa Linn. Plant Cell Rpt. 2:198-200.

47. Nair S, PK Gupta, MV Shirgurkar, AF Mascarenhas: 1984. In vitro organogenesis from leaf explants of Annona squamosa Linn. Plant Cell Tissue Organ Culture 3:29-40.

48. Navarro L, J Juarez: 1977. Tissue culture techniques used in Spain to recover virus-free citrus plants. Acta Hort. 78:425-435.

49. Navarro L, JM Ortiz, J Juarez: 1985. Aberrant citrus plants obtained by somatic embryogenesis of nucelli cultured in vitro. HortScience 20:214-215.

50. O'Hair SK, MP Asokan, RE Litz, HH Bryan: 1983. Fluid drilling of somatic potato embryos as a means of crop establishment, pp. 144-146. In: WJ Hooker (ed). Research for potato in the year 2000. Centro Internacional de Papas, Lima, Peru.

51. Patel GK, VA Bapat, PS Rao: 1983. In vitro culture of organ explants of Morus indica: plant regeneration and fruit formation in axillary bud culture. Z. Pflanzenphysiol. 111:465-468.

52. Rangan TS, T Murashige, WP Bitters: 1968. In vitro initiation of nucellar embryos in monoembryonic Citrus. HortScience 3:226-227.

53. Roca WM: Cassava, pp. 269-301. 1984. In: WR Sharp, DA Evans, PV Ammirato, Y Yamada (eds). Handbook of plant cell culture. Vol. 2. Crop species. Macmillan Publishing Co., New York.

54. Singh LB: 1960. The mango. Leonard Hill, Ltd., London.

55. Staritsky G: 1974. Xanthosoma brasiliense Eng. propagated virus-free in vitro. Tropical Root and Tuber Crops Newsletter 7:38-39.

56. Strauss MS, J Arditti: 1980. Plantlet regeneration from shoot tip cultures of Xanthosoma caracu. Ann. Bot. 45:209-212.

57. Tisserat B, T Murashige: 1977. Probable identity of substances in Citrus that repress asexual embryogenesis. In Vitro 13:785-789.

58. Tisserat B, T Murashige: 1977. Repression of asexual embryogenesis *in vitro* by some plant growth substances. In Vitro 13:799-805.

59. Wakasa K: 1979. Variation in the plants differentiated from the tissue culture of pineapple. Japan. J. Breed. 29:13-22.

60. Yie ST, SI Liaw: 1977. Plant regeneration from shoot tips and callus of papaya. In Vitro 13:564-568.

61. Zatyko JM, I Simon, CS Szabo: 1975. Induction of polyembryony in cultivated ovules of red currant. Plant Sci. Lett. 4:822-825.

COMMERCIAL *IN VITRO* PROPAGATION AND PLANTATION CROPS

RAYMOND P. OGLESBY AND JOHN L. GRIFFIS, JR.

Oglesby Plant Labs, Hollywood, FL 33023, U.S.A.

For several thousand years man has been domesticating numerous plants for a variety of uses including food, fiber, pharmaceuticals, and aesthetics. Certain of these crops, usually perennials that are planted over large areas and maintained for many growing seasons and many harvests, are referred to as "plantation crops." Among these are coffee, tea, coconut, banana, pineapple, cacao, oil palm, rubber, various spices, and forest trees. For a variety of reasons, including both geographic distribution of the plants and the success of conventional propagation methods for these crops, *in vitro* propagation techniques have not until recently received the research efforts that have gone into herbaceous temperate and greenhouse crops. Today, however there are active research programs dealing with *in vitro* propagation of all these crops going on around the world. At Oglesby Plant Labs our ongoing research programs deal with some of these "plantation crops" including: pineapple, banana, oil palm, coconut, clove, allspice, cinnamon, and some coniferous forest species. Our greatest contributions have been in the development of practical schemes for commercial production of both banana and pineapple using *in vitro* techniques. Our largest project to date has involved the *in vitro* production of over 1.5 million banana plants for a large banana plantation in the Caribbean.

Of all the plantation crops listed above, bananas were probably one of the earliest cultivated by man. This fact may be attributed to many factors including the rapidity with which the plant produces a crop and produces vegetative offsets, the natural occurrence of sterile, seedless triploid fruits, and the ease with which the edible portion may be picked and eaten. Because the plant was herbaceous rather than woody, it was also one of the first plants in this group to attract the attention of plant tissue culturists. The production of virus-free bananas from meristem culture was reported more than 10 years ago (1). Since that time, a number

Zimmerman, R.H. et al. (eds), Tissue Culture as a Plant Production System for Horticultural Crops.
 ISBN 90-247-3378-2.

of others have done additional work with dessert bananas, plantains, and abaca (2, 3, 5, 6). Oglesby Plant Labs has been involved in the development of commercially viable in vitro propagation schemes of various bananas since 1979.

Our current project has involved the large-scale production of the sterile triploid banana AAA 'Grande Naine'. At this time, Oglesby produces 15,000 banana plantlets in vitro for shipment to the Caribbean every week. The project had the same beginnings as any other several years ago when the first explants were placed into sterile culture. Our original techniques were very similar to those reported by Krikorian and Cronauer (5). A summary of the process is included in Table 1.

TABLE 1. Shoot apex isolation and in vitro multiplication of 'Grande Naine'.

1. Obtain one or more stock plants and cut off the top of the plant.
2. Remove all of the soil and roots. Wash corm thoroughly.
3. Remove the outer sheathing leaf bases one at a time by hand.
4. Remove a large portion of the corm with a knife, rinse thoroughly.
5. Remove more leaf bases and wash thoroughly again.
6. Sterilize remaining explant in recommended concentration of Alcide, 20 min.
7. Rinse under laminar flow hood with sterile distilled water 1 or 2 times.
8. Remove remaining leaves with scalpel under dissecting microscope.
9. Remove shoot apex and small amount of basal corm with scalpel.
10. Sterilize shoot apex in Alcide for 1-2 minutes, do not rinse.
11. Place explant on solid stage I explant medium[1].
12. Observe for 3 weeks; watch for explant enlargement and color development.
13. Transfer to fresh medium every 3 weeks as explant forms plantlet.
14. Plantlet 3 to 6 cm in height is ready for multiplication.
15. Plantlet is removed and transverse cut separates leaves from base & apex.
16. Remove any darkened or necrotic tissue from basal piece of plantlet.
17. Cut basal piece in half, through the shoot apex.
18. Place both halves in fresh stage II culture medium.
19. After 4 weeks, remove new shoots formed on original explant.
20. Repeat Steps 15-18 until desired number of cultures is obtained.
21. Remove individual shoots from multiple shoot cultures with scalpel.
22. Place shoots in vessels containing stage III rooting medium with charcoal.
23. Roots are formed after 14-21 days and vessels are shipped to grower.

[1] Culture Medium, stage I: MS minimal organics salts, sucrose 0.12 M, thiamine HCl 2.97 µM, inositol 5.55 mM, CW 15% v/v, agar 0.8%; pH 5.7; heat sterilized.

Of course, every step in the in vitro process was important. However, the key to success at Oglesby was in step 23, for without the grower to purchase the end product, the entire process would have remained academic rather than commercial.

The use of in vitro propagation techniques to produce banana plants gave the grower several advantages over conventional propagation. The cultivar AAA 'Grande Naine' was new to some of the islands, and there would be considerable difficulty in importing enough clean planting stock to plant an acreage of any size. Tissue-cultured plants, however, could be imported in large quantities and could be easily tested and certified disease-free. Additionally, banana "seed" corms were bulky and expensive to ship and they could be held by agricultural inspectors for extended periods of time to determine if diseases or pests were present. TC plants were relatively inexpensive to ship because of their size and weight and they readily passed international inspection since they were "packaged" in a sterile environment.

The original project expanded as the grower became more familiar in dealing with the cultured plantlets. Initially, only 50 acres of land (30,000 plants) were to be planted with bananas from tissue culture. These were to provide "seed" stock plants for conventional propagation for the rest of the acreage. However, upon completion of the initial planting, the grower expanded the project to 500 acres (300,000 plants). This proved to be so successful that the acreage was eventually expanded to include 2000 acres of tissue cultured banana plants.

The project itself was begun to reintroduce a commercially viable plantation crop for export from the Caribbean islands to Europe. The success or failure of the project would determine whether or not bananas, especially AAA 'Grande Naine', could again become a significant commercial crop throughout the region. All the criteria for planting, growing, harvesting, and shipping of bananas would be established with this project.

The site selection for the banana project itself was somewhat unusual. The land had no recent commercial usage and had only 12 inches of soil above the water table. Bananas require about four feet of soil, so a large amount of drainage pipe had to be installed and numerous drainage canals constructed to divert the water from the field. Additionally, the soil had to be tilled and prepared for planting, irrigation pipe installed (95 miles of pipe for each 500 acres) and a structure had to be built to acclimatize the TC bananas before they could be planted in the fields. This acclimatization process is summarized in Table 2.

TABLE 2. Tissue cultured bananas from laboratory to harvest.

1. Plants are received from Oglesby Plant Labs in plastic shipping vessels.
2. Plants are removed from vessels and placed in tubs of water to facilitate removal of the agar medium from the tender roots.
3. Plants are planted into soil mix in liner trays and placed under 70% shade.
4. After 2 weeks, plants are transferred to 30% shade.
5. After 2 more weeks, plants are transferred to full sun.
6. After 3 to 4 more weeks, plants are about 12 inches tall and are ready to be planted in the field. Six laborers can plant 15,000 liners in the field in only 2 days. Conventionally propagated material would take much longer.
7. With proper fertilization and watering, the plants reach maturity and fruit harvest begins after only 10 to 12 months in the field. Harvest of conventionally propagated bananas begins after 12 to 14 months in the field.

Additionally, all new harvesting equipment also had to be installed. Today, overhead conveyors transport the stalks of bananas to packing sheds, where the fruit is removed, graded, de-latexed, weighed, and packaged for shipment. The overwhelming success of this project is now assured!

At Oglesby Plant Labs and other labs around the world, tissue culture techniques are being constantly improved. There is no doubt that a bright future exists for the relationship between plantation crops and in vitro propagation.

REFERENCES

1. Berg LA, M Bustamante: 1974. Heat treatment and meristem culture for the production of virus-free bananas. Phytopathology 64:320-322.

2. Cronauer SA, AD Krikorian: 1984. Rapid multiplication of bananas and plantings by in vitro shoot tip culture. HortScience 19:234-235.

3. Hwang SC, CL Chen, JC Lin, HL Lin: 1984. Cultivation of banana using plantlets from meristem culture. HortScience 19:231-233.

4. Jarrett RL, W Rodriguez, R Fernandez: 1985. Evaluation, tissue culture propagation, and dissemination of 'Saba' and 'Pelipita' plantains in Costa Rica. Scientia Hort. 25:137-147.

5. Krikorian AD, SA Cronauer: 1984. Banana, pp. 327-348. In: 1984. WR Sharp, DA Evans, PV Ammirato, Y Yamada (eds). Handbook of plant cell culture. Vol. 2. Crop species. Macmillan Publishing Co., New York.

6. Mante S, HB Tepper: 1983. Propagation of *Musa textilis* Née plants from apical meristem slices *in vitro*. Plant Cell Tissue Organ Culture 2:151-159.

USE OF TISSUE CULTURE FOR MICROPROPAGATION OF VEGETABLE CROPS[1]

TIMOTHY J. NG

Department of Horticulture, University of Maryland, College Park, MD 20742, U.S.A.

1. INTRODUCTION

The use of tissue culture with vegetable crops has an extensive history dating back to the classic studies of White (52) and Caplin and Steward (8). As a result of the efforts of many investigators in this area, techniques for micropropagation of most major species of vegetables have been developed. Most of these techniques involve the use of vegetative meristematic tissue; however, methods have also been developed in many crops for regeneration of plants from callus and from protoplasts. Due to the sheer number of plant species that are categorized as vegetables, a comprehensive literature review is not possible within the scope of this article. In addition, techniques for micropropagation of many of the major vegetable crops have recently been reviewed (1, 14, 46). Thus, this article will concentrate on describing situations where production of vegetable plants through tissue culture represents an economically viable prospect for the vegetable industry.

For most vegetables, the use of plants derived from tissue culture for field production of an edible product is not a feasible alternative when compared to current plant production systems. This is due in part to the fact that most vegetable crops are seed-propagated. Seeds of these vegetables are usually produced in abundance and are therefore relatively inexpensive to produce when compared to the costs of plants from micropropagation; in addition, seeds are durable, compact, self-sufficient, easily handled and relatively stable genetically. Even in vegetables where asexual propagation is the established mode of propagation (e.g. potato,

---

[1] Scientific Article 7289, Contribution A4300, of the Maryland Agricultural Experiment Station (Department of Horticulture).

Zimmerman, R.H. et al. (eds), Tissue Culture as a Plant Production System for Horticultural Crops.
 ISBN 90-247-3378-2.

sweet potato), micropropagation does not represent an economic alternative because of the ease of obtaining a large number of plants through bedding or the use of seed pieces. Where micropropagation of vegetable crops may show the most promise at the moment are situations where it serves as an adjunct to existing seed technology (e.g. in the production of specific parental genotypes for hybrid seed production) and in specific cases where difficulties are involved in traditional propagation methods for a given species.

2. PARENTS FOR HYBRID SEED PRODUCTION

2.1. Genotypes incapable of self-pollination

Many vegetable seed companies have recently established tissue culture facilities at their main field installations. While these in part were built to interface with programs currently under way in affiliated biotechnological companies, these laboratories are also being used to investigate the feasibility of producing parental plants for hybrid seed production. This approach appears to be most feasible in situations where the parental genotype is incapable of self-pollination, whether through male sterility, gynoecy or self-incompatibility, or in cases where the quality of resulting seed is extremely poor.

Male sterility and gynoecy are useful traits in hybrid production in that costs concomitant with hand emasculation are avoided. Also, in species where the natural mode of pollination is through natural vectors, these phenotypes eliminate the costs involved in hand pollination. Where micropropagation is most useful in producing male sterile plants is in species where an established cytoplasmic sterility system does not exist. For example, in tomato (Lycopersicon esculentum) a number of nuclear genes exist that convey male sterility (54). Since a cytoplasmic male sterile factor has not yet been identified, these male sterile lines must be maintained in a heterozygous situation and male fertile segregants must be rogued out each generation in order to insure true hybrid seed production from the male steriles. Tissue culture production of tomato plants (30, 39, 54) allows for rapid propagation of these genotypes with fidelity to the phenotypes.

Another interesting situation exists in asparagus (Asparagus officinalis), a dioecious vegetable where male plants exhibit superior survival and establishment in the field and also tend have superior vigor and yield. Maleness is determined by a single dominant allele; males are normally heterozygous at that locus and females homozygous recessive. In situations where this crop is seed-propagated, half of the field population is male and the other half female, resulting in yields that may be less than those potentially possible with an all male planting. Dore (11) and Reuther (44) have reported that "allmale" genotypes (homozygous for the dominant male allele) have been produced through doubled haploids and that micropropagation of these "allmale" lines is possible. Use of these lines as pollen sources result in the production of asparagus seed that produce populations consisting solely of male plants.

Genetic gynoecy is common in the cucurbit family (5). In cucumber (Cucumis sativus) the production of strictly pistillate flowers allows for complete cross-pollination and strictly hybrid seeds from plants that are consistent in this phenotype. However, maintaining these lines involves the production of staminate flowers from these genotypes, an involved process that includes rooting 3-4 cuttings from a given genotype and treating at least 1 cutting with gibberellin or an ethylene inhibitor such as silver ion. Handley and Chambliss (18) demonstrated that up to 5 or 6 axillary buds could be obtained from each cutting through tissue culture, thereby increasing the amount of plant material available from a given plant. Wehner and Locy (50) demonstrated that a number of genotypes of cucumber were capable of regenerating shoots and roots in culture.

Self-incompatibility can also pose a problem in breeding and seed production programs (23). Procedures for self-pollination of these lines utilize bud pollination techniques which are labor-intensive, particular in the case of vegetables like broccoli (Brassica oleracea) where flower size is small. Seed derived from bud pollinations also tend to be poor quality and low in vigor. Anderson and Carstens (2) demonstrated that broccoli plantlets could be obtained from flower bud explants; stem, leaf and leaf rib explants also appear to be suitable material for culture (22). When relative costs are compared, micropropagation can represent an economically viable alternative to bud pollination (3).

2.2. Polyploid genotypes

In certain vegetables, genotypes with chromosome complements other than the standard diploid level can be desirable in certain situations. In cases where these genotypes represent autopolyploids, line maintenance can be a problem because of irregularities resulting from abnormal chromosome pairing (e.g., trivalent and quadrivalent formation) during meiosis, although these irregularities may form the basis for desirability of these genotypes in field production (20).

An example would be the case of the triploid watermelon (Citrullus lanatus), a popular development because of the seedless nature of the fruits from these plants (24). Triploid watermelons have existed for many years but few widely-grown cultivars exist that exploit this genetic combination because of the difficulties involved in breeding and seed production. As triploids contain only vestigial traces of aborted seeds, triploid seeds are usually produced by transferring pollen from diploid plants to tetraploid ones. Triploid plantlets from micropropagation could be produced for field production but would represent an expensive alternative to triploid seed production. Where micropropagation may show the most promise in triploid production is through cloning of the tetraploid parent. Sexual propagation of tetraploid watermelon by seed is not without problems: aneuploidy can result from abnormal chromosome pairing; tetraploid plants tend to produce few seeds; and seeds from tetraploid plants tend to be low in vigor and viability (37). Thus, techniques developed for watermelon tissue culture (5, 6) could prove useful in maintaining the parental tetraploid lines.

3. PROPAGATION AND PRESERVATION OF SPECIFIC GENOTYPES

3.1. Rapid propagation

In certain instances, tissue culture production represents the most rapid means of obtaining large numbers of a specific gene combination, particularly when gene combination is highly heterozygous or when the species requires an extensive amount of time to go from generation to generation (12, 27, 31, 32). This can be the case in many crops where a specific genotype is not easily duplicable although use of these techniques commercially tend to be somewhat limited because of cost. However, there are specific instances where it is cost effective to use tissue culture for producing plants for field production. For instance, in many cases asparagus is propagated by crown division and by cuttings; these procedures are slow and yield few progeny. Although micropropagation by callus and cell cultures tends to be unreliable because of genetic instability (53), techniques using vegetative tissue have proven to be reliable and useful for micropropagation (10, 17, 19, 35).

In potato (Solanum tuberosum), seed tubers are generally planted for propagation and the annual multiplication rate is approximately 6- to 10-fold. This slow rate can represent a major impediment to the introduction and release of a new cultivar. With micropropagation techniques, it is possible to get as high as an 80-fold increase in potato plants within 40 days (31).

3.2. Virus elimination

Perhaps one of the most established commercial uses of tissue culture in vegetables is for elimination of viruses in vegetative tissue, particularly in crops that are asexually propagated such as potato and sweet potato. Elimination of viral contamination allows for distribution of disease-free stock to growers as well as guaranteeing pure plant tissue for germplasm storage. The procedure typically involves excision of meristems, culture on a suitable medium at high temperature, plantlet regeneration and rooting, and virus-indexing (28). However, the possibility exists that heritable changes in the genetic material may occur during the viral elimination process (34).

3.3. Germplasm preservation

As breeding efforts continue in the vegetable area and the number of available genotypes for a given crop proliferate, archival storage of useful germplasm is becoming an important problem (51). Numerous studies have been conducted in using low temperature storage of plant cell and tissue cultures as alternatives to traditional procedures. These appear most feasible in cases where the crop is either vegetatively propagated and storage facilities are at a minimum, or where the seeds are relatively short-lived.

In order to successfully store cultures, growth must be slowed to a minimum; this can be accomplished through various techniques including minimal media, low temperature, mineral oil, partial dehydration or hypobaric conditions (4). Once cell growth and division have been effectively suppressed, cultures can be transferred to low (4-6°C) or cryogenic (-196°C) temperatures for storage. In cases where cryogenic storage is used, slow freezing and hardening with abscisic acid appear to aid survival of the cultures (9). These cultures can be thawed at 40°C after cryopreservation and washed with hormone free media; although specific lesions may occur during freezing, there is sufficient cell survival to allow proliferation.

4. FIELD ESTABLISHMENT OF PLANTS FROM MICROPROPAGATION

4.1. Direct establishment of plantlets

One of the problems involved with using tissue culture to propagate plants for field production is the high cost involved. On a per plantlet basis, most of these costs are incurred in the post-culture process of rooting and hardening prior to transplanting. This process generally requires fairly precise temperature and humidity control within a greenhouse or growth chamber environment as well as the labor input. Once the plantlets are established, techniques and equipment developed for field establishment of transplants grown from seed (e.g., seedling plugs, transplanting machinery) are suitable for their transfer to the field.

Currently there are a number of investigators working with the possibility of directly transferring plantlets to the field. For instance, Levy (31) transferred potato plantlets from proliferated shoots directly from glass tubes to the field with subsequent irrigation. Many of the plants survived, indicating that further refinement of this procedure may eventually eliminate the need for post-culture hardening of the plantlets. Useful treatments might include the use of antitranspirants to decrease water loss from the plants during the critical period of field establishment, although the concentration of these compounds will probably have to be somewhat lower than those recommended for use with normal vegetative tissue (49).

4.2. Encapsulation of somatic embryos

A relatively new and interesting development in the area of micropropagation of vegetables is the prospect of encapsulating somatic embryos to create synthetic seeds (25). Somatic embryos have been reported to proliferate in callus and tissue cultures of a number of plants (21, 25, 33, 40, 45, 47); free-floating embryos are also sometimes found in liquid media cultures (7). These embryos are capable of undergoing differentiation and growth to yield rooted plantlets when transferred to media free of growth regulators.

Kitto and Janick (25, 26) obtained somatic embryos of carrots (*Daucus carota*) by culturing 14-day old stationary phase cell suspensions on a hormone-free basal medium and encapsulated them with 5% polyethylene oxide. Initially only 3% of the embryos survived encapsulation and drying to a constant weight; this survival rate increased when the cultures were pretreated with ABA during embryo induction.

In studies with carrot and pumpkin somatic embryos, Smith and Murashige (47) emphasized the need for synchronization of embryo initiation and the production of normal and uniform embryos. They reported that fractionation of cell cultures through a series of sieves followed by centrifugation and resuspension in sucrose suppressed precocious germination and increased the proportion of carrot embryos in earlier developmental stages. Redenbaugh et al. (43) coated celery (*Apium graveolens*) embryos with sodium alginate and immersed them in a calcium chloride solution. The capsule immediately hardened and, when subsequently planted *in vitro*, exhibited 90-100% survival. However, survival was much lower when the capsules were planted under greenhouse conditions.

Obviously, although appropriate techniques need to be worked out to increase survival and performance of the encapsulated embryos, the process of encapsulating somatic embryos shows great potential. A U.S. patent has recently been awarded for a method of producing synthetic seeds by encapsulating embryos in a biodegradable polymer (16). The prospect also exists that encapsulation materials may be able to incorporate growth regulators, pesticides, fertilizers and field amendments (e.g., bacteria) to increase cropping efficiency. Normal seeds often germinate poorly under adverse conditions (38) or when subjected to inadequate storage conditions (41); high quality synthetic seeds with appropriate amendments may provide better field emergence in these situations. In addition, if somatic embryos can be produced inexpensively and uniformly in the future, techniques such as those developed for fluid drilling of pregerminated seeds may also be adaptable for transfer of the somatic embryos to the field.

5. DIFFICULTIES WITH CURRENT MICROPROPAGATION SYSTEMS

5.1. Genetic stability of cultures

Besides the high cost of micropropagation, there are a number of problems that need to be overcome prior to any widespread use of tissue culture in producing vegetable plants. One of these problems is the genetic instability often observed in long-term cultures, particularly those maintained as callus or cell suspensions (42). Problems may even arise with short-term cultures (48). Genetic anomalies that can occur range from point mutations (13) to chromosome aberrations including aneuploidy and polyploidy (36, 40, 48). The resulting genetic changes may result in a range of desirable and undesirable characters including loss of regeneration ability (13, 40).

5.2. Differential genotype effects

Another major problem currently faced is genetic differences within a species for the ability to regenerate plants from callus or differentiated tissue. Ziv et al. (54) reported that callus induction and shoot formation in tomato cultures were promoted by the presence of a male sterile allele. Also in tomato, Kurtz and Lineberger (29) reported that different genotypes showed little difference in callus formation over different concentrations of indoleacetic acid but differed considerably in shoot regeneration ability. Other differences in tissue culture performance among tomato genotypes have been reported (30, 39). Wehner and Locy (50) investigated 85 genotypes of cucumber and reported differences in abilities of hypocotyl and cotyledon explants to form shoots and roots. The fact that proliferation and differentiation ability in culture can be a heritable trait (15) indicates a potential for incorporating screening for tissue culture performance into a breeding program to complement current screening procedures for more traditional characters.

6. SUMMARY

Micropropagation has not yet become a standard propagation practice in most facets of the vegetable industry. Start-up costs are high for initiating a program and labor costs can remain high even after the establishment of a laboratory. However, it does show promise in certain areas of vegetable plant production. In certain situations, it can serve as a useful adjunct for seed production technology in some crops by providing a means of propagating difficult-to-propagate parental material. It is also useful for increasing the propagation rate of some clonally propagated vegetables.

REFERENCES

1. Ammirato PV, DA Evans, WR Sharp, Y Yamada (eds): 1984. Handbook of plant cell culture. Volume 3, Crop species. Macmillan Publishing Co., Inc., New York.

2. Anderson WC, JB Carstens: 1977. Tissue culture propagation of broccoli, Brassica oleracea (Italica Group), for use in F_1 hybrid seed production. J. Amer. Soc. Hort. Sci. 102:69-73.

3. Anderson WC, GW Meagher, AG Nelson: 1977. Cost of propagating broccoli plants through tissue culture. HortScience 12:543-544.

4. Bajaj YPS: 1983. Cryopreservation and international exchange of germplasm, pp. 19-41. In: SK Sen, KL Giles (eds). Plant cell culture in crop improvement. Plenum Press, New York and London.

5. Barnes LR, FD Cochran, RL Mott, WR Henderson: 1978. Potential uses of micropropagation for cucurbits. Cucurbit Genet. Coop. 1:21.

6. Blackmon WJ, BD Reynolds: 1982. In vitro shoot regeneration of Hibiscus acetosella, muskmelon, watermelon, and winged bean. HortScience 17:588-589.

7. Bragdo-Aas M: 1977. Regeneration of plants from callus of potato tubers. Acta Hort. 78:133-137.

8. Caplin SM, FC Steward: 1949. A technique for the controlled growth of excised plant tissue in liquid media under aseptic conditions. Nature 163:920-924.

9. Chen THH, KK Kartha, LV Gusta: 1985. Cryopreservation of wheat suspension culture and regenerable callus. Plant Cell Tissue Organ Culture 4:101-109.

10. Chin CK: 1982. Promotion of shoot and root formation in asparagus in vitro by ancymidol. HortScience 17:590-591.

11. Dore C: 1977. In vitro techniques as an efficient tool in asparagus breeding. Acta Hort. 78:89-93.

12. Dunstan DI, KC Short: 1977. In vitro studies on organogenesis and growth in Allium cepa tissue cultures. Acta Hort. 78:139-145.

13. Evans DA, WR Sharp: 1983. Single gene mutations in tomato plants regenerated from tissue culture. Science 221:949-951.

14. Evans DA, WR Sharp, PV Ammirato, Y Yamada (eds): 1983. Handbook of plant cell culture. Volume 1, Techniques for propagation and breeding. Macmillan Publishing Co., Inc., New York.

15. Frankenberger EA, PM Hasegawa, EC Tigchelaar: 1981. Diallel analysis of shoot-forming capacity among selected tomato genotypes. Z. Pflanzenphysiol. 102:233-242.

16. Gebhart F: 1985. Plant Genetics gets synthetic seed patent. Genet. Engin. News, Oct. 1985. p. 18.

17. Guri A, S Izhar: 1984. Improved efficiency of plant regeneration from protoplasts of eggplant (Solanum melongena L.). Plant Cell Reports 3:247-249.

18. Handley LW, OL Chambliss: 1979. In vitro propagation of Cucumis sativus L. HortScience 14:22-23.

19. Hasegawa PM, T Murashige, FH Takatori: 1973. Propagation of asparagus through shoot apex culture. II. Light and temperature requirements, transplantability of plants, and cyto-histological characteristics. J. Amer. Soc. Hort. Sci. 98:143-148.

20. Holl FB: 1975. Innovative approaches to genetics in agriculture. Can. J. Genet. Cytol. 17:517-524.

21. Jarrett RL, S Salazar, R Fernandez Z: 1984. Somatic embryogenesis in sweet potato. HortScience 19:397-398.

22. Johnson BB, ED Mitchell Jr: 1978. In vitro propagation of broccoli from stem, leaf, and leaf rib explants. HortScience 13:246-247.

23. Keller WA, T Rajhathy, J Lacapra: 1975. In vitro production of plants from pollen in Brassica campestris. Can. J. Genet. Cytol. 17:655-666.

24. Kihara H: 1951. Triploid watermelons. Proc. Amer. Soc. Hort. Sci. 58:217-230.

25. Kitto SL, J Janick: 1985. Production of synthetic seeds by encapsulating asexual embryos of carrot. J. Amer. Soc. Hort. Sci. 110:277-282.

26. Kitto SL, J Janick: 1985. Hardening treatments increase survival of synthetically-coated asexual embryos of carrot. J. Amer. Soc. Hort. Sci. 110:283-286.

27. Klimaszewska K, WA Keller: 1985. High frequency plant regeneration from thin cell layer explants of Brassica napus. Plant Cell Tissue Organ Culture 4:183-197.

28. Kuo CG, BJ Shen, MJ Shen, SK Green, DR Lee: 1985. Virus-free sweet potato storage roots derived from meristem tips and leaf cuttings. Scientia Hort. 26:231-240.

29. Kurtz SM, RD Lineberger: 1983. Genotypic differences in morphogenic capacity of cultured leaf explants of tomato. J. Amer. Soc. Hort. Sci. 108:710-714.

30. Kut SA, DA Evans: 1982. Plant regeneration from cultured leaf explants of eight wild tomato species and two related Solanum species. In Vitro 18:593-598.

31. Levy D: 1985. Propagation of potato by direct transfer of in vitro proliferated shoot cuttings into the field. Scientia Hort. 26:105-109.

32. Litz RE, RA Conover: 1978. In vitro propagation of sweet potato. HortScience 13:659-660.

33. Liu JR, DJ Cantliffe: 1984. Somatic embryogenesis and plant regeneration in tissue cultures of sweet potato (Ipomea batatas Poir.) Plant Cell Reports 3:112-115.

34. Moyer JW, WW Collins: 1983. 'Scarlet' sweet potato. HortScience 18:111-112.

35. Murashige T, MN Shabde, PM Hasegawa, FH Takatori, JB Jones: 1972. Propagation of asparagus through shoot apex culture. I. Nutrient medium for formation of plantlets. J. Amer. Soc. Hort. Sci. 97:158-161.

36. Murata M, TJ Orton: 1983. Chromosome structural changes in cultured celery cells. In Vitro 19:83-89.

37. Nerson H, HS Paris, Z Karchi, M Sachs: 1985. Seed treatments for improved germination of tetraploid watermelon. HortScience 20:897-899.

38. Ng TJ, EC Tigchelaar: 1973. Inheritance of low temperature seed sprouting ability in tomato. J. Amer. Soc. Hort. Sci. 98:314-316.

39. Ohki S, C Bigot, J Mousseau: 1978. Analysis of shoot-forming capacity in vitro in two lines of tomato (Lycopersicon esculentum Mill.). Plant & Cell Physiol. 19:27-42.

40. Orton TJ: 1985. Genetic instability during embryogenic cloning of celery. Plant Cell Tissue Organ Culture 4:159-169.

41. Pesis E, TJ Ng: 1984. The role of anaerobic respiration in germinating muskmelon seeds. I. In relation to seed lot quality. J. Expt. Bot. 35:356-365.

42. Phillips GC, JF Hubstenberger: 1985. Organogenesis in pepper tissue cultures. Plant Cell Tissue & Organ Culture 4:261-269.

43. Redenbaugh K, J Nichol, ME Kossler, B Paasch: 1984. Encapsulation of somatic embryos for artificial seed production. In Vitro 20:256. (Abstr.).

44. Reuther G: 1977. Adventitious organ formation and somatic embryogenesis in callus of asparagus and iris and its possible application. Acta Hort. 78:217-224.

45. Ronchi VN, MA Caligo, M Nozzolini, G Luccarini: 1984. Stimulation of carrot embryogenesis by proline and serine. Plant Cell Reports 3:210-214.

46. Sharp WR, DA Evans, PV Ammirato, Y Yamada (eds): 1984. Handbook of plant cell culture. Volume 2, Crop species. Macmillan Publishing Co., Inc., New York.

47. Smith WA, T Murashige: 1984. A method for quantitative study of development of somatic embryos in plant cell culture. In Vitro 20:256. (Abstr.).

48. Sree Ramulu K, P Dijkhuis, S Roest, GS Bokelmann, B DeGroot: 1984. Early occurrence of genetic instability in protoplast cultures of potato. Plant Sci. Letters 36:79-86.

49. Sutter EG, M Hutzell: 1984. Use of humidity tents and anti-transpirants in the acclimatization of tissue-cultured plants to the greenhouse. Scientia Hort. 23:303-312.

50. Wehner TD, RD Locy: 1981. In vitro adventitious shoot and root formation of cultivars and lines of Cucumis sativus L. HortScience 16:759-760.

51. Westcott RJ, GG Henshaw, BWW Grout, WW Roca: 1977. Tissue culture methods and germplasm storage in potato. Acta Hort. 78:45-49.

52. White PR: 1934. Potentially unlimited growth of excised tomato root tips in a liquid medium. Plant Physiol. 9:585-600.

53. Yang Hsu-Jen: 1977. Tissue culture technique developed for asparagus propagation. HortScience 12:140-141.

54. Ziv M, D Hadary, N Kedar, G Ladizinsky: 1984. Lycopersicon esculentum: Trifoliate plants recovered from anther cultures of heterozygous tf tf plants. Plant Cell Reports 3:10-13.

TISSUE CULTURE TECHNIQUES AND PLANT INTRODUCTION/QUARANTINE PROCEDURES

BRUCE J. PARLIMAN

U.S. Department of Agriculture, Agricultural Research Service, Plant Introduction Station, 11601 George Palmer Highway, Glenn Dale, MD 20769, U.S.A.

1. INTRODUCTION

Two U.S. Department of Agriculture agencies, the Agricultural Research Service (ARS) and the Animal and Plant Health Inspection Service - Plant Protection and Quarantine (APHIS-PPQ), are responsible for introduction/ quarantine aspects of foreign plant germplasm importation. Personnel from ARS in cooperation with APHIS-PPQ operate the U.S. Plant Introduction and Quarantine Station (USPI/QS). ARS personnel receive, establish, maintain in quarantine, index, conduct virus[1] therapy, multiply and distribute clonally and sexually propagated foreign plant germplasm introductions. They also maintain temporary collections of quarantine-released fruit crop cultivars (31, 38). The research objectives of the ARS scientists at USPI/QS are 1) to develop and apply methodology for producing disease-free-germplasm and for multiplying rare and/or hard-to-propagate germplasm, and 2) to develop improved methods of virus detection (graft and mechanical transmission, serology, dsRNA detection, nucleic acid hybridization and electron microscopy) and study the etiology of new viruses found in foreign introductions. The ARS has the sole responsibility for plant materials from importation through quarantine, indexing, propagation and distribution. APHIS-PPQ personnel are responsible for establishing import and quarantine regulations, specifying safeguards, conducting inspections and overseeing virus indexing and therapy of quarantined plants. Only APHIS-PPQ personnel have the authority

---

1 Unless otherwise specified, the use of the term virus in this text will include true viruses, viroids and mycoplasmalike organisms (MLO) etc. It is understood that different treatments and procedures may be required to detect and eliminate these distinct organisms.

Zimmerman, R.H. et al. (eds), Tissue Culture as a Plant Production System for Horticultural Crops.

ISBN 90-247-3378-2.

to place plant items in quarantine and later release that same germplasm from quarantine.

2. ROLE OF PLANT QUARANTINES

Plant quarantines are designed to restrict or prevent the establishment of exotic plant pests introduced by man and thereby protect agriculture and the environment from avoidable damage by hazardous organisms (16). Specific plant quarantines are usually established on the basis of pest risks that are assessed through a careful analysis of biological variables as related to applicable import safeguards (18, 20, 31). Federal regulations specify whether or not imported plant germplasm is listed in the prohibited or restricted (including post entry) quarantine categories (3). Plants within the prohibited category have a very high pest risk, because they may have a high frequency of latent virus/MLO infections, they may harbor new and/or hard-to-detect mites and races of fungi or bacteria, or they may originate in countries where high pest risks exist (16, 21, 35). For the United States, seeds of some major crops and tissues of many economically important vegetatively propagated plants fall within the prohibited plant quarantine category. A sample list of prohibited genera is presented in Table 1. Plants and plant parts listed in the prohibited quarantine category are usually imported into the United States for research purposes by ARS and are sent to the USPI/QS for quarantine. Other public or private research scientists may hold prohibited category import permits if they have appropriate expertise and adequate quarantine facilities, but this happens infrequently.

Pest risk components for particular plants can be reduced through the use of *in vitro* cultures (15, 16, 17, 19, 24). *In vitro* cultured tissues rank among the most appropriate forms/sources for importing quarantined plant germplasm as 1) passage through a quarantine station, 2) pest-free certification by an exporting country, 3) *in vitro* cultures, 4) plants grown in greenhouses under specified conditions (3), 5) plants from regulated field plots, 6) plants from commercial nurseries, 7) plants from farms, 8) dormant tubers, roots, etc. collected from a marketplace, and 9) plants and plant parts collected from the wild.

TABLE 1. Examples[1] of vegetatively propagated crop plants in the "PROHIBITED" (3, 31) category.

Crop	World source	Pest
Acer spp. (maple)	Many countries	Xanthomonas leaf disease and virus
Chrysanthemum spp.	Most countries	White Rust
Citrus spp. (orange)	All[2]	Viruses
Fragaria spp. (strawberry)	Many countries	Red Stele
Fraxinus spp. (ash)	Europe	Canker and dwarfing disease
Gladiolus spp.	Africa & other	Rusts
Ipomoea spp. (sweet potato)	All except Canada	A range of diseases including viruses
Malus spp. (apple)	All[2]	Viruses
Prunus spp.[3] (peach etc.)	All[2]	A range of diseases including viruses
Pyrus spp. (pear)	All[2]	A range of diseases including viruses
Rubus spp. (raspberry etc.)	All[2]	A range of diseases including viruses
Solanum spp.[3] (potato)	All except Canada	A range of diseases including viruses

[1] List is intended to show examples only, and is not inclusive. Close relatives of the named crops are usually included under the same restrictions (3).

[2] All except a few countries where special certification is issued and for which U.S. permits are granted.

[3] Seed of Prunus spp. susceptible to plum pox disease and being imported from Europe without an appropriate phytosanitary certificate, are prohibited. Seed of Solanum spp. have been placed in the prohibited category. Seedlings are tested in quarantine for the presence of Andean potato latent virus (APLV), potato virus T, and potato spindle tuber viroid (PSTV)(3).

3. ADVANTAGES OF TISSUE CULTURE

Quarantine-related advantages to tissue culture techniques include:

3.1. The quantity of plant tissue and the size of the consignment shipped can be be considerably smaller when materials are imported *in vitro* (20).

3.2. Many pests can be detected as contaminants in an *in vitro* system. Snails, insects and mites can be detected through the glass and the culture media frequently support the growth of nonobligate, nonfastidious bacteria and fungi and may show saprophytes that are carried into the system by obscure pests such as nematodes and mites (19).

3.3. Tissues are buffered against the environment during shipping.

3.4. Tissues are not likely to need surface sanitization at the time of port-of-entry inspection.

3.5. Meristem-tip cultures may be free of tested viruses (2, 29, 33, 40).

3.6. All plants used in introduction/quarantine activities can be derived from the same meristem-tip culture, minimizing the need for cumbersome subclone identifications during the quarantine period. Under these conditions, all tissues from a given genotype can be considered to have the same virus status (7, 9, 31).

3.7. Virus- or pest-indexed tissues can be shipped, stored and maintained with little opportunity for reinfestation.

3.8. If virus-infected tissues are shipped into the quarantine system _in vitro_, there is little opportunity for other quarantined genotypes of the same genus to become cross-contaminated while in quarantine.

3.9. Potentially, large collections of genotypes of a particular genus or species (such as _Ipomoea_, _Manihot_, or _Solanum_) can be accurately and effectively maintained in several replicated sites and still not require much space or maintenance effort.

3.10. Tissues from _in vitro_ cultured shipments can be placed directly into a pest therapy program (thermotherapy and/or meristem-tip or micrograft), while explants of the same material are being established in greenhouse/field index blocks. Streamlining the system shortens the quarantine period. Examples include _Ipomoea_ and _Solanum_.

3.11. Tissue culture systems, especially for _Prunus_ genotypes, can be a source of uniform (size, age and genotype), virus-free rootstocks and own-rooted indicator stocks, which are used in greenhouse/field indexing.

For some index tests, candidate tissues (buds from imported genotypes) are budded onto rootstock genotypes that act as indicators themselves. In other tests, candidate and indicator tissues are budded simultaneously onto standard rootstocks (6, 8, 10, 31, 36).

3.12. Solanum viruses can be maintained in tissue cultured plantlets, but symptoms are infrequently expressed (28). If a system could be standardized, tissues of genotypes of this and other genera could be taken from tissue cultured plantlets and used directly in virus-indexing procedures. This would constitute a very controlled system and would eliminate the need for greenhouse- grown stock plants.

3.13. Virus-indexed germplasm that is released from quarantine is distributed only from the indexed parent plant and its clonal progeny. Tissue from parent plants can be held in *in vitro* cultures (see 3.6. above) during greenhouse/field indexing. This saves space, decreases maintenance and nearly eliminates any opportunity for introduction of additional plant pests.

3.14. Plants grown in culture characteristically can be multiplied rapidly. By maintaining parent plant lines in culture, the USPI/QS has a head start on the historic problems of rapid cloning quarantine-released germplasm prior to national distribution. Associated distribution/shipping problems can be alleviated.

3.15. Germplasm that has been released from quarantine, *in vitro*, has the potential to be moved to its appropriate repository where portions of the material could be established in the field while other portions are being moved directly into a long-term preservation system (4, 22, 34, 39). This would be useful with germplasm in genera such as *Rubus*, *Prunus*, *Vitis*, etc., where field-grown collections are prone to pathogen reinfestation.

3.16. Under certain circumstances, an explorer/collector/importer may have the opportunity to establish and ship freshly collected germplasm via *in vitro* cultures. Rarely would this be feasible but for all the above reasons it would be very useful.

3.17. There is potential, as yet relatively unexplored, to use tissue culture systems to test for the presence of deleterious and obscure pests. Under certain circumstances, microenvironments might be created that would allow deleterious microorganisms to flourish and would not inhibit the active growth of the candidate plant germplasm. Under present circumstances, consignments of certain types of seed or small shipments of vegetative tissue are actually destroyed by testing and/or clean-up procedures.

4. DISADVANTAGES OF TISSUE CULTURE

Tissue culture procedures do have disadvantages or limitations in facilitating introduction/quarantine activities.

4.1. Tissue culture procedures do not necessarily eliminate obligate parasites such as certain fungi, rusts, mildews, chytrids and viruses (19, 23, 24).

4.2. To date, tissue cultures of prohibited items as prepared and shipped from laboratories of other countries have not received regulatory approval as being pest free. Regulations are not likely to change until in vitro culture importation methodology has been demonstrated to be effective and consistent. Therefore, all prohibited category germplasm, shipped in vitro, is tested for pathogens before being released from US quarantines.

4.3. Currently, there are no in vitro virus/pathogen indexing procedures in use. If in vitro-cultured germplasm arrives at the USPI/QS, samples of each genotype must be removed from culture, rooted, acclimated and grown under greenhouse/screenhouse conditions to a size sufficient to take samples for virus indexing. Tissue from the original parent culture may be subcultured and held in vitro until in vivo grown plants are released from quarantine but all testing must be conducted under greenhouse/field or laboratory conditions.

4.4. Tissue culture media are not universally useful across genotypes. In general germplasm exchanges are based on the premise that new, unusual, rare and or untested genotypes of known or unknown genera are being imported. Thus, the development or adaptation of media to unknown plant material, frequently must precede importation.

4.5. Standard introduction/quarantine procedures are effective.

4.5.1. The development of new, more effective and more universal virus identification and testing technologies is moving more rapidly than are changes in tissue culture procedures. Thus, virus detection systems may continue to surpass the need to develop new *in vitro* testing procedures.

4.5.2. To a large extent, the movement of plant germplasm in the form of properly collected, packed, shipped, handled, and established (propagated) tissues is still an efficient, effective and convenient process. Shipping small consignments of rare, hard-to-propagate germplasm via traditional methods (if done correctly) may still be the most efficient and effective method for today's international movement of valuable plant germplasm.

4.6. Facilities may not always be available. Surface sanitization, etc., procedures may be quite difficult when collecting wild or even cultivated material.

4.7. The generation of variant genotype forms may occur in genera such as *Ipomoea*, etc.

4.8. Collecting germplasm out of season cannot be helped by tissue culture procedures. Frequently germplasm explorations are conducted at times of optimum fruit ripening or times for disease appearance etc., but frequently not at optimal times for clonal propagation or seed collection.

4.9. Tissue culture and virus indexing etc. procedures require a high degree of technical background and skill. There is a continuous need to upgrade and increase the technical force handling quarantined plant germplasm. The increased application of modern tissue culture and virology techniques to plant introduction/quarantine activities is continually being slowed by a need for retraining and by a need to increase the number of qualified personnel working in the area.

Tissue culture techniques are used in conjunction with the introduction and quarantine of plants of several genera (Table 2). At present the prime use is for virus elimination including meristem-tip and micrograft therapy. Currently, only *in vivo* thermotherapy is used, but there is potential for *in vitro* heat treatment (25). The use of *in vitro* procedures for temporary storage of subclones of genotypes being tested in quarantine,

TABLE 2. Tissue culture techniques used to facilitate the release of clonally propagated plant germplasm from quarantine.

Genus	Tissues	Treatments/activities
Asparagus spp.	METP[1], PLNT	SHPT, TH[2] (15)
Citrus spp.	METP, PLNT	TH, ST
	etiolated seedling	micrograft therapy (29, 30)
Fragaria spp.	METP, PLNT	SHPT, TH, ST, PROP (27)
Ipomoea spp.	METP, PLNT	SHPT, TH, ST, PROP (2)
Manihot spp.	METP, PLNT	SHPT, TH, ST, PROP (1, 22)
Prunus spp.	PLNT	virus-tested rootstocks and own-rooted indicator-stocks
Saccharum spp.	dormant bud	TH (11, 37)
Solanum spp.	METP, PLNT	SHPT, TH, ST, PROP
		heat therapy (5, 25, 26, 32)
Vitis spp.	METP, PLNT	TH
		Micrograft therapy (12, 13, 14, 33)

[1] METP=meristem-tip, PLNT=plantlets and plant tissues suitable for propagation/storage/shipment/therapy activities, SHPT=shipment, TH=therapy, ST=storage, PROP=propagation/multiplication.

[2] Therapy procedures may require heat treatment in conjunction with meristem-tip, etc. cultures.

is rapidly becoming a critical component in the USPI/QS system. *In vitro* storage of large shoot pieces increases indexing/therapy efficiency, reduces whole-plant space and maintenance requirements, and reduces the incidence of vector transmission of viruses. If techniques have been developed for the above, then media, etc., are usually applicable to propagation and shipping. Tissue culture procedures are being used only infrequently in conjunction with the introduction/quarantine activities of plants in genera such as *Malus*, *Pyrus*, *Rubus*, *Acer*, vegetatively propagated grasses etc. For many genotypes within these genera, tissue culture technology is available (40), but either the frequency of importation is too low to warrant the establishment of a system or traditional importation/ quarantine methods are acceptable.

By integrating *in vitro* techniques into the introduction/quarantine system, the process becomes more efficient and effective for both recipients and the USPI/QS. For example, *Ipomoea* spp. germplasm has historically been imported into the United States as root pieces. From these roots, plants were grown in greenhouses and herbaceous index tests were conducted on subclones of each genotype over several seasons. Shoots from original parent plants were released and distributed to recipients if tests were negative. Historically, more than 50% of the candidate genotypes showed positive results in index tests and had to be meristem-tip cultured, reestablished in the greenhouse and retested. This procedure took several years to complete, did not guarantee that vectors would not cross-contaminate greenhouse-grown plants, and consumed time and space, etc., for maintenance, etc., of subclones as whole plants. At present many _Ipomoea_ introductions arrive at the USPI/QS _in vitro_ and those that arrive as roots are placed in culture as soon as new shoots can be forced.

Ipomoea germplasm is no longer held in quarantine in greenhouses, but is held as tissue-cultured shoots grown in incubators. Meristem-tip cultures are established from tissue-cultured shoots. Clonal progeny derived from the meristem-tip cultures are maintained in culture and established in the greenhouse. Greenhouse-grown plants are tested for viruses with herbaceous and ELISA tests through two cycles. When released from quarantine, only tissue-cultured progeny from the original meristem-tip cultures are distributed. All other specimens of the genotype are destroyed. This procedure should reduce the standard quarantine period for _Ipomoea_ by reducing establishment, maintenance, and therapy time, but will not necessarily reduce indexing time.

In conclusion, tissue culture techniques have not been used extensively in shipping quarantined germplasm countries or to establish, maintain, propagate or test quarantined germplasm. Only recently has it been used as a support system to provide virus-tested rootstocks and indicator stocks. However, tissue culture procedures have been used regularly for virus therapy. In the future, as procedures improve and become adapted, tissue culture will be accepted and used more extensively and will continue to make plant introduction/quarantine activities more efficient and effective.

REFERENCES

1. Adejare GO, RHA Coutts: 1981. Eradication of cassava mosaic disease from Nigerian cassava clones by meristem-tip culture. Plant Cell Tissue Organ Culture 1:25-32.

2. Alconero R, AG Santiago, F Morales, F Rodriguez: 1975. Meristem tip culture and virus indexing of sweet potatoes. Phytopathology 65:769-773.

3. Anonymous: 1980. Foreign quarantine notices, Title 7 - Agriculture. Chapter 111 - Animal and Plant Health Inspection Service. Part 319:37:1-26. Federal Register 45(94):31572-31597. (Current editions available through USDA, Regulatory Support Staff, APHIS-PPQ, Hyattsville, MD 20782).

4. Barlass M, KGM Skene: 1983. Long term storage of grape in vitro, pp. 19-21. In: JT Williams, NM Anishetty (eds). The plant genetic resources newsletter. Bull. 53. Int. Board for Plant Genetics Resources, Plant Prod. and Protection Div. of FAO, Rome, Italy.

5. Facioli G, C Rubies-Autonell: 1982. PVX and PVY distribution in potato meristem tips and their eradication by the use of thermotherapy and meristem-tip culture. Phytopath. Z. 103:66-76.

6. Fridlund PR: 1967. The relationship of inocular-receptor contact period to the rate of graft transmission of twelve Prunus viruses. Phytopathology 57:1296-1299.

7. Fridlund PR: 1983. Distribution of chlorotic leaf spot virus in apple budsticks. Plant Dis. Rptr. 57:865-869.

8. Fridlund PR: 1979. Intrageneric graft compatibility in relation to graft transmission of Prunus viruses. Lucrari Stiintifice (Seria Agronomie) 23:5-10 (Institutul Agronomic, Ion Ionescu de La Brad, IASI).

9. Fridlund PR: 1979. Distribution of apple stem pitting and apple mosaic viruses in apple budsticks. Lucrari Stiintifice (Seria Horticulture) 23:5-8 (Institutul Agronomic, Ion Ionescu de La Brad, IASI).

10. Fridlund PR: 1980. Glasshouse indexing for fruit tree viruses. Acta Phytopath. Acad. Sci. Hung. 15:153-158.

11. Gillaspie AG, CC McKnew: 1975. Improved sugarcane quarantine facilities and procedures at Beltsville, MD. Sugar J. 38:40-43.

12. Goheen AC: 1977. Virus and viruslike diseases of grapes. HortScience 12:465-469.

13. Goheen AC: 1980. The California clean grape stock program. Cal. Agr. 34:15-16.

14. Harris RE, JH Stevenson: 1979. Virus elimination and rapid propagation of grapes in vitro. Comb. Proc. Intl. Plant Prop. Soc. 29:95-107.

15. Kahn RP: 1976. Aseptic plantlet culture to improve the phytosanitary aspects of plant introduction for asparagus. Plant Dis. Rptr. 60:459-461.

16. Kahn RP: 1977. Plant quarantine: principles, methodology and suggested approaches, pp. 289-307. In: WB Hewitt, L Chiarappa (eds). Plant health and quarantine in international transfer of genetic resources. CRC Press, Cleveland.

17. Kahn RP: 1978. International exchange of genetic stocks, pp. 233-245. In: KW Hughes, R Henke, M Constantin (eds). Propagation of higher plants through tissue culture, a bridge between research and application. Knoxville. April 16-19. Pub. Tech. Inf. Center, U.S. Dept. of Energy, Oak Ridge, TN.

18. Kahn RP: 1979. A concept of pest risk analysis. EPPO Bull. 9:119-130.

19. Kahn RP: 1979. Tissue culture applications for plant quarantine, pp. 186-201. In: K Maramorosch, H Hirumi (eds). Practical tissue culture applications. Academic Press, New York.

20. Kahn RP: 1984. Safeguarding the international exchange of plant germ, pp. 866-871. In: Proc. 10th Intl. Congr. Plant Protection, Vol. 2. Brighton, United Kingdom, 1983.

21. Kahn RP, HE Waterworth, AG Gillaspie Jr, JA Foster, AC Goheen, RL Monroe, WL Povish, RG Mock, CF Luhn, EC Calavan, CN Roistacher: 1979. Detection of viruses or viruslike agents in vegetatively propagated plant importations under quarantine in the United States: 1968-1978. Plant Dis. Rptr. 63:775-779.

22. Kartha KK: 1984. Tissue culture techniques for virus elimination and germplasm preservation. Plant Breed. Rev. 2:265-282.

23. Klocke JA, P Myers: 1984. Chemical control of thrips on cultured Simmondsia chinesis (Jojoba) shoots. HortScience 19:400.

24. Langhans RW, RK Horst, ED Earle: 1977. Disease-free plants via tissue culture propagation. HortScience 12:149-150.

25. Lozoya-Saldana H, WO Dawson: 1982. The use of constant and alternating temperature regimes and tissue culture to obtain PVS-free potato plants. Amer. Potato J. 59:221-230.

26. Lozoya-Saldana H, O Merlin-Lara: 1984. Thermotherapy and tissue culture elimination of potato virus X (PVX) in Mexican potato cultivars resistant to late blight. Amer. Potato J. 61:735-739.

27. McGrew JR: 1980. Meristem culture for production of virus-free strawberries, pp. 80-85. In: Proc. conf. on nursery production of fruit plants through tissue culture--applications and feasibility. USDA, Sci. & Educ. Admin., Agr. Res. Results, ARR-NE-11.

28. McMorran JP, TC Allen: 1983. Maintenance, symptoms and distribution of potato viruses X, S, Y, A, M and leafroll in potato tissue culture plantlets. Amer. Potato J. 60:839-847.

29. Nauer EM, CN Roistacher, TL Carson, T Murashige: 1983. In vitro shoot-tip grafting to eliminate citrus viruses and virus-like pathogens produces uniform budlines. HortScience 18:308-309.

30. Navarro L, CN Roistacher, T Murashige: 1975. Improvement of shoot tip grafting in vitro for virus-free citrus. J. Amer. Soc. Hort. Sci. 100:471-479.

31. Parliman BJ, GA White: 1985. The plant introduction and quarantine system of the United States. Plant Breed. Rev. 3:361-431.

32. Roca WM: 1978. Introduction de germoplasma como cultvio de meristemas. Proc. II Latin American Bot. Congr. Tissue Culture Symp., Univ. of Brasilia, Brazil.

33. Sasahra H, K Tada, M Iri, T Takezawa, M Tazaki: 1981. Regeneration of plantlets by meristeming tip culture for virus-free grapevine. J. Japan. Soc. Hort. Sci. 50:169-175.

34. Silva GH: 1985. In vitro storage of potato tuber explants and subsequent plant regeneration. HortScience 20:139-140.

35. Waterworth HE: 1981. Control of plant diseases by exclusion: quarantine and disease-free stocks, pp. 269-296. In: D Pimentel (ed). CRC handbook of pest management in agriculture. Vol. 1. CRC Press, Boca Raton, FL.

36. Waterworth HE, WR Povish: 1980. Indexing quarantined apple and pear cultivars for viruses and mycoplasma-like organisms. Indian Phytopath. 33:275-281.

37. Waterworth P, RP Kahn: 1978. Thermotherapy and aseptic bud culture of sugarcane to facilitate the exchange of germplasm and passage through quarantine. Plant Dis. Rptr. 62:772-776.

38. Wilkes G: 1983. Current status of crop plant germplasm. CRC Critical Rev. Plant Sci. 1(2):133-181.

39. Withers LA: 1980. Preservation of germplasm, pp. 101-136. In: IK Vasil (ed). Perspectives in plant cell and tissue culture. Suppl. 11B, Intl. Rev. Cytol., Academic Press, New York.

40. Zimmerman RH: 1983. Tissue culture, pp. 124-135. In: JN Moore, J Janick (eds). Methods in fruit breeding. Purdue Univ. Press, West Lafayette, IN.

TISSUE CULTURE PROPAGATION OF ORNAMENTAL CROPS: AN OVERVIEW

PAUL E. READ AND MARY A. HOSIER

Department of Horticultural Science and Landscape Architecture, University of Minnesota, St. Paul, MN 55108, U.S.A.

1. INTRODUCTION

In recent years, propagation of numerous ornamental plants by tissue culture has become accepted commercial practice. Great progress has been made since Haberlandt (29) described the basic principles of plant tissue culture, followed by successful culture of plant callus by Nobecourt and Gautheret in France and by White in the U.S. (79). The rapid strides achieved in the field of micropropagation of ornamental plants have been made possible by the research accomplishments of myriad dedicated scientists. It is impossible to list all of these accomplishments in a brief overview, but some of the more significant contributions will be included in the following categories of ornamentals: orchids, ferns, foliage plants, bulbous species, herbaceous ornamentals, and trees and shrubs.

2. ORCHIDS

No discussion of tissue culture propagation of ornamental plants can be presented without giving credit to the work of Morel, often (and justifiably) referred to as the "Father of Micropropagation." Morel's successful culture of apical and axillary buds of Cymbidium and Cattleya paved the way for rapid development of these techniques as a means of cloning desirable orchid plants (53, 54, 55). Rao and Avadhani (63) demonstrated that Vanda seeds could be used as a source of clonal material, but other workers have applied Morel's techniques to Phalaenopsis (41), Cymbidium (77), Dendrobium (69) and Cattleya (40, 67). Arditti (7) has compiled a review on clonal propagation of orchids, which gives details of appropriate media, environmental requirements and explant stage and source, together with an extensive list of references. It is readily apparent that

Zimmerman, R.H. et al. (eds), Tissue Culture as a Plant Production System for Horticultural Crops.
 ISBN 90-247-3378-2.

direct application of these technologies has become a standard practice in the orchid propagation industry.

3. FERNS

Ferns as well as orchids were explored at an early stage as candidates for propagation in vitro. In 1949, Wetmore and Morel (78) successfully employed apical meristems of Adiantum for propagation purposes and Bristow (16) was able to differentiate both gametophytic and sporophytic tissues from callus of Pteris cretica. Rhizome segments or tips have been used as explant sources for Ampelopteris (50), Nephrolepis (2, 57) and Rumohra (18), while leaf tissue could be used as an explant source for Microgramma (35) and Rumohra (18). Although some difficulties, especially appearance of unusual and off-type plants, were encountered in early commercialization for Nephrolepis, tissue culture is now commonplace in propagation of many fern species.

4. FOLIAGE PLANTS

Perhaps the most rapid adoption of micropropagation as a commercial practice has taken place in the foliage plant industry. A wide range of species are routinely propagated by tissue culture methods. A plethora of publications could be listed in this area of specialization, such as the report by Jones and Murashige (44) of in vitro propagation of Aechmea and other bromeliads. Others of interest include the success of Debergh and DeWael with Ficus lyrata (24), and the development of micropropagation techniques for Peperomia (32) and Spathiphyllum (27). Indeed, micropropagation of Spathiphyllum was the primary profit maker for several companies that entered the tissue culture business. Work on Dieffenbachia then provided guidelines for disease-free propagation (45). As mentioned earlier for ferns, some problems of trueness to type may occur, such as that reported by Chua, et al. (20) for Dracaena marginata 'Tricolor', which lost its tricolor pattern when propagated in vitro. In general, however, a large number of foliage plant species are now being propagated via tissue culture with little or no problems of variability.

5. BULBOUS SPECIES

Numerous species with fleshy subterranean organs, i.e., bulbs, corms, fleshy rhizomes, etc., have been investigated for propagation by tissue culture. Robb's early work with *Lilium speciosum* essentially recognized that the common practice of scaling could be more efficient when performed *in vitro* (68). Cormous species such as *Gladiolus* (81) and *Freesia* (13, 14, 22, 37) were successfully cultured in the early 1970's. Several members of the Liliaceae and Iridaceae were reported to be propagated by tissue culture by the mid-1970's. Hyacinth (38, 60, 61, 62), tulip (10), lilies (30, 59, 75, 76), narcissus (38, 39, 71), bearded iris (51) and grape hyacinth (37, 38) have been cultured with varying degrees of success. Despite Bancilhon's paper on tulip tissue culture published in 1974 (10), tulip has remained recalcitrant. Wright and Alderson (80) and Nishiuchi (58) have made some strides toward *in vitro* propagation of tulip, but only a few cultivars have been responsive and their performance has been inconsistent and difficult to reproduce. Furthermore, conflicting results for hyacinth have also been published (60, 62), suggesting that a great deal of work is yet to be accomplished with the bulbous species. It is likely that more definitive investigation of bulb physiology and environmental factors will be required. For example, Stimart and Ascher (75) found that temperature and dormancy factors profoundly influenced tissue culture of *Lilium longiflorum*.

6. HERBACEOUS ORNAMENTALS

As with bulbs, some of the early strides in tissue culture of herbaceous ornamentals resulted from application of known techniques of leaf cuttage to the *in vitro* approach. For example, Schraudolf and Reinert (70) employed leaf discs of *Begonia rex* and Heide (31) used leaves of *Begonia* X *hiemalis* in early studies. Other plant parts were soon used as explants and numerous other herbaceous species were successfully cultured *in vitro*. Shoot tips were used for chrysanthemum (11) and fuchsia (23) and various leaf portions were employed for *Saintpaulia* (12, 46), *Streptocarpus* (6) and *Episcia* (43). Portions of petals and other flower parts have also been used (33), with one recent report employing flower parts to propagate chimeral *Saintpaulia* (47). It is apparent that a wide range of species can be cultured *in vitro* and many are now commercially propagated by tissue culture methods.

7. TREES AND SHRUBS

In contrast to the herbaceous species, woody plants have been more resistant to tissue culture attempts and only recently have they been successfully propagated in vitro. Ball's work in 1950 with Sequoia (8) encouraged researchers, but little further success was achieved until several workers began to employ media containing salt levels reduced from the high levels of the traditional Murashige and Skoog type of medium commonly used for herbaceous species (3, 5, 48, 56). That development, together with the use of direct rooting championed by Anderson (4), led to a succession of reports of in vitro propagation of numerous woody angiosperms. Rhododendron (3, 4, 5, 49), Eucalyptus (25), Daphne (21), Rosa (42, 73), Syringa (34, 52), Alnus (28), Kalmia (48) and Malus (72) are a few examples. However, coniferous species have been more difficult to culture in vitro. Aitken et al. (1) and Horgan and Aitken (36) have published several reports of regeneration of plants from Pinus radiata cultures. Ball (9) reported clonal propagation of Sequoia and Bonga (17), Cheng (19) and Sommer and Brown (74) have made progress with in vitro propagation of several other coniferous genera. Briggs and McCulloch (15) recently presented a summary of woody plant micropropagation in the United States and western Canada in which they point out that a large number of important woody species are yet to be cultured successfully, especially the conifers. It appears that much research is yet required to micropropagate these recalcitrant woody plants.

8. CHALLENGES AND OPPORTUNITIES

What will be some of the scientific advances necessary to the economic application of tissue culture technology to a wide range of species? Some recent research suggests that manipulation of the stock plant environment (26, 64), use of new methods of delivery of currently used and new plant growth regulating chemicals (65), and gaining a better understanding of the requirements for acclimation of tissue culture-derived plants (66) show promise in improving tissue culture methods. As these and other emerging technologies become further investigated and adapted, it is likely that micropropagation will become a practical propagation method for an ever-increasing list of species.

REFERENCES

1. Aitken J, KJ Horgan, TA Thorpe: 1981. Influence of explant selection on the shoot-forming capacity of juvenile tissue of *Pinus radiata*. Can. J. For. Res. 11:112-117.

2. Albrecht CN: 1977. Tissue culture propagation of Boston fern, *Nephrolepis exaltata* 'Bostoniensis'. M.S. Thesis. Washington State University.

3. Anderson WC: 1975. Propagation of rhododendrons by tissue culture: Part 1. Development of a culture medium for multiplication of shoots. Comb. Proc. Intl. Plant Prop. Soc. 25:129-135.

4. Anderson WC: 1978. Rooting of tissue cultured rhododendrons. Comb. Proc. Intl. Plant Prop. Soc. 28:135-139.

5. Anderson WC: 1984. A revised tissue culture medium for shoot multiplication of *Rhododendron*. J. Amer. Soc. Hort. Sci. 109:343-347.

6. Appelgren M, O Heide: 1972. Regeneration in *Streptocarpus* leaf discs and its regulation by temperature and growth substances. Physiol. Plant. 27:417-423.

7. Arditti J: 1977. Clonal propagation of orchids by means of tissue culture--a manual, pp. 203-293. In: J Arditti (ed). Orchid biology, reviews and perspectives, I. Cornell University Press, Ithaca, N.Y.

8. Ball EA: 1950. Differentiation in a callus culture of *Sequoia sempervirens*. Growth 14:295-325.

9. Ball EA: 1978. Cloning in vitro of *Sequoia sempervirens*, p. 259 (abstr.). In: KW Hughes, R Henke, M Constantin (eds). Propagation of higher plants through tissue culture, a bridge between research and application. Washington, D.C. Technical Information Center, U.S. Dept. of Energy Conf. 7804111.

10. Bancilhon L: 1974. First experiments on vegetative propagation in vitro of *Tulipa gesneriana* L. variety Paul Richter. C. R. Acad. Sci. Ser. D. 279:983-986.

11. Ben-Jaacov J, RW Langhans: 1972. Rapid multiplication of chrysanthemum plants by stem-tip proliferation. HortScience 7:289-290.

12. Bilkey PC, BH McCown, AC Hildebrandt: 1978. Micropropagation of African violet from petiole cross-sections. HortScience 13:37-38.

13. Brants DH, H Vermeulen: 1965. Production of virus-free Freesia's by means of meristem culture. Neth. J. Pl. Path. 71:25-27.

14. Brants DH: 1968. A revised medium for Freesia meristem cultures. Neth. J. Pl. Path. 74:120-121.

15. Briggs BA, SM McCulloch: 1983. Progress in micropropagation of woody plants in the United States and western Canada. Comb. Proc. Intl. Plant Prop. Soc. 33:239-248.

16. Bristow JM: 1962. The controlled in vitro differentiation of callus derived from a fern, Pteris Cretica L., into gametophytic or sporophytic tissues. Developmental Biology 4:361-375.

17. Bonga JM: 1981. Organogenesis in vitro of tissues from mature conifers. In Vitro 17:511-518.

18. Chen SY, PE Read: 1983. Micropropagation of leather leaf fern (Rumohra adiantiformis). Proc. Fla. State Hort. Soc. 96:266-269.

19. Cheng T: 1975. Adventitious bud formation in culture of Douglas fir [Pseudotsuga menziesii (Mirb) (Franco)]. Plant Sci. Lett. 5:97-102.

20. Chua BU, JT Kunisaki, Y Sagawa: 1981. In vitro propagation of Dracaena marginata 'Tricolor'. HortScience 16:494.

21. Cohen D, PM LeGal: 1976. Micropropagation of Daphne X Burkwoodii Turrill. Comb. Proc. Intl. Plant Prop. Soc. 26:330-333.

22. Davies DR: 1972. Speeding up the commercial propagation of freesias. Grower 77:711.

23. Dabin P, D Choisez-Givron: 1982. Culture in vitro de deux variétés de Fuchsia ('Constance' et 'Symphony') et comparaison avec les méthodes traditionelles. Bull. Soc. Roy. Bot. Belg. 115:27-32.

24. Debergh P, J DeWael: 1977. Mass propagation of Ficus lyrata. Acta Hort. 78:361-364.

25. DeFossard RA, RA Bourne: 1976. Vegetative propagation of Eucalyptus ficifolia F. Muell by nodal culture in vitro. Comb. Proc. Intl. Plant Prop. Soc. 26:373-388.

26. Economou AS, PE Read: 1986. Influence of light duration and irradiance on micropropagation of a hardy deciduous azalea. J. Amer. Soc. Hort. Sci. 11:146-149.

27. Fonnesbech M, A Fonnesbech: 1979. In vitro propagation of spathiphyllum. Scientia Hort. 10:21-25.

28. Garton S, MA Hosier, PE Read, RS Farnham: 1981. In vitro propagation of Alnus glutinosa Gaertn. HortScience 16:758-759.

29. Haberlandt G: 1902. Kulturversuche mit isolierten Pflanzenzellsn. Sitzungsber. Akad. der Wiss. Wien, Math.-Naturwiss. Kl.111:69-92.

30. Hackett WP: 1969. Aseptic multiplication of lily bulblets from bulb scales. Comb. Proc. Intl. Plant Prop. Soc. 19:105-108.

31. Heide OM: 1968. Auxin level and regeneration of Begonia leaves. Planta 81:153-159.

32. Henny RJ: 1978. In vitro propagation of Peperomia 'Red Ripple' from leaf discs. HortScience 13:150-151.

33. Heuser CW, DA Apps: 1976. In vitro plantlet formation from flower petal explants of Hemerocallis 'Chipper Cherry'. Can. J. Bot. 54:616-618.

34. Hildebrandt V, PM Harney: 1983. In vitro propagation of Syringa vulgaris 'Vesper'. HortScience 18:432-434.

35. Hirsch AM: 1975. The effect of sucrose on the differentiation of excised fern leaf tissue into either gametophytes or sporophytes. Plant Physiol. 56:390-393.

36. Horgan KJ, J Aitken: 1981. Reliable plantlet formatlon from embryos and seedling shoot tips of radiata pine. Physiol. Plant. 53:170-175.

37. Hussey G: 1975. Totipotency in tissue explants and callus of some members of the Liliaceae, Iridaceae, and Amaryllidaceae. J. Exp. Bot. 26:253-262.

38. Hussey G: 1977. In vitro propagation of some members of Liliaceae, Iridaceae and Amaryllidaceae. Acta Hort. 78:303-309.

39. Hussey G: 1982. In vitro propagation of Narcissus. Ann. Bot. 49:707-719.

40. Ichihashi S, S Kako: 1973. Studies on clonal propagation of Cattleya through tissue culture method. I. Factors affecting survival and growth of shoot meristem of Cattleya in vitro. J. Jap. Soc. Hort. Sci. 42:264-270.

41. Intowong O, Y Sagawa: 1974. Clonal propagation of Phalaenopsis by shoot tip culture. Amer. Orchid Soc. Bull. 43:893-895.

42. Jacobs G, P Allan, CH Bornman: 1969. Tissue culture studies on rose: Use of shoot tip explants. 1. Auxin: cytokinin effects. Agroplantae 1:179-188.

43. Johnson BB: 1978. In vitro propagation of Episcia cupreata. HortScience 13:596.

44. Jones JB, T Murashige: 1974. Tissue culture propagation of Aechmea fasciata Baker and other bromeliads. Comb. Proc. Intl. Plant Prop. Soc. 24:117-126.

45. Knauss JF: 1976. A tissue culture method for producing Dieffenbachia picta cv. Perfection free of fungi and bacteria. Proc. Florida State Hort. Soc. 89:293-296.

46. Kukulczanka K, G Suszynska: 1972. Regenerative properties of Saintpaulia ionantha Wendl. leaves cultured in vitro. Acta Soc. Bot. Pol. 41:503-509.

47. Lineberger RD, M Druckenbrod: 1985. Micropropagation of chimeral African violets. Ohio Agr. Res. and Dev. Center Res. Circ. 284:25-28.

48. Lloyd GB, BH McCown: 1980. Commercially feasible micropropagation of mountain laurel, Kalmia latifolia, by use of shoot tip culture. Comb. Proc. Intl. Plant Prop. Soc. 30:421-427.

49. McCown BH, GB Lloyd: 1983. A survey of the response of Rhododendron to in vitro culture. Plant Cell Tissue Organ Culture 2:77-85.

50. Mehra PN, DS Sulklyan: 1969. In vitro studies on apogamy, apospory and controlled differentiation of rhizome segments of the fern, Ampelopteris prolifera (Retz.) Copel. J. Linn. Soc. London (Bot) 62:431-443.

51. Meyer MM, LH Fuchigami, AN Roberts: 1975. Propagation of tall bearded irises by tissue culture. HortScience 10:479-480.

52. Minocha SC: 1980. Lilacs in test tubes: Potential for cloning of lilacs by cell and tisue culture. Lilacs 8:12-19.

53. Morel GM: 1948. Recherches sur la culture associée de parasites obligatoires et de tissues végétaux. Ann. Epiphyties 14:123-234.

54. Morel GM: 1960. Producing virus-free Cymbidium. Amer. Orchid Soc. Bull. 29:495-497.

55. Morel GM: 1964. Tissue culture: a new means of clonal propagation of orchids. Amer. Orchid Soc. Bull. 33:473-478.

56. Murashige T, F Skoog: 1962. A revised medium for rapid growth and bioassays with tobacco tissue cultures. Physiol. Plant. 15:473-497.

57. Murashige T: 1974. Plant propagation through tissue culture. Annu. Rev. Plant Physiol. 25:135-166.

58. Nishiuchi Y: 1980. Studies on vegetative propagation of tulips. IV. Regeneration of bulblets in bulb scale segments cultured in vitro. J. Japan. Soc. Hort. Sci. 49:235-240.

59. Novak FJ, Eva Petru: 1981. Tissue culture propagation of Lilium hybrids. Scientia Hort. 14:191-199.

60. Pierik RLM, MA Ruibing: 1973. Regeneration of bulblets on bulb scale segments of hyacinth in vitro. Neth. J. Agric. Sci. 21:129-138.

61. Pierik RLM, HHM Steegmans: 1975. Effect of auxins, cytokinins, gibberellins, abscisic acid and ethephon on regeneration and growth of bulblets on excised bulb scale segments of hyacinth. Physiol. Plant. 34:14-17.

62. Pierik RLM, J Woets: 1971. Regeneration of isolated bulb scale segments of hyacinth. Acta Hort. 23(2):423-428.

63. Rao AN, PN Avadhani: 1963. Some aspects of in vitro culture of Vanda seeds, pp. 194-202. In: Proc. Fourth World Orchid Conf., Singapore.

64. Read PE, AS Economou: 1982. Supplemental lighting in the propagation of deciduous azaleas. Comb. Proc. Intl. Plant Prop. Soc. 32:639-645.

65. Read PE, Q Yang, MA Hosier, SW Lee: 1985. Forcing solutions: a novel plant growth regulator delivery method for in vitro culture of woody species. HortScience 20:592.

66. Read PE, CD Fellman: 1985. Accelerating acclimation of in vitro propagated woody ornamentals. Acta Hort. 166:15-20.

67. Reinert RA, HC Mohr: 1967. Propagation of Cattleya by tissue culture of lateral bud meristems. Proc. Amer. Soc. Hort. Sci. 91:664-671.

68. Robb SM: 1957. The culture of excised tissue from bulb scales of Lilium speciosum Thumb. J. Exp. Bot. 8:348-352.

69. Sagawa Y, T Shoji, T Shoji: 1967. Clonal propagation of Dendrobiums through meristem culture. Amer. Orchid Soc. Bull. 36:856-859.

70. Schraudolf H, J Reinert: 1959. Interaction of plant growth regulators in regeneration processes. Nature 184:465-466.

71. Seabrook JEA, BC Cumming, LA Dionne: 1976. The in vitro induction of adventitious shoot and root apices on Narcissus (daffodil and narcissus) cultivar tissue. Can. J. Bot. 54:814-819.

72. Singha S: 1982. In vitro propagation of crabapple cultivars. HortScience 17:191-192.

73. Skirvin RM, MC Chu: 1979. In vitro propagation of 'Forever Yours' rose. HortScience 14:608-610.

74. Sommer HE, CL Brown: 1974. Plantlet formation in pine tissue cultures. Amer. J. Botany 61:11.

75. Stimart DP, PD Ascher: 1981. Foliar emergence from bulblets of Lilium longiflorum Thunb. as related to in vitro generation temperatures. J. Amer. Soc. Hort. Sci. 106:446-450.

76. van Aartrijk J, GJ Blom-Barnhoorn: 1980. Effects of sucrose, mineral salts, and some organic substances on the adventitious regeneration in vitro of plantlets from bulb-scale tissue of Lilium speciosum 'Rubrum'. Acta Hort. 109:297-304.

77. Wimber DE: 1963. Clonal multiplication of cymbidiums through tissue culture of the shoot meristem. Amer. Orchid Soc. Bull. 32:105-107.

78. Wetmore RH, G Morel: 1949. The growth and development of Adiantum pedatum L. on nutrient agar. Amer. J. Bot. 36:805-806.

79. White PR: 1963. The cultivation of animal and plant cells (2d ed). Ronald Press, New York.

80. Wright NA, PG Alderson: 1980. The growth of tulip in vitro. Acta Hort. 109:263-269.

81. Ziv M, AH Halevy, R Shilo: 1970. Organs and plantlets regeneration of Gladiolus through tissue culture. Ann. Bot. 34:671-675.

TISSUE CULTURE AS A PLANT PRODUCTION SYSTEM FOR FOLIAGE PLANTS

R.D. HARTMAN AND F.W. ZETTLER

Hartman's Plants, Inc., Palmdale, Florida 33944 and Department of Plant Pathology, University of Florida, Gainesville, FL 32611 USA, respectively

Boston fern (Nephrolepsis exaltata bostoniensis) was the first foliage plant to be commercially propagated in vitro, about 15-20 years ago. Others were soon added to the list, and today the majority of boston ferns sold, as well as the popular aroids, Dieffenbachia, Syngonium, Spathiphyllum, and Philodendron (excluding P. selloum and P. oxycardium), are tissue culture derivatives. Considering that the first successful in vitro propagation of any aroid was only in 1974 (6), it is obvious that the deployment of tissue culture technology has been rapid indeed and has ameliorated the tremendous explosion of the foliage plant industry during the 1970's from a 28 to 300 million dollar industry (14). The decade of the 70's was also marked by the genesis of numerous commercial tissue culture laboratories with investment capitals ranging from 20,000 to 250,000 dollars (7). At one time, over 100 such operations were in existence, many of which failed shortly after their inception. While the number of such private laboratories has declined in recent years, most of those that survived have expanded their operations exponentially. Recent investments in excess of 1,000,000 dollars are not unusual.

During the past 18 months, lab expansions in the United States have raised the potential output from 20-30 million plants per year to over 100 million plants per year. Though current production levels are meeting less than 10% of the potential foliage market (the market for micropropagation companies is wholesale foliage liner producers and finishers), this level exceeds the developed market. Many foliage growers are still not convinced that their starting materials need to be derived from tissue culture. This excessive production capability, over developed market needs, has had the recent effect of gluts and even price wars in certain unprotected product lines. Projecting this type of problem many labs have taken measures to protect their cultivars. This has generally taken the form of trademarks.

Zimmerman, R.H. et al. (eds), Tissue Culture as a Plant Production System for Horticultural Crops.
 ISBN 90-247-3378-2.

Since trademarks only protect the name, not the product, new names are simply assigned to these trademarked products by the competition, causing some confusion in the industry as several plant cultivars are being sold under as many as three names. This problem may be solved in the future with more widespread use of plant patents. Even here, however, there will be numerous problems in adequately characterizing a cultivar to assure the defensibility of the patent.

Since Florida leads the nation in foliage plant production, it is not surprising that most of the larger tissue culture operations are located here. The three largest are Hartman's Plants, Inc., Oglesby Labs, Inc., and Weyerhaeuser, Inc. Other large operations are located in Utah (Native Plants, Inc.), Texas (HyClone, Inc.), California (Twyford) and Tennessee (Phyton Technologies, Inc.). In other countries, major tissue culture laboratories are located in England, Holland, France, Belgium, and Israel as well as most other industrialized countries of the world. Certain eastern block countries also have labs including USSR with 6 and Poland with over 20. The growing commercial importance of tissue culture to the foliage industry is obvious, especially when one considers that the current wholesale market value of a typical foliage explant ready for transfer to soil is 15-25 cents.

The current price levels, though high in relation to other economic plant species, especially those that are seed propagated, are in many cases insufficient to generate the profit margins required to sustain the high level of research and development needed to realize all the benefits this technology has to offer. The benefits tissue culture propagation has had on the foliage industry that are already being realized include:

1) Provides large numbers of plants free of known pathogens

2) Makes available large numbers of selected types of plants in a relatively short period of time

3) Allows producers to devote 100% of their bench space to salable products

4) Adds incentive to breeders and hybridizers to increase their efforts because of the rapid introduction and use of their progeny

5) Allows free exchange of plant material between countries

6) Allows sustained year-round production. For example, it is not subject to season, disease or insect outbreaks, droughts, etc.

7) Allows growers with high quality facilities and growing expertise to realize the full genetic potential of a given plant

8) Gives growers the opportunity to truly program the use of their facilities

9) Has had an impact in overall quality improvement in the industry

10) Improved habit and other horticultural characteristics have caused buyers to specify habits that are only possible from tissue culture derived plants

11) Because of ease of shipping it is simple to supply sample quantities of product lines

However, many other potential benefits await realization. Many important foliage plants such as palms, aglaonema and crotons, await commercial _in vitro_ exploitation. Furthermore, such areas as mechanization that have great potential in reducing unit cost, require substantial up-front research investments with no guaranteed return. Problems such as mutation, vitrification, habituation and contamination still plague the industry. Strong price competition that ignores profit margins in order to gain market share can thwart many research and development activities and leave the industry as a whole weaker.

Among the benefits of micropropagation, none has had the significance that elimination of phytopathogens has. The value of tissue culture as a means to propagate plants free of virus and other pathogens has been recognized for about 20 years (13), and already tissue culture has made a significant impact on the foliage industry as a means of controlling such troublesome diseases as dasheen mosaic virus (6, 8). Once common, if not ubiquitous, in certain cultivars of _Dieffenbachia_ (1, 17), this virus is now rarely, if ever, encountered in nurseries availing themselves of tissue culture-derived stock. Nevertheless, many foliage plant propagators today perceive this technology solely from the standpoint of rapid micropropagation. Boston ferns, for example, are relatively free of disease problems to begin with, and there would be little justification for going to the initial expense of tissue culture were this to be the only advantage to be gained in so doing. The intrinsic value of disease control through tissue culture also tends to be underestimated because aseptic culture is a necessary adjunct to this technology. Microbial contaminations of any kind, regardless of whether or not pathogens are involved, are usually fatal to the developing explant in culture and thereby must be eliminated as a matter of course. Fortunately, most microbial contaminations, especially fungal, can readily be detected *in vitro* by visual inspection, even by relatively unskilled personnel.

Other pathogens, particularly viruses and certain bacteria, are not as readily detected *in vitro*, however, and thus may pose special threats to the tissue culture industry. It is important to recognize that, by its very nature as a means to mass-propagate plants, tissue culture technology can render the industry uniquely vulnerable to certain types of disease problems. Orchid growers were among the first to realize this. Those who neglected to take the precaution of indexing their plants for virus while having them processed for micropropagation actually increased disease incidence in their operations, and many suffered severe financial losses accordingly. The potential for disease problems can reach enormous proportions in the foliage industry, especially for such crops as *Dieffenbachia* which comprises 5% of the plants sold in Florida (15). Through tissue culture it is theoretically possible to produce up to 70,000 *Dieffenbachia* cuttings from a single shoot tip within a year (9). Obviously then, if the plants being increased are latently infected with such debilitating pathogens as dasheen mosaic virus and/or *Erwinia chrysanthemi*, catastrophic losses can be expected when the diseased plants are ultimately prepared for release to customers (11, 17).

Since the full market potential of tissue culture laboratories has by no means yet been realized, many of the larger operations are continuing to expand in order to keep abreast of increasing consumer demands. As these demands are met in future years, however, the importance of disease control will be given proportionately greater attention. Of all crops cultivated, the foliage ornamental probably has the most severe qualitative constraints, since all plant parts, without exception, must be grown to near perfection if the product is to be marketable. With this in mind, it is obvious that the competitive edge will favor those tissue culture operations with a sustained reputation for producing quality, disease-free plants on a regularly scheduled basis.

Recognizing this, many companies have already taken definitive steps to protect the integrity of their product. These steps have included the employment of pathologists and/or microbiologists to conduct various certification programs including that recommended by Knauss (11). Ultimately, the specific disease control strategy employed will depend upon the crop in question, but in principle will probably emulate the "mother-block" systems described many years before the tissue culture revolution (4, 5, 16). In most instances, however, it is probable that indexing procedures will not need to be as frequently applied because of the tremendous acceleration of propagation rates in vitro and because of the remote prospect of re-infections while plants are in culture.

The diagnostic procedures used for each foliage plant species will vary according to the pathogen involved. Indexing for most bacterial or fungal diseases is likely to be relatively straightforward. However, there seems to be an increasing number of bacterial contaminants being discovered that are either (a) heat tolerant in that they form endospores and can often survive autoclaving or (b) are slow growing fastidious organisms or (c) are anaerobic and therefore can't grow at all or are slow growing with routine indexing procedures. All of these organisms require special media and/or techniques for detection. Special techniques are also necessary for many viral (or viroid) diseases, especially in instances where symptom expression is unreliable. For certain ones, the highly sensitive ELISA (Enzyme-linked immunosorbent assay) procedure (3) might apply. ELISA methods have repeatedly proven their worth for such crops as potatoes, which are grown on a large-scale basis. However, for many ornamentals that are grown in limited quantities and/or are susceptible to viruses (or viroids) for which no detecting antiserum is available, alternative methods may be necessary, such as the light microscopic methods described by Christie and Edwardson (2) or the methods of Jordan et al. (10) for detecting double-stranded RNA in infected plants. It is possible that specific diagnostic facilities will eventually become an integral part of most tissue culture operations, particularly the larger ones that can afford them. For smaller operations, it is hoped that such services will become available by arrangements made through privately owned or institutional laboratories.

In conclusion, tissue culture as a factor in controlling foliage plant diseases should be considered in conceptual terms, and this can perhaps best be illustrated by the following example. Bamboo Nurseries of Orlando, Florida, has been a supplier for many years of hybrid vining and self-heading Philodendrons (12). Only about 25% of the total greenhouse area could be put into direct use for the plants being marketed; the remaining space was committed to the maintenance of stock plants, which required considerable care and handling. Despite the routine use of fungicides and nematicides, 20% losses to diseases could be expected, and the plants being readied for sale required manual rinsing of the leaves to eliminate the objectionable spray residues. Recently, the organization has converted its operation to the use of tissue culture-derived plants. Select hybrids are forwarded, on a contractual basis, to a major tissue culture facility for increase, and are later received as microcuttings in sealed, plastic envelopes. The grower receives about 10,000 such microcuttings each week, and these are planted into sterilized soil mix by a single employee. Fertilization and watering are accomplished automatically by a remote system, and the plants are allowed to develop on the premises for three months prior to being sold as liners. Under this regimen, a product of exceptional quality and uniformity is produced, the use of pesticides is obviated, and disease losses are no longer experienced (H.N. Miller, personal communication). Such operations in the future will become the rule rather than the exception. Without doubt, tissue culture has major potential as control of foliage plant diseases.

There are, however, many other additional benefits still to be realized and problems to be overcome. These will require extensive research and development with their related costs. Short-term gluts that result in price cutting stall the progress. The laboratory management teams must be exceedingly innovative in their approach so that new products, presented in new forms and developed in new ways, are the sources of competition that have the overall benefits of advancing the technology and industry and allow the real value of the products to be realized.

REFERENCES

1. Chase AR, FW Zettler: 1982. Dasheen mosaic virus infections of dieffenbachia cultivars. Plant Disease 66:891-893.

2. Christie RG, JR Edwardson: 1977. Light and electron microscopy of plant virus inclusions. Florida Agr. Expt. Sta. Monograph Series No. 9. 150 pp.

3. Clark MF: 1981. Immunosorbent assays in plant pathology. Annu. Rev. Phytopathol. 19:83-106.

4. Dimock AW: Disease control. 1941. Florist's Review 87(2252):19.

5. Dimock AW: 1962. Obtaining pathogen-free stock by cultured cutting techniques. Phytopathology 52:1239-1241.

6. Hartman RD: 1974. Dasheen mosaic virus of cultivated aroids and its control by seed propagation and culture of shoot tips. Ph. D. dissertation, University of Florida, Gainesville. 80 pp.

7. Hartman RD: 1979. Plant tissue culture from a propagator's viewpoint. Foliage Digest 11(3):2-8.

8. Hartman RD: 1974. Dasheen mosaic virus and other phytopathogens eliminated from caladium, taro, and cocoyam by culture of shoot tips. Phytopathology 64:237-240.

9. Henny RJ, JF Knauss, A Donnan Jr: 1981. Foliage plant tissue culture, pp. 137-178. In: JN Joiner (ed). Foliage plant production handbook. Prentice-Hall Publ., New York.

10. Jordan RL, JA Dodds, HD Ohr: 1983. Evidence for viruslike agents in avocado. Phytopathology 73:1130-1135.

11. Knauss JF: 1976. A tissue culture method for producing Dieffenbachia picta cv. Perfection free of fungi and bacteria. Proc. Florida State Hort. Soc. 89:293-296.

12. McColley RH, HN Miller: 1956. Philodendron improvement through hybridization. Proc. Florida State Hort. Soc. 78:409-415.

13. Morel GM: 1960. Producing virus-free cymbidiums. American Orchid Soc. Bull. 29:495-497.

14. Smith CN, Scarborough EF: 1981. Status and development of foliage plant industries, pp. 1-39. In: JN Joiner (ed). Foliage plant production handbook. Prentice-Hall Publ. Co., New York.

15. Smith CN, JR Strain: 1976. Market outlets and product mix for Florida foliage plants. Proc. Florida State Hort. Soc. 89:274-278.

16. Tammen J, RR Baker, RE Skiver: 1959. Control of vascular wilt diseases of carnation by culture-indexing. Phytopathology 50:356-360.

17. Zettler FW, JF Knauss, ME Taylor-Knauss, RD Hartman, AR Chase: 1980. Evaluation of Dieffenbachia maculate 'Perfection' plants free of dasheen mosaic virus. Acta Hort. 110:259-263.

COMMERCIAL MICROPROPAGATION OF FLORIST FLOWER CROPS

DENNIS P. STIMART

Department of Horticulture, University of Maryland, College Park, MD 20742 U.S.A.

1. INTRODUCTION

The micropropagation of florist flower crops has been substantial and as a group they are easy to propagate *in vitro*. Maximum productivity is achieved by determining optimal environmental conditions and nutrient medium composition for shoot multiplication and rooting. A review of the scientific literature indicates procedures exist to micropropagate numerous plant materials; however, it is difficult to assess commercial feasibility and subsequent problems from this information. The material presented within this paper is an attempt to review techniques and problems associated with commercial micropropagation of florist flower crops. The information presented was obtained from interviews with commercial laboratories operating in the United States. The scientific literature was not referred to in gathering information. A reluctance of some laboratories to give proprietary information is reflected by lack of detail in the text.

2. FLORIST FLOWER CROPS COMMERCIALLY MICROPROPAGATED

The range of plant materials commercially micropropagated as florist items include herbaceous annuals, biennials and perennials, and woody plants. The wide diversity is shown in Table 1 listing 75 genera from 36 botanical families. Plants listed have been aseptically isolated (Stage 1), multiplied (Stage 2), rooted (Stage 3), and transplanted (Stage 4). The list is not intended to be all inclusive but rather to be representative of the diversity available at the laboratories interviewed.

Scientific Article No. A-4271, Contribution No. 7260 of the Maryland Agricultural Experiment Station.

Zimmerman, R.H. et al. (eds), Tissue Culture as a Plant Production System for Horticultural Crops.

Printed in The Netherlands. ISBN 90-247-3378-2.

TABLE 1. Floriculture crops for which commercial micropropagation systems are available[1,2].

Family	Species	Explant	Regeneration/ multiplication system
Acanthaceae	Aphelandra sp.	ST	ASP
	Crossandra	ST	ASP
Agavaceae	Yucca	ST	ASP
Alstroemeriaceae	Alstroemeria sp.	ST	AS
Amaryllidaceae	Agapanthus	S	AS
	Haemanthus	S	AS
	Hippeastrum sp.	S	AS
Apocynaceae	Amsonia tabernaemontana	ST	ASP
Araceae	Anthurium	ST,L	ASP
	Caladium	ST	ASP
	Spathiphyllum	ST	ASP
	Zantedeschia	ST	ASP
Asclepiadaceae	Asclepias tuberosa	ST	Embryoids from callus
Begoniaceae	Begonia x hiemalis	P	Shoot regeneration
	B. tuberhybrida	ST,L	
Boraginaceae	Brunnera macrophylla	ST,FB,L,R	Adventitious shoots from callus
Campanulaceae	Campanula elatines	ST	ASP
	C. glomerata	ST	ASP
	C. linifolia	ST	ASP
	C. rotundifolia	ST	ASP
	Platycodon grandiflorus	NA	NA
Caryophyllaceae	Dianthus caryophyllus	ST	ASP
	D. gratianopolitanus	ST	ASP
	Gypsophila paniculata	ST,FB	ASP
	Herniaria glabra	ST	ASP
	Silene vulgaris	ST	ASP
	Petrorhagia saxifraga	ST	ASP
Compositae	Aster novi-belgii	ST,FB	ASP
	Chrysanthemum coccineum	ST	ASP
	C. maximum	ST	ASP

TABLE 1. (Continued)

Family	Species	Explant	Regeneration/ multiplication system
Compositae (Continued)	*C.* x *morifolium*	ST	ASP
	Coreopsis sp.	ST,FB	ASP
	Gerbera jamesonii	ST	ASP
	Inula ensifolia	ST,L	ASP
	Liatris aspera	ST	ASP
	Ratibida pinnata	ST	ASP
	Rudbeckia fulgida	ST	ASP
	Stokesia laevis	NA	NA
	Tagetes erecta	ST	ASP
Crassulaceae	*Kalanchoe blossfeldiana*	ST	ASP
Euphorbiaceae	*Euphorbia pulcherrima*	NA	NA
Fumariaceae	*Dicentra spectabilis*	ST	ASP
Gentianaceae	*Gentiana* sp.	ST	NA
Geraniaceae	*Pelargonium*	ST	ASP
Gesneriaceae	*Gloxinia*	ST,L	ASP
	Saintpaulia	L	ASP
Heliconiaceae	*Heliconia*	NA	NA
Hypericaceae	*Hypericum calycinum*	ST	ASP
Iridaceae	*Crocosmia*	S	AS
	Freesia	S	NA
	Iris sp.	NA	NA
Labiatae	*Ajuga reptans*	ST	ASP
	Lamiastrum galeobdolon	ST	ASP
	Nepeta cataria	ST	ASP
	Salvia	NA	NA
Leguminosae	*Thermopsis caroliniana*	ST	ASP
Liliaceae	*Hemerocallis* sp.	FB	AS,ASP
	Hosta fortunei	S,DVB	ASP,AS
	H. lancifolia	S	AS,ASP
	H. plantaginea	S	AS,ASP
	H. undulata	DVB	ASP
	Lilium sp.	FB	Bulblets

TABLE 1. (Continued)

Family	Species	Explant	Regeneration/ multiplication system
Liliaceae	Liriope sp.	ST	ASP
(Continued)	Maianthemum canadense	ST	ASP
Lobeliaceae	Lobelia cardinalis	ST	ASP
Orchidaceae	Cypripedium sp.	seed	seed
Plumbaginaceae	Armeria	ST,F	NA
	Limonium	ST,F	NA
Polemoniaceae	Phlox glaberrima	ST	ASP
	P. paniculata	ST	ASP
	P. subulata	ST	ASP
Primulaceae	Dodecatheon meadia	ST	ASP
	D. jeffreyi	ST	ASP
Ranunculaceae	Aquilegia caerulea	ST	ASP
	Clematis	ST	NA
	Helleborus sp.	ST	NA
	Thalictrum aquilegifolium	ST	ASP
Rosaceae	Geum triflorum	ST	ASP
	Rosa hybrida	ST	ASP
Rubiaceae	Bouvardia	ST	ASP
Rutaceae	Dictamnus albus	ST	ASP
Saxifragicaceae	Astilbe x arendsii	ST	ASP
	Bergenia cordifolia	ST	ASP
	B. ciliata	ST	ASP
	B. stracheyi	ST	ASP
	Hydrangea macrophylla	ST	ASP
Scrophulariaceae	Erinus alpinus	ST	ASP
	Mimulus ringens	ST	ASP
Solanaceae	Petunia sp.	ST,L	ASP,AS
	Salpiglossis	ST,L	ASP,AS

[1]Abbreviations: AS, adventitious shoots; ASP, axillary shoot proliferation; DVB, dormant vegetative buds; FB, flower buds; L, leaves; NA, information not available; P, petals; R, roots; S, scapes; ST, shoot tip.
[2]Information derived from (1) and sources indicated in Acknowledgments.

3. MEDIUM

The media most commonly cited for micropropagation of floricultural crops were formulations devised by Murashige-Skoog, Schenk-Hildebrandt, Gamborg's B5, White's, and Heller's. Many other types are used; however, the Murashige-Skoog medium was used most frequently.

The selection process of determining a medium for micropropagation varied between laboratories. Previous research in the literature or experience in the laboratory on related species, genera or family members was used to give an indication of the most likely media for initial testing. It was recognized that certain types of media work better than others and these should be selected first and tested. Another approach was based on knowledge of the native habitat of the plant. Plants growing outdoors in an acid environment would be screened in vitro on media used previously for acid-thriving plants or low salt requiring plants would not be cultured on high salt media. Some laboratories found mass screening of media formulations currently on hand (3 to 20 different types) an efficient way of determination. Modifications of the basal medium consisted mainly of dilution of basal salts, changing to liquid or semi-solid medium or alteration of plant growth regulators. Manipulation of basal medium major and minor elements and vitamins was done as a last resort.

4. PLANT GROWTH REGULATORS

Auxin, cytokinin, and gibberellin are used to control morphogenesis during micropropagation of floricultural plants. Most commonly used auxins (in order of preference) include indole-3-acetic acid, indole-3-butyric acid, and naphthaleneacetic acid; cytokinins (in order of preference) include benzyladenine, kinetin, and isopentenyladenine. The concentration range used is 10^{-5} to 10^{-7} M.

Growth regulators are used primarily during Stage 2 and 3 of micropropagation to facilitate increases in shoots and roots, respectively. Factorial comparisons of auxin to cytokinin ratios are employed to determine optimum root and shoot morphogenesis. Multiplication of herbaceous plants requires combinations of auxins and cytokinins. Routinely, addition of two auxins and two cytokinins at different concentrations are used totalling four growth regulators. As plants became less succulent and more woody, cytokinin predominated as the principle growth regulator and is used at low concentrations.

The specific growth regulator used and the relative concentration of each required for morphogenesis varies among taxa and among cultivars. Hosta cultivars have similar requirements for plant growth regulators while Hemerocallis have specific requirements for each cultivar. Most present day Hemerocallis are derived from a narrow genetic base of 2 to 3 species and would be expected to respond uniformly. However, a broader genetic base (30-35 species) in present-day Hosta cultivars have similar requirements.

Changes in plant requirements for growth regulators occur during micropropagation. During rapid shoot multiplication, it may be necessary to alter concentration or type used every 2 months. Most frequent adjustments are required when establishing Stage 2 cultures. Once rapid shoot proliferation occurs, changes are required less frequently. Certain genotypes of Gerbera and Hosta require high cytokinin at the outset, but once multiplication starts, lower cytokinins are required. A decrease in multiplication may occur later and require temporary reduction of cytokinin by one-half the present concentration or complete cytokinin removal for 1 or 2 subcultures before returning to normal cytokinin level. Thus, it appears that cytokinin initially promotes shoot growth but later suppresses development, perhaps from excessive cytokinin accumulation.

Stunted shoot growth during Stage 2 micropropagation of Aster and Chrysanthemum was related to cytokinin. Prolonged subculture on Stage 2 medium containing cytokinin resulted in abnormally compact shoot growth. If cytokinin pulses are used but separated by 1 or 2 subcultures on medium without cytokinin, non-stunted shoots develop. If removal of cytokinin is delayed, plants may never revert to normal growth. Gibberellic acid applied to Aster and Chrysanthemum cultures partially overcame this response and caused stem elongation, but did not affect other species.

Maintenance of cultures of some cultivars on high cytokinin levels can result in cytokinin habituation; shoot proliferation continues indefinitely once cytokinin is removed. Habituation results in excessive shoot proliferation and production of small plants with poor survival in Stage 4. The specific cytokinin and the concentration required to cause habituation on plants are determined from experimentation and Stage 2 medium would utilize a lower concentration.

Excessive shoot elongation during multiplication occurs on some species and is undesirable. Application of the growth retardants ancymidol [α-cyclopropyl-α(p-methoxyphenyl)-5-pyrimidinemethanol] or daminozide [(butanedioic acid mono (2,2-dimethylhydrazide)] reportedly shortens shoots.

Rooting success can be associated with previous cytokinin treatment. Long-term cultures on Stage 2 cytokinin concentrations reduced Stage 3 rooting of *Gerbera* and *Rosa* cultivars. A reduction in cytokinin concentration during Stage 2 for 1 or 2 subculture cycles immediately before Stage 3 counteracted this response. Presumably, residual cytokinin in plant tissue overrides the root promoting effect of auxin. Subculturing in the absence of cytokinin reduced this response.

Survival post-*in vitro* can depend on the form of cytokinin used during multiplication. Benzyladenine provided to *Liriope* in Stage 2 resulted in 10 to 20 percent survival in Stage 4, while use of kinetin resulted in 80 percent or more survival. Rooting differences were not observed to be affected by the form of cytokinin under similar environmental conditions.

5. EXPLANT

Many sources of explants are used to micropropagate florist flower crops including actively growing shoot tips, flower buds and scapes, leaf blades and petioles, dormant buds, embryos, and roots (Table 1). Shoots have been obtained directly from all types of explants. The explant source used varies depending on the plant propagated, the availability of explants, and genetic stability. The easiest and preferred route to production of shoot multiplication utilizes shoot tips from which axillary shoots proliferate. This technique maintains the highest level of genetic stability since few adventitious shoots develop. Eighty-three percent of the plants listed in Table 1 utilized shoot tips as the explant source. Flower buds from *Aster*, *Coreopsis*, *Gypsophila*, *Hosta*, and *Rudbeckia*, and scapes from *Hosta* regenerate adventitious shoots rapidly with little or no callus, if taken when immature. Shoots can arise from petioles of recently expanded leaves of *Begonia* x *hiemalis*, and leaves or leaf sections from *Anthurium*, *Begonia*, *Hosta*, *Inula*, *Phlox*, *Rudbeckia*, *Saintpaulia*, and *Bergenia*. Leaf explants of *Anthurium* are induced to produce enough callus to form adventitious shoots that are used as explants for axillary shoot proliferation. Direct shoot formation from roots is rare; however, roots from *Brunnera* and *Phlox subulata* have regenerated adventitious shoots. Roots of herbaceous perennials develop callus.

Plants that produce frequent spontaneous mutations are avoided in micropropagation since they are suspected of instability. However, micropropagation of periclinal *Hosta* cultivars are stable, if tissue cultures are initiated from active shoot tips from which axillary shoots proliferate. Adventitious shoots derived from flower buds or scape explants result in instability.

Indexing is done commonly for bacteria and fungi and occasionally for viruses, depending on the plant species. Normally, a plant is not placed into production until indexing is completed. However, several laboratories reported conducting indexing and production simultaneously. A variation is indexing followed by production with indexing conducted during shoot multiplication, rooting, and after transplanting. A common practice is to rapidly proliferate several lines of a cultivar concurrently, each started from separately indexed stock at separate times and maintained as distinct lines. The culture initiation may be spaced apart by weeks or months. If bacteria, fungi or viruses are detected in a line, it is removed from production. Lines are retained 1 year in production, moved through and out of the system, and the cycle is repeated.

Preconditioning of stock plants facilitates aseptic isolation and establishment of rapid shoot development *in vitro*. Stock plants maintained vegetatively as vigorously growing plants under greenhouse conditions or in conditioned rooms with reduced humidity and temperature allow easier aseptic explant isolation and quicker shoot proliferation. Explants derived from species grown outdoors are more contamination prone. Dormant plants are avoided unless vigorous growth can be induced after satisfying rest or by cytokinin application. However, explants from species including *Hosta* can be taken year-round. Plants requiring vegetative shoot tips as explants when reproductive are not taken until after flowering and vigorous vegetative growth resumes. The role of gas, light, moisture, nutrients, and temperature treatments given to stock plants used as explant sources has been poorly studied and seldom utilized by industry.

6. SHOOT MULTIPLICATION

Establishment time required to achieve optimum shoot multiplication depends on the plant genotype. A lag from explant isolation until shoot multiplication can be 2 to 3 weeks in *Rosa hybrida* 'Del Mator,' 10 weeks for *R. hybrida* 'Crystal' to over 3 years for some herbaceous perennials.

An average lag time is 4 to 6 months. Frequent subcultures are required during this period that decreases once proliferation commences. For Gypsophila, the initial subculture duration is every 2 weeks replaced by every 3 to 4 weeks during rapid shoot multiplication, while for R. hybrida 'Crystal' subculture every 10 days initially is replaced by every 3 weeks after rapid shoot proliferation.

Maintenance of shoot proliferation is done by frequent and timely subculture. Deviation from a well-managed subculture routine reduces multiplication rate, yield and quality of shoots, rooting, and subsequent growth. These parameters and leaf yellowing or necrosis are used as measures to determine critical time spans between subcultures. Once the time is determined, subculture frequency can be established. Requirements of cultivars are specific and should be adhered to.

Subculture technique during Stage 2 usually involves division down to a single shoot. Division to clumps of shoots during multiplication are used less frequently. Subculture technique can affect shoot multiplication. Orientation of the shoot in a vertical or horizontal position can be important. Removal or damage of the shoot apex can enhance or reduce shoot multiplication. Crown damage on Gerbera during subculturing yields blind shoots that fail to proliferate. In Rosa hybrida, there is an optimal size of proliferated shoots necessary before subculture. If too small or if an excessive amount of old wood is removed, poor multiplication results. Drying during separation of proliferated shoots was considered a problem of some plants and a major problem for certain R. hybrida cultivars.

Size of proliferated shoots reportedly causes difficulty during multiplication. Excessive shoot stretching can be controlled by growth retardants, as mentioned previously. Increased light and cytokinin levels during proliferation have been used to shorten shoots. Compact shoots with a rosette type of growth periodically occur and are rogued before subculturing since they usually fail to revert to normal growth. However, shoot growth that typically is short and compact can be partially overcome by gibberellin application in the Stage 2 medium, by submerging proliferated shoots halfway in liquid medium, or by light reduction. Also, excessive proliferation creates more compact shoots; reduction in shoot multiplication by lowering cytokinin levels or frequency of subculture ended this undesirable response.

Rhythmic patterns of shoot multiplication are experienced throughout the year with *Rosa hybrida*. Certain cultivars in September proliferate one and one-half to two and one-half times more shoots than when cultured in January, February, and March.

Vitreous shoots were reported on many of the floricultural crops during rapid multiplication, and are considered a problem, but not understood. Generally, vitrification is believed to be a symptom of environmental stress. The water-soaked appearance of Stage 2 shoots is more apparent in liquid cultures than on semi-solid medium. Certain types of agar and high concentrations of certain cytokinins were cited as contributing to vitrification. Increased concentration of agar and sucrose, air movement into culture vessels, and incorporation of charcoal in Stage 2 reduced vitreous shoot formation. Changing agar types was also beneficial. Micropropagated Caryophyllaceous species were considered to show the greatest tendency toward vitrification.

Stock renewal is an integral part of successful micropropagation. Failure to renew explants of *Dianthus*, *Gerbera*, *Gypsophila*, and *Rosa* result in loss of shoot multiplication, reduced plant size *in vitro*, misshappen leaves, failure of leaf unfolding, and reduced vigor after greenhouse acclimation. A minimum of 1 year and a maximum of 2 years of rapid multiplication are allowed on these 4 genera before renewal. In other plants, constant reinitiation circumvents contamination from slow-growing bacteria within explants or high mutation frequencies from adventitious shoots from leaf petiole explants of *Begonia* x *hiemalis* and callus cultures of *Caladium* and *Spathiphyllum*. In contrast, *Hemerocallis* and *Hosta* are not renewed typically. Shoot multiplication from axillary buds maintains stability.

7. ROOTING

Rooting of microcuttings of floriculture crops is not difficult. Most species root quickly by producing a satisfactory number of well-branched primary roots. It is not uncommon to have 90-95% success of rooting. Not all plants require a separate rooting stage since shoots may root *in vitro* after multiplication.

Microcuttings or clumps of proliferated shoots requiring a rooting stage can be rooted *in vitro* or in the greenhouse. Preference toward *in vitro* or greenhouse rooting was dependent on the species of plant and laboratory technique. *In vitro* rooting of *Gerbera jamesonii* was considered essential to attain rapid shoot growth in the greenhouse by one laboratory while rooting in the greenhouse was deemed satisfactory by another laboratory. A combination of *in vitro* and greenhouse rooting also occurred. For *Begonia* x *hiemalis*, root induction and development until appearance of root primordia occurred *in vitro*. Plants were subsequently transplanted into a sterilized soil-based medium and kept axenic for one month prior to complete greenhouse acclimation.

Rooting in the greenhouse required high humidity during all seasons from intermittent mist or fog and supplemental bottom heat during the winter from electric heating pads or water circulating through piping placed beneath rooting plants. Very soft succulent foliage developed under fog was followed by acclimation problems. Excessive moisture from intermittent mist caused nutrient leaching and tissue rot. However, both methods were recommended by laboratories. Clear tray covers placed over rooting plants was an alternative; however, seasonal increases in intensity and duration of solar radiation caused heat build-up and tray temperatures must be monitored and vented as needed.

8. IN VITRO ENVIRONMENT

The manipulation of environment during micropropagation has not been studied extensively but is considered important. Some light and temperature manipulation was reported as beneficial to responses *in vitro* and post-*in vitro*. Complete darkness for shoot formation and multiplication of *Brunnera* is absolute. No shoots develop or proliferate in light. Explants from scapes of *Hosta* form callus that forms adventitious shoots in complete darkness. If scape explants are cultured in light, callus fails to develop and no shoots form. *In vitro* photoperiod treatments given to *Dianthus caryophyllus* affect the phase change from vegetative to reproductive. Under 16 hours of light, flowering occurs on shoots. By use of 12 hours of light, most shoots are vegetative and few flower.

Light intensity reportedly affects shoot proliferation of several genera. For *Dicentra spectabilis*, shoot multiplication is increased under reduced light compared with higher levels. High light intensity increases shoot multiplication and quality of *Aster*, *Chrysanthemum*, and *Gypsophila*. When light intensity is reduced on *Aster* and *Chrysanthemum* fewer shoots form and these are of reduced quality due to vitrification and extreme succulence. Most plants in the Caryophyllaceae express poor shoot multiplication under reduced light. Once rapid shoot multiplication of *Hosta* has started, no light intensity requirement exists and plants do equally well under varied light intensities.

Temperature dramatically affects *in vitro* and post-*in vitro* responses. In *Dicentra spectabilis*, 20°C enhances rapid shoot multiplication and 27°C results in poor explant growth characterized by a suppression of shoot multiplication. In *Lilium*, bulblets generated *in vitro* at 25°C are in deeper dormancy post-*in vitro* than those generated at 30°C, based on time until leaf emergence. Axis elongation from bulblets developed *in vitro* at 30°C is quicker than those generated at 25°C. Generally, plants that are more tropical are cultured *in vitro* at 28°C and those that are more temperate at 22°C.

9. ACCLIMATION

Acclimation of micropropagated plants is complex and different for each crop. Seasonal changes require additional adjustments. An essential component is providing high humidity from fog systems or low volume intermittent mist. Survival post-*in vitro* and onset of vigorous plant growth of some species depends on subculture frequency. Delayed subculture during Stage 2 reduces survival while scheduled and more frequent subculture resulted in excellent survival.

10. POST-IN VITRO ABERRATIONS

Temporary and permanent changes in vegetative and reproductive growth habit of micropropagated plants are not uncommon. Some of these alterations are beneficial while others are destructive. Some of the more harmful changes can be reduced or eliminated by *in vitro* manipulations. The greater the proliferation rate, the greater the chance of abnormal type recovery. Reduction of proliferation in *Hosta* reduced aberrations. Also, the longer the time span between stock renewal and the use of a callus phase, the greater the number of aberrant type recovery. Plant genotype influences stability.

Shoot multiplication of micropropagated *Aster*, *Chrysanthemum* and *Hosta* under field conditions commonly exceeds conventionally propagated stock, and is beneficial. Ten divisions per plant of field-grown *Aster* and

Chrysanthemum with conventional propagation are common whereas micropropagated plants produce around 100 divisions, a tenfold increase over conventional methods. This response generally does not persist more than one season. Similarly, *Begonia* x *hiemalis* plants post-*in vitro* produce a 2- to 3-branched plant considered more marketable. In *Rosa hybrida*, more flowering canes with more branches per cane develop initially after micropropagation, creating fuller plants. However, in *Gerbera* and *Saintpaulia*, post-*in vitro* branching is undesirable.

Stem diameter of micropropagated *Chrysanthemum*, *Dianthus*, *Gypsophila*, and *Kalanchoe* can be small during shoot multiplication and remain small post-*in vitro* resulting in inadequate thickness and plant lodging. Micropropagated *Gypsophila* grown as field cut flowers twist off at the soil line as a result of a thin stem and being wind blown. Small-calipered stems post-*in vitro* can be partially eliminated by reducing the *in vitro* multiplication rate and temperature, and increasing light intensity.

Leaf color and shape aberrations on micropropagated plants exist. In *Hosta*, many beneficial leaf color mutants have been selected and released. Many sectorial and mericlinal chimeras of *Hosta* develop in micropropagated plants that are unstable, regardless of propagation technique. Also, nonvariegated *Hosta* can be recovered from originally variegated plants. However, variegation can gradually return causing a delay in marketing. Alterations in leaf shape of *Hosta*, such as twisting and lanceolate forms rather than cordate, are temporary responses. Reflexing and cupping of *Gerbera* leaves commonly occurs and is a temporary post-*in vitro* change.

Reluctance to flower and lack of uniform flowering of some floricultural crops were considered major limiting factors for micropropagated plants. Reluctance to flower was reported in *Gypsophila* and *Kalanchoe*. Only a few plants flowered of micropropagated *Gypsophila* in a very large cut flower planting. Clones were selected that would flower following micropropagation. Certain clones of *Kalanchoe* will not flower after micropropagation, a response they fail to outgrow. Results imply that reversion to juvenility occurs and occasionally is irreversible, although genetic variability controlling reversion exists. Lack of uniform flowering in micropropagated *Chrysanthemum*, *Dianthus*, and *Kalanchoe* makes accurate crop scheduling difficult. This currently limits commercial acceptance of micropropagated cultivars of these genera.

Petal number on the first flowers to develop after micropropagation of some double-flowered Rosa hybrida cultivars approaches 12 per flower. After 2 to 3 additional flower flushes, petalage returns to the normal number of 35 per flower. This change in flower petal number closely resembles the transitional changes associated with Rosa seedlings, suggesting reversion to juvenility during micropropagation.

Permanent flower color changes of Chrysanthemum and Dianthus were reported. Temporary changes in Rosa reverted after several flushes of flowers. In yellow cultivars, the brightness of color is decreased on the first flowers to bloom.

Polyploidy and aneuploidy were reported in almost all crops but the frequency was low and not considered a major problem if adventitious shoot development from callus was avoided.

11. MARKETING

Marketing of micropropagated florist crops in the United States currently has a limited distribution. Between seventy and eighty-five percent of the plants generated by laboratories were sold within the immediate state and the remainder sold and shipped to surrounding states. Customer acceptance of micropropagated plants has been slow because of plant appearance, not having the physical facility to acclimate plants and the knowledge of how to work with micropropagated plants. Also, a reluctance to accept changes in cultural practices has limited distribution. A marketing problem of micropropagated Rosa is to convince growers to utilize their greenhouse space differently. Conventionally, old Rosa are replaced during the winter (January) by new plants. Convincing growers to plant during summer (June and July) in addition to winter would create a better flow of quality Rosa from the laboratory to the market.

Plant accumulation is desirable and practiced to meet production limitations and market requirements. An approach is to store under cool conditions and under low light intensity for extended time. Plants that are extremely vigorous and placed into cold storage become necrotic. Those species sensitive to chilling injury are stored at 10°C or higher and insensitive species are stored at less than 10°C. The most commonly used low temperature for storage ranges from 2.2° to 6.7°C.

The most expensive aspect of micropropagation of an established laboratory was labor. It was estimated that 50 to 85 percent of the cost of micropropagated plants was labor related. Interest in mechanization was considered a priority item; however, no laboratory reported active research in this area. Large scale embryogenesis was considered as an alternative to reducing labor costs although the concern of genetic instability with the use of callus was expressed.

ACKNOWLEDGMENTS

The following persons and firms supplied information used in preparing this paper:

MR Burke, Ceres 2000, Winter Haven, Florida, USA;
L Casey, Hyclone, Inc., Conroe, Texas, USA;
Dai-Ichi Seed Company, Tokyo, Japan;
A Donnan, Plant-Tech Inc., Apopka, Florida, USA;
D Donohoe, Sunny Border Nurseries, Inc., Kensington, Connecticut, USA;
D Egger, Twyford Plant Laboratories, Inc., Santa Paula, California, USA;
C Falstad, Walters Gardens, Inc., Zeeland, Michigan, USA;
Fukukaen Nurseries and Bulb Company, Ltd., Nagoya, Japan;
S Garton, Native Plants, Inc., Salt Lake City, Utah, USA;
JJ Goodwin, Native Plants, Inc., Salt Lake City, Utah, USA;
RD Hartman, Hartman's Plants, Inc., Palmdale, Florida, USA;
G Hennen, Ogelsby Plant Laboratories, Altha, Florida, USA;
D Mikkelsen, Mikkelsen's Inc., Ashtabula, Ohio, USA;
Miyoshi and Company, Ltd., Tokyo, Japan;
PJ Myers, Oakdell, Inc., Apopka, Florida, USA;
M Roh, Florist and Nursery Crops Laboratory, U.S. Department of Agriculture, Beltsville, Maryland, USA;
R Strode, Agri-Start's, Inc., Apopka, Florida, USA;
J Will, Armstrong Roses, Inc., Somis, California, USA;
M Zilis, T and Z Nursery, Inc., Winfield, Illinois, USA.

REFERENCES

1. Zilis M, D Zwagerman, D Lamberts, L Kurtz: 1979. Commercial propagation of herbaceous perennials by tissue culture. Comb. Proc. Intl. Plant Prop. Soc. 29:404-414.

IN VITRO PROPAGATION OF FLOWER-BULB CROPS

J. VAN AARTRIJK AND P.C.G. VAN DER LINDE

Bulb Research Centre, Lisse, The Netherlands

1. INTRODUCTION

The research on and the application of plant-tissue culture techniques have been extended rapidly in the past c. 15 years, and bulbous crops have certainly shared in this development. Although meristem culture, embryo culture, and other *in vitro* techniques have gained the well-deserved interest of many investigators, by far the most attention has been paid to the development of methods for rapid vegetative propagation, probably because of the low natural multiplication rate of many bulbous crops. Such methods, in combination with other propagation techniques, e.g., scaling, chipping or scooping, can be valuable tools to build up rapidly commercial stocks of selected plants with desired characters, viz., new genotypes or disease-free plants. The availability of such methods will probably increase breeding and selection efforts, which in turn may lead to an increase in the average quality of the bulbs produced.

Ever since the first report by Robb (50), the number of reports in the literature dealing with micropropagation of flower-bulb crops has increased. Although the major part of these reports refer to a small number of crops, especially *Lilium*, *Hyacinthus*, and *Tulipa*, many crops belonging to the *Liliaceae*, *Iridaceae* or *Amaryllidaceae* have been studied to some extent (Table 1). Many of these reports do not include quantitative data but merely give an indication that *in vitro* propagation is feasible. Reports in literature with quantitative data have served as the main basis for this review.

We shall restrict ourselves to some main lines and deal with the subject by discussing five major aspects of almost any propagation scheme, viz., explant type, induction of plantlets, (bulblet) growth, subculture stage(s), and transfer to soil. Although for the commercial application all five aspects are equally important, in literature most attention has been paid to the first three aspects.

Zimmerman, R.H. et al. (eds), Tissue Culture as a Plant Production System for Horticultural Crops.
 ISBN 90-247-3378-2.

TABLE 1. Ornamental bulbous crops shown to be capable of being propagated *in vitro*.

Family	Genus	References
Liliaceae	*Lilium*	(7, 59, 66)
	Tulipa	(8, 36)
	Hyacinthus	(14, 29, 45)
	Ornithogalum	(24, 34)
	Muscari	(15, 51)
	Fritillaria	(67)
	Lachenalia	(35)
	Alstroemeria	(71)
Iridaceae	*Iris*	(18, 25)
	Gladiolus	(26, 32)
	Freesia	(46, 60)
	Crocus	(17)
Amaryllidaceae	*Narcissus*	(28, 56)
	Hippeastrum	(33, 42, 69)
	Nerine	(30, 47)
	Haemanthis	(21)
	Zephyranthes	(19)
	Eucharis	(48)

TABLE 2. Explant types used for *in vitro* propagation studies of various ornamental bulbous crops.

Explant	*Lilium*	*Hyacinth*	*Gladiolus*	*Narcissus*	*Tulipa*	*Iris*	*Hippeastrum*	*Nerine*
Bulb scale	X	X		X	X	X	X	X
Leaf	X	X		X				
Stem	X	X	X	X	X	X	X	X
Bud	X		X		X	X		
Flower petal		X	X		X			
Cotyledon	X							
Root	X							

2. EXPLANT TYPES

Explants of many types of tissue have been shown to be able to regenerate plantlets. In Table 2 data are listed for a number of important ornamental bulbous plants. It is easily seen that the best suited explants are from (a) bulb scales, especially (some *Liliaceae*) or exclusively (*Iridaceae*, *Amaryllidaceae*) from the basal part near the meristematic zone at the junction of scales and basal plate (19, 20, 28, 48, 50, 57, 69; Fig. 1); (b) stems, especially thin slices from young, unelongated stems (8, 10, 11, 23, 25, 42, 43, 70; Fig. 2); and (c) buds (1, 18, 26, 27, 32; Fig. 3).

FIGURE 1 (left). Adventitiously formed bulblets from the zone at the junction of bulb-scale and basal plate of *Iris hollandica* 'Prof. Blaauw'.
FIGURE 2 (right). Adventitious regeneration of shoots on slices of young stem of tulip 'Merry Widow'.

FIGURE 3. Regeneration of shoots on the swollen base of an axillary bud isolated from a corm of *Gladiolus* 'Peter Pears'.

3. INDUCTION OF PLANTLETS

The majority of the described propagation methods is based on adventitious formation of plantlets, which originate from one or a few cells within an explant. Plantlet induction thus starts with the initiation of cell divisions in more or less specialized cells. For commercial propagation it is important to be able to influence the number of plantlets induced per explant. Many factors have been shown to affect the regeneration process. Only a few important ones will be discussed here.

3.1. Factors related to the plant tissue

3.1.1. Size of explants. This can probably be interpreted as the relative amount of wound-damage to the tissue (4). Wounding is generally considered to be the initial event that puts into motion the processes that ultimately lead to cell divisions. Both inhibitory and stimulatory effects of using smaller-sized explants have been described (4, 44, 68).

3.1.2. Position effects. Except for position effects of explants within a plant or a bulb (28, 50, 57, 68, 69), position effects may occur within explants, viz., basipetal polarity (4, 28, 43, 48, 69), and lateral polarity (20, 31, 69). Many of these effects can often be interpreted as age effects. These positional effects generally lead to special requirements with regard to orientation of the explant on the media.

3.1.3. Pretreatment of plant tissue may profoundly influence the regeneration behavior. Both an increase and a decrease of plantlet induction have been observed after cold storage of parent-bulb tissue (2-9°C; 3, 14, 28, 52), the effect being dependent on storage time (3, 14, 58), auxin concentration in the medium (3), and culture temperature (58). For tulip (38) and iris, a pretreatment of bulbs at 30°C is stimulatory.

In general, it can be said that cell divisions and plantlet induction preferably take place in (unelongated) cells of young, non-dormant, suitably wounded and well-oriented explants.

3.2. Factors related to the nutrient medium

However regenerative an explant may be, regeneration generally does not occur without a proper nutrient environment containing salts, sugar, some vitamins, and plant growth regulators (PGRs).

3.2.1. PGRs. In bulbous crops low concentrations ($< 10^{-6}$ M) of auxin appear to be important to increase the induction of cell divisions and plantlets. Their role has been established in explants of, for instance, *Narcissus* (28), *Lilium* (3, 57), *Ornithogalum* (24), *Hyacinthus* (11, 45), *Zephyranthes* (19), and *Eucharis* (48). Effects of various other factors, viz., degree of wounding, temperature, ethylene, and triiodobenzoic acid have been shown (5, 7, 9) to depend on the presence of auxin. Auxins generally act according to an optimum curve. Many reports describe a stimulatory effect of cytokinins on the number of plantlets formed (23, 26, 29, 68), although these compounds do not seem to be necessary in all cases, especially not when explants from bulbs are used (3, 19, 24, 28, 57). This group of growth regulators also generally act according to an optimum curve. Abscisic acid (ABA) appears to inhibit plantlet induction (41, 45), whereas varying results have been obtained with gibberellic acid (GA_3) (24, 41, 45) and ethylene or ethephon (7, 45).

3.2.2. Other medium factors. Relatively few data on the effects of other medium constituents are available. A pH of c. 6.0 (14, 18, 62, 68) and the presence of mineral salts (generally a normal or double concentration of MS salts, 2, 62) and some vitamins (2) appear to be satisfactory. Sucrose generally acted according to an optimum curve, with the optimum between 3 and 10% (w/v; 2, 14, 62), which is rather high when compared with the concentrations used with other crops.

3.3. Environmental factors

3.3.1. Temperature. Within the range of c. 10°-30°C, an increase in culture temperature results in either an increase in the number of plantlets per explant (4, 58, 69) or a decrease (43, 62), or it acts according to an optimum curve (37, 53). A temperature of 35°C is clearly supra-optimal (58, 69). When literature data are compared, a more or less chaotic picture emerges, which may be clarified to some extent by assuming that higher temperatures, within limits, stimulates cell divisions, but inhibits the subsequent differentiation into plantlets.

3.3.2. Light. Real confusion arises when the reported effects of light on plantlet induction are studied. Light, generally cool-white fluorescent light at low intensities, has been reported to stimulate (31) or to inhibit (13, 56, 71) plantlet induction, to be without effect (62, 69) or to depend on other parameters (29, 40). One should keep in mind, however, that in these experiments day- and wave-length in the lighted environment may have differed considerably.

3.3.3. It can be concluded that many factors influence the process of plantlet induction. They frequently act according to optimum curves and cause varying and sometimes even conflicting effects. The existence of interactions between the effects of various factors may explain the variations in the reported effects. Indeed, many interactions between the effects of various factors have been reported in or can be deduced from the literature. With *Lilium* explants, experiments have been performed to test the hypothesis that the action of various factors on the induction of cell divisions and plantlet induction is exerted through one biochemical process. It was found that the number of plantlets per explant after a 10-week culture period, as affected by wounding, temperature, auxin and aminoethoxyvinyl glycin, an inhibitor of ethylene biosynthesis, was highly correlated with the relative production by the explants of ethane and ethylene in the first few days of the culture period (6, 7).

Considering these results and data in the literature, for instance, on the synthesis of ethane and ethylene, we suggest that the effects of various extracellular signals may be translated into an intracellularly comprehensible form, as in animal cells, by changes in transmembrane Ca^{2+}-fluxes, leading to a modulation of Ca^{2+}-dependent processes in the cytoplasm.

4. BULBLET GROWTH

Bulblets or cormlets have generally been regarded as the final product of propagation schemes, and hence, investigations have been directed to the regulation of bulblet growth *in vitro*. The experimental results on the regulation of bulblet growth of induced plantlets appear to be more uniform than those on the regulation of plantlet induction.

Once vigorously growing shoots are obtained, they tend to form a bulblet or cormlet sooner or later in the culture period, this bulbing process being generally thought to coincide with a gradual deepening of the dormant state. Various factors have been reported to stimulate bulbing of shoots, generally at the expense of further leaflet formation. These factors are: higher culture temperature (within the range c. 15-30°C; 36, 58), darkness (13, 31, 56, 57, 72), high sucrose levels (> 5% w/v; 36, 49, 63, 72), relatively low pH (c. 5.0; 13, 36), and the presence of a low concentration of auxin ($< 5 \cdot 10^{-6}$ M; 13, 23, 36, 57). On the contrary, light (13, 31, 57), cytokinins (3, 26, 36, 57, 64), low sucrose (63, 72), and low(er) temperature (30, 58) have been shown to promote shoot or leaflet growth or to inhibit or retard the onset of bulbing.

Shoots of several crops, unfortunately, do not grow very vigorously, but tend to stop growing in an early stage of development. Examples are tulip and iris.

To obtain bulblets in these systems, efforts should be directed to the induction of the bulbing process. Several treatments have been reported in the literature to cause such a bulbing response, viz., a well-timed treatment for 2-10 weeks at c. 5°C (36, 49), a treatment with GA_3, either applied as a soak or added to the medium, in combination with a cold treatment or not (49), and ABA or capric acid, in combination with a cold treatment (39).

5. FURTHER PROPAGATION

Micropropagation methods generally include one or more subculture stages following the initiation stage. For bulbous crops three different approaches may be distinguished:

- the approach that may be successful for, for instance, *Lilium* (3, 62), hyacinth (12, 14), *Ornithogalum* (24), and *Muscari* (34) (Fig. 4). This approach can roughly be characterized by the following: (a) in the initiation stage and subculture stages similar explants (leaf, bulb scale, etc.) are used, generally with similar culture requirements,

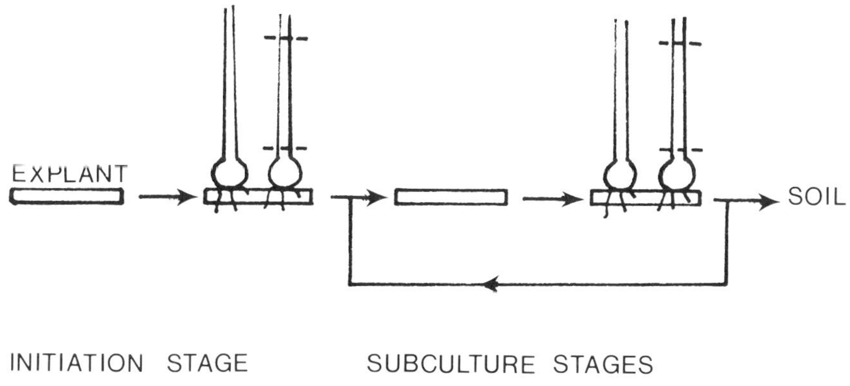

FIGURE 4. Scheme for mass propagation of, for instance, *Lilium*, *Hyacinthus*, *Ornithogalum*, and *Muscari*.

(b) shoots tend to grow vigorously, (c) the subculture periods are relatively long, and (d) a relatively low input of labor is necessary per produced plantlet.

- the approach as used for Narcissus (28), Gladiolus (26, 32), Nerine (30), and iris (van der Linde and van Aartrijk, unpublished results; Fig. 5). Some general characteristics of such a scheme are:

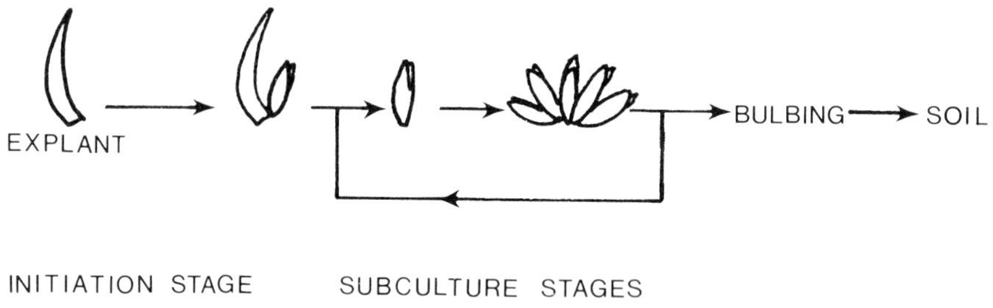

FIGURE 5. Scheme for mass propagation of, for instance, Narcissus, Gladiolus, Iris, and Nerine.

(a) different types of explants are often used in initiation and subculture stages; generally trimmed and split developing shoots are used in the subculture stages, (b) the subculture periods are relatively short, (c) shoots tend to stop growth, (d) subculture demands are often different from the ones in the initiation stage, and (e) a relatively high input of labor is necessary per plantlet or bulblet produced.

- the 'callus' approach: proliferation of callus with subsequent regeneration of plantlets or bulblets. Because of the well-known tendency for callus to differentiate into genetically aberrant plants, its use for commercial propagation schemes is limited. Crops able to produce genetically stable callus, such as Lilium (54, 55) and Freesia (16, 60), may be propagated in this way (Fig. 6).

6. TRANSFER TO SOIL

Once produced in vitro, plantlets sooner or later have to be transferred to soil or other substrates. Although the amount of research devoted to

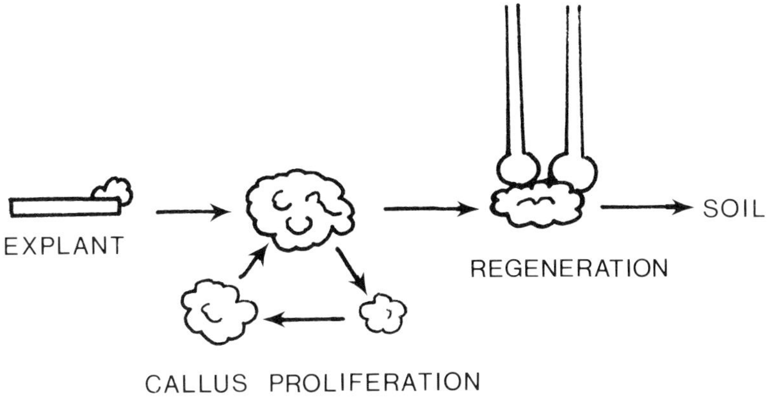

FIGURE 6. Scheme for mass propagation by callus proliferation with subsequent regeneration of plantlets or bulblets.

this important aspect is rather limited, several phenomena have been observed, although not all of them have been described yet.

6.1. Dormancy

Bulblets or plantlets may have entered a more or less dormant state during the *in vitro* culture period. This will be the case especially when bulbing treatments have been given, such as, a high sucrose level in the medium (65). Pretreatments of the bulblets or plantlets before planting may be necessary.

Such treatments are:

- cold treatment (c. 0-9°C for 2-10 weeks) given within or outside the culture vessels. This has been promotive to, for instance, *Lilium* (65, 71), *Narcissus* (28) and *Gladiolus* (22).
- high temperature treatment. A treatment of 1 hour at 45°C given to bulblets of *Lilium longiflorum* breaks their dormancy, especially when the bulblets are cultured before at 30°C when compared with 25°C (61). Iris bulblets stored at 30°C for 4-8 weeks will germinate more rapidly and synchronized than bulblets without such a treatment (van der Linde, unpublished results).
- a red-light treatment (2 weeks, 16 hours per day) given at the end of the culture period releases bulblets of *Lilium longiflorum* from dormancy (61).

6.2. Other phenomena

Sometimes, plantlets of iris and hyacinth have been observed that, after transfer to soil, continued to form new leaves. These plantlets produced neither bulbs nor flowers and appeared to have 'lost' the normal periodicity. Pretreatments with low or high temperature restored the 'normal' situation (van der Linde, unpublished results).

Lily bulblets produced in vitro, when stored at low temperature (c. 0-5°C), may develop a soft rot starting from the vegetative apex. The incidence of the disease depended on storage temperature, duration of storage period, and previous culture conditions, viz., the auxin concentration in the nutrient medium. The phenomenon seems to have characteristics in common with 'chilling' injury in chilling-sensitive plants.

In general, plantlets of ornamental bulbous crops produced in vitro appear to be genetically identical to the parent plants, if care is taken to avoid callus.

Lilium has even been named the most stable species in tissue culture (54, 55), although minor abnormalities have been observed (9). Although callus of freesia has been described to be genetically stable in culture (16, 60), freesia plantlets derived from in vitro cultured flower-buds frequently bear abnormal flowers (Fig. 7).

FIGURE 7. Genetically aberrant flower of Freesia plant derived in vitro from a flower bud (courtesy of Prof. Dr. R. L. M. Pierik). At the right: control. At the left: flower of in vitro cultured plant.

REFERENCES

1. Aartrijk J van, GJ Blom-Barnhoorn: 1979. Some influences of 1-naphthylacetic acid on the differentiation of meristems of Lilium speciosum 'Rubrum' nr. 10 in vitro. Acta Hort. 91:269-279.

2. Aartrijk J van, GJ Blom-Barnhoorn: 1980. Effects of sucrose, mineral salts, and some organic substances on the adventitious regeneration in vitro of plantlets from bulb-scale tissue of Lilium speciosum Rubrum. Acta Hort. 109:297-302.

3. Aartrijk J van, GJ Blom-Barnhoorn: 1981. Growth regulator requirements for adventitious regeneration from Lilium bulb-scale tissue in vitro, in relation to duration of bulb storage and cultivar. Scientia Hort. 14:261-268.

4. Aartrijk J van, GJ Blom-Barnhoorn: 1983. Adventitious bud formation from bulb-scale explants of Lilium speciosum Thunb. in vitro. Effects of wounding, TIBA and temperature. Z. Pflanzenphysiol. 110:355-363.

5. Aartrijk J van, GJ Blom-Barnhoorn: 1984. Adventitious bud formation from bulb-scale explants of Lilium speciosum Thunb. in vitro. Interacting effects of NAA, TIBA, wounding, and temperature. J. Plant Physiol. 116:409-416.

6. Aartrijk J van, GJ Blom-Barnhoorn, J Bruinsma: 1985. Adventitious bud formation from bulb-scale explants of Lilium speciosum Thunb. in vitro. Production of ethane and ethylene. J. Plant Physiol. 117:411-422.

7. Aartrijk J van, GJ Blom-Barnhoorn, J Bruinsma: 1985. Adventitious bud formation from bulb-scale explants of Lilium speciosum Thunb. in vitro. Effects of aminoethoxyvinyl glycin, 1-aminocyclopropane-1-carboxylic acid, and ethylene. J. Plant Physiol. 117:401-410.

8. Alderson PG, RD Rice, NA Wright: 1983. Towards the propagation of tulip in vitro. Acta Hort. 131:39-47.

9. Bennici A: 1979. Cytological chimeras in plants regenerated from Lilium longiflorum tissues grown in vitro. Z. Pflanzenzüchtung 82:349-353.

10. Bigot C: 1974. Comparaison des aptitudes pour le bourgeonnement de tissus superficiels et de tissus profonds cultivés in vitro. Cas de la tige d'un lis hybride, cv. Enchantment. Compt. Rend. Acad. Sci. Paris 278D:1027-1030.

11. Chung JD, CK Chun, YK Suh, EM Rhee: 1981. In vitro propagation of Hyacinthus orientalis L. Effect of auxins on callus formation and bulblet differentiation from flower organ and flower stem tissue. J. Kor. Soc. Hort. Sci. 22:146-151.

12. Chung JD, CK Chun, YK Suh, EM Rhee: 1983. In vitro propagation of Hyacinthus orientalis L. Organ redifferentiation through the subculture of bulblet scale and leaf-tissue segments originating from flower stalk tissue. J. Kor. Soc. Hort. Sci. 24:162-168.

13. Chug JD, CK Chun, YK Suh, EM Rhee: 1983. In vitro propagation of Hyacinthus orientalis L. Effect of auxin concentration and light or dark incubation on organogenesis from flower-stalk tissue, and comparison of totipotency between bulb-scale and flower-stalk tissue. J. Kor. Soc. Hort. 24:76-85.

14. Chung JD, CK Chun, EM Rhee: 1984. In vitro propagation of Hyacinthus orientalis L. Effect of sucrose concentration, pH value, bulblet age, and chilling treatment of bulblet on organogenesis from bulblet scale and leaf tissue. J. Kor. Soc. Hort. Sci. 25:72-79.

15. Cumming BG, DE Peck: 1984. Tissue culture of grape hyacinth Muscari spp. HortScience 19:723-724.

16. Davies DR: 1972. Speeding up the commercial propagation of freesias. The Grower 77:711.

17. Ding BZ, SH Bai, Y Wu, XP Fan: 1981. Induction of callus and regeneration of plantlets from the corm of Crocus sativus. Acta Bot. Sin. 23:419-420.

18. Fujino M, T Fujiura, K Hamada: 1972. Multiplication of Dutch iris (Iris hollandica) by organ culture. Effect of growth regulators, culture media, pH, peptone, and agar on the growth and differentiation of excised lateral buds. J. Jap. Soc. Hort. Sci. 41:66-71.

19. Furmanova M, H Oledka: 1981. Plant regeneration from excised bulb-scale segments of Zephyranthes robusta Baker. Acta Soc. Bot. Poloniae 50:399-404.

20. Grootaarts H, JHN Schel, RLM Pierik: 1981. The origin of bulblets formed on excised twin scales of Nerine bowdenii. Plant Cell Tissue Organ Culture 1:39-46.

21. Hall JH, PB Tomlinson: 1973. Natural callus-like tissue produces vegetative buds on the leaves of the monocotyledon Haemanthis cinnabarinum Decaisne (Amaryllidaceae). Nature New Biol. 243:252-255.

22. Hildebrandt AC: 1970. Growth and differentiation of single plant cells and tissues. pp. 71-93. Proceedings of 2nd Intl. Conf. Plant Tissue Culture, Strasbourg, France.

23. Hosoki T, T Asahira: 1980. In vitro propagation of Narcissus. HortScience 15:602-603.

24. Hussey G: 1976. Plantlet regeneration from callus and parent tissue in Ornithogalum thyrsoides. J. Exp. Bot. 27:375-382.

25. Hussey G: 1976. Propagation of Dutch Iris by tissue culture. Scientia Hort. 4:163-165.

26. Hussey G: 1977. In vitro propagation of Gladiolus by precocious cillary shoot formation. Scientia Hort. 6:287-296.

27. Hussey G: 1980. Propagation of some members of Liliaceae, Iridaceae, and Amaryllidaceae by tissue culture. In: Petaloid Monocotyledons. Linn. Soc. Symp. Ser. 8:33-42.

28. Hussey G: 1982. In vitro propagation of Narcissus. Ann. Bot. 49:707-719.

29. Kim YJ, PM Hasegawa, RA Bressan: 1981. In vitro propagation of hyacinth. HortScience 16:645-647.

30. Leffring L: 1983. Vegetatieve vermeerdering van Nerine bowdenii. Vakbl. Bloem. 40:33-34.

31. Leshem B, H Liliën-Kipnis, B Steinitz: 1982. The effect of light and of explant orientation in the regeneration and subsequent growth of bulb-scale sections cultured in vitro. Scientia Hort. 17:129-136.

32. Logan AE, FW Zettler: 1985. Rapid in vitro propagation of virus-indexed gladiolus. Acta Hort. 164:169-180.

33. Mii M, T Mori, N Iwase: 1974. Organ formation from the excised bulb scale of Hippeastrum hybridum in vitro. J. Hort. Sci. 49:241-244.

34. Nel D: 1981. Rapid propagation of Ornithogalum hybrid in vitro. Agroplantae 13:83-84.

35. Nel D: 1983. Rapid propagation of Lachenalia hybrids in vitro. S. Afr. J. Bot. 2:245-246.

36. Nishiuchi Y: 1980. Studies on the vegetative propagation of tulip. Regeneration of bulblets in bulb-scale segments cultured in vitro. J. Jap. Soc. Hort. Sci. 49:235-240.

37. Nishiuchi Y: 1980. Studies on the vegetative propagation of tulip. Effects of temperature on the adventitious bud formation of excised bulb scales cultured in vitro. J. Hokkaido Univ. Educ. 30:189-194.

38. Nishiuchi Y: 1982. Studies on multiplication of tulip bulb by tissue culture method. J. Hokkaido Univ. Educ. 33:49-65.

39. Nishiuchi Y: 1983. Studies on vegetative propagation of tulip. Effect of growth regulators on the bulb formation of adventitious buds cultured in vitro. J. Hokkaido Univ. Educ. 34:9-15.

40. Novak FJ, E Petri: 1981. Tissue culture propagation of Lilium hybrids. Scientia Hort. 14:191-199.

41. Paek KY, H Sung: 1983. Factors affecting regeneration ability and physiology of dormancy in the mature bulbil segments of Lilium lancifolium in vitro. J. Kor. Soc. Hort. Sci. 24:149-157.

42. Pajerski L, PD Ascher: 1977. Propagation of Amaryllis (Hippeastrum) from cultured scape tissue. HortScience 12:293.

43. Pierik RLM, J Woets: 1971. Regeneration of isolated bulb-scale segments of hyacinth. Acta Hort. 23:423-428.

44. Pierik RLM, AJM Post: 1975. Rapid vegetative propagation of Hyacinthus orientalis L. in vitro. Scientia Hort. 3:293-297.

45. Pierik RLM, HHM Steegmans: 1975. Effects of auxins, cytokinins, gibberellins, abscisic acid, and ethephon on regeneration and growth of bulblets on excised bulb-scale segments of hyacinth. Physiol. Plant. 34:14-17.

46. Pierik RLM, HHM Steegmans: 1976. Vegetative propagation of freesia through the isolation of shoots in vitro. Neth. J. Agric. Sci. 24:274-277.

47. Pierik RLM: 1981. Vegetatieve vermeerdering van Nerine in kweekbuizen. Vakbl. Bloem. 38:28-31.

48. Pierik RLM, PA Sprenkels, J van Bragt: 1983. Rapid vegetative propagation of Eucharis grandiflora in vitro. Acta Hort. 147:179-186.

49. Rice RD, PG Alderson, NA Wright: 1983. Induction of bulbing of tulip shoots in vitro. Scientia Hort. 20:377-390.

50. Robb SM: 1957. The culture of excised tissue from bulb scales of Lilium speciosum Thunb. J. Exp. Bot. 8:348-352.

51. Roest S, GS Bokelman: 1981. Vegetatieve vermeerdering van Muscari in kweekbuizen. Vakbl. Bloem. 36:134-135.

52. Seabrook JEA, BG Cumming, LA Dionne: 1976. The in vitro induction of adventitious shoot and root apices on Narcissus (daffodil and narcissus) cultivar tissue. Can. J. Bot. 54:814-819.

53. Seabrook JEA, BG Cumming: 1982. In vitro morphogenesis and growth of Narcissus in response to temperature. Scientia Hort. 16:185-190.

54. Sheridan WF: Plant regeneration and chromosome stability in tissue cultures, pp. 263-295. 1974. In: L Ledoux (ed). Genetic manipulation with plant material. Plenum Publ. Corp., New York.

55. Simmonds JA, BG Cumming: 1976. Propagation of Lilium hybrids. II. Production of plantlets from bulb-scale callus cultures for increased propagation rates. Scientia Hort. 5:161-170.

56. Steinitz B, H Yahel: 1982. In vitro propagation of Narcissus tazetta. HortScience 17:333-334.

57. Stimart DP, PD Ascher: 1978. Tissue culture of bulb-scale sections for asexual propagation of Lilium longiflorum Thunb. J. Amer. Soc. Hort. Sci. 103:182-184.

58. Stimart DP, PD Ascher: 1981. Developmental responses of Lilium longiflorum bulblets to constant or alternating temperatures in vitro. J. Amer. Soc. Hort. Sci. 106:450-454.

59. Stimart DP, PD Ascher, HF Wilkins: 1982. Overcoming dormancy in Lilium longiflorum bulblets produced in tissue culture. J. Amer. Soc. Hort. Sci. 107:1004-1007.

60. Stimart DP, PD Ascher: 1982. Plantlet regeneration and stability from callus culture of Freesia x hybrida Bailey cultivar 'Royal'. Scientia Hort. 17:153-157.

61. Stimart DP, PD Ascher, HF Wilkins: 1983. Axis elongation from tissue-culture-generated bulblets of Lilium longiflorum Thunb. J. Amer. Soc. Hort. Sci. 108:99-101.

62. Takayama S, M Misawa: 1979. Differentiation in Lilium bulbscales grown in vitro. Effects of various cultural conditions. Physiol. Plant. 46:184-190.

63. Takayama S, M. Misawa: 1980. Differentiation in Lilium bulbscales grown in vitro. Effects of activated charcoal, physiological age of bulb, and sucrose concentrations on differentiation and scale-leaf formation in vitro. Physiol. Plant. 48:121-125.

64. Takayama S, M Misawa: 1982. Regulation of organ formation by cytokinin and auxin in Lilium bulbscales grown in vitro. Plant Cell Physiol. 23:67-74.

65. Takayama S, M Misawa, Y Takashige, H Tsumori, K. Ohkawa: 1982. Cultivation of in vitro-propagation Lilium bulbs in soil. J. Amer. Soc. Hort. Sci. 107:830-834.

66. Takayama S, M Misawa: 1983. A scheme for mass propagation of Lilium in vitro. Scientia Hort. 18:353-362.

67. Wietsma WA: 1982. Fritillaria in kweekbuizen te vermeerderen. Vakbl. Bloem. 37:35.

68. Wright NA, PG Alderson: 1980. The growth of tulip tissues in vitro. Acta Hort. 109:263-270.

69. Yanagawa T, Y Sakanishi: 1977. Regeneration of bulblets on Hippeastrum bulb segments excised from various parts of a parent bulb. J. Jap. Soc. Hort. Sci. 46:250-260.

70. Ziv M, AH Halevy, R Shilo: 1970. Organs and plantlet regeneration of Gladiolus through tissue culture. Ann. Bot. 34:671-676.

71. Ziv M: 1973. Vegetative propagation of Alstroemeria in vitro. Scientia Hort. 1:271-277.

72. Ziv M: 1979. Transplanting Gladiolus plants propagated in vitro. Scientia Hort. 11:257-260.

WOODY ORNAMENTALS, SHADE TREES, AND CONIFERS

B.H. MCCOWN

Department of Horticulture, University of Wisconsin, Madison, WI 53706, U.S.A.

1. INTRODUCTION

The previous presentations in this symposium have amply documented the application of microculture in the production of numerous herbaceous crops. However, the number of woody shrubs, trees, and conifers now in commercial micropropagation is much more limited. One significant factor leading to this difference is that there has been proportionately much less research and time spent in applying tissue culture techniques to woody perennial crop plants than to their herbaceous counterparts. When significant time has been devoted to micropropagating woody perennial plants (as with tree fruits), good progress has been observed. However, there appears to be fundamental problems based on the biology of these plants that also accounts for the relative difficulty in their in vitro culture.

This talk will focus on these more recalcitrant plants and attempt to pinpoint some of the reasons why utilization of microculture technology for these crops has lagged behind the application to herbaceous crops.

2. APPLICABLE MICROPROPAGATION SYSTEMS

2.1. Shoot cultures

In no system of in vitro culture is the general lack of understanding of woody perennial plant physiology more apparent than with shoot cultures. Nevertheless, shoot cultures remain the preeminent in vitro method of clonal multiplication for these plants. The basic reason is that shoot cultures can be manipulated with some success without understanding the basic underlying reasons why the system responds. Problems arise when the empirically-derived techniques fail and new approaches must be developed. Such development is strongly hindered by the lack of basic knowledge of what controls the growth dynamics of these plants.

Zimmerman, R.H. et al. (eds), Tissue Culture as a Plant Production System for Horticultural Crops.

ISBN 90-247-3378-2.

A micropropagation system can be considered as consisting of two periods--the culture establishment period and the propagule production period. Woody perennial plants can show problems in both periods; however, the greatest difficulties are usually encountered in the culture establishment period.

The culture establishment period itself consists of three phases--the isolation, stabilization and shoot production phases (5). The isolation protocol for tissues from woody plants does not differ greatly from that used for most herbaceous plants. However, one recurring problem with woody perennial plants is a lack of complete decontamination of the tissue. We and many other labs have continually observed bacterial associates growing within and on the cultured shoots. In some cases, even screening the tissues on bacteriological media does not show definitive evidence of the contaminants since their growth in the absence of viable plant tissue may be slow or nil. That this problem would be greater with woody perennial plants than with herbaceous annuals is logical since such close associations would have a greater likelihood of evolving with plant species typified by long life spans and long-term complex relationships with the environment. The exact effect of such contamination on the growth and responsiveness of cultured shoots is unknown in most cases. We have seen effects that range from inhibition to no apparent influence.

The greatest differences between woody perennial and herbaceous plants are observed during the stabilization phase of the culture establishment period. This phase is typified by a change in the growth characteristics of the tissue from the unpredictable and often abnormal shoot development observed with new isolates to the highly predictable and uniform growth seen with commercially-utilizable cultures. With most annuals and herbaceous perennials, this phase takes a relatively short time and can be accomplished in a small number of subcultures after the initial isolation. With woody perennials, this phase can require relatively little time (as with many shrubs), may require several years (as with mature deciduous trees), or may not ever be completed (as with some north temperate climax forestry species).

An overwhelmingly important factor influencing the stabilization of a shoot culture is the phase-state of the original stock plants. In short, the more juvenile a stock plant is, the more readily it is cultured. This becomes particularly apparent when isolates are taken from different portions of the same plant such as young shoots originating from the collar region versus buds from apical portions of well-established branches. With the elepidote rhododendrons, collar-shoots establish readily and have a stabilization phase of less than six months; shoots from flowering stems may not stabilize at all or may require more than one year.

A second factor is the type of shoot growth dynamics that the plant normally displays in its seasonal cycle. In general, plants that show a pattern of continuous shoot growth throughout the season (such as Potentilla, Populus, and Thuja) are much easier to stabilize in culture than are those species denoted by episodic or flushing growth (such as Syringa, Quercus, and northern Picea). A major problem observed with the episodic species is that the flushing shoot growth is maintained in culture; thus the resultant cultures are never truly predictable or uniform and are not usually of commercial value. Interestingly, the episodic pattern can be reversed in some plants, e.g., Rhododendron, and continuous-growing shoot cultures produced. The propagules from such cultures also display a non-flushing habit for a period after removal from culture, a characteristic that is highly advantageous to the grower. This may be an indication of the rejuvenation that has occurred in culture since, for such species as the elepidote rhododendrons, juvenile plants (seedlings, collar-generated shoots) also show non-flushing growth habits.

The relationship between seasonal growth habits and culturability is apparent with the conifers (Table 1). In addition, some conifers, notably the pines, have axillary buds that are highly specialized and diminutive (fascicular buds); these are even more recalcitrant in culture when shoot culture-type systems are attempted. Recently however, some progress has been made with these plants (3).

The basic obstacle to solving each of the above problems is our lack of understanding of biological controls of juvenility and episodic shoot cycles. Until we have more fundamental theories on which to base a design for a culture system, our approaches will remain rather empirical and thus may remain only marginally effective.

Besides the major problems of phase-change and seasonal growth dynamics, a number of other phenomena are observed during the stabilization phase that create cultural problems in a commercial setting. Some species (Acer, Ulmus, Cornus) have a tendency to produce heavy callus at the base of the shoot cultures. This not only directly inhibits shoot growth but limits

TABLE 1. An ordering of various conifer genera as to their predicted culturability as shoot cultures on the basis of their seasonal shoot growth dynamics and axillary bud characteristics. (From 1)

GROUP I. Readily cultured. Generally display continuous (or nearly so) seasonal shoot growth and have fully-developed axillary buds.

Examples: Chamaecyparis, Metasequoia, Sequoia, Thuja

GROUP II. Cultured with some difficulty. Characterized by multiple growth flushes in a season and have well-developed axillary buds.

Examples: Juniperus, Taxus, Tsuga

GROUP III. Very difficult to culture. Show highly episodic shoot growth with usually only one flush each season but have well-developed axillary buds.

Examples: Picea, Pseudotsuga

GROUP IV. Very difficult to culture. Show highly episodic growth and do not have well-developed axillary buds.

Examples: Abies, Pinus (northern species are the most recalcitrant)

---

the effectiveness of the medium in stimulating axillary bud development (since the buds are no longer in direct contact with the medium) and thus multiplication. Such over-callusing is usually stimulated by high cytokinin levels. Thus a trade-off often must be made where the producer lowers the cytokinin levels several orders of magnitude below what is probably optimal for multiplication, but at the same time reduces the over-callusing problem to a manageable level.

Production of phenolics as evidenced by browning of the medium 1 to 2 weeks after subculture is a problem with some species (Magnolia, Paeonia, many conifers). Such browning often can be seen to originate from the dead leaf bases or from the newly cut surfaces at the base of the shoot cultures. In any case, such cultures can become autocatalytic where the exudates create more injury, which in turn produces more exudate, the final result being the degeneration of the culture before the normal subculture lifespan. Charcoal-based media help to alleviate the toxicity problem, but also strongly interfere with the effectiveness of any hormonal additions to the medium.

Some woody plants have not shown any of the symptoms noted above, but continue to produce weak growth with yellowish and early-senescing leaves (*Celtis*). The causes are unknown.

Finally, vitreous growth is not uncommon in woody perennial plant cultures. The factors that contribute to the onset of this malady are apparent and include high cytokinin levels, high humidity in the culture vessel, tightly sealed vessels, and liquid/Gelrite supports. The condition is highly genotype dependent. Again, as with over-callusing, the rate of shoot growth/multiplication is sacrificed by a reduction in the cytokinin levels so that vitreous growth can be kept under control.

In the production phase of the culture establishment period, the aim is the reproducible and infinite generation of shoot cultures of uniform and predictable habit. This is usually obtainable as long as the stabilization phase has been fully achieved. Unfortunately, many attempts at commercialization of a micropropagation system have failed because the tissue was not yet stabilized and thus the upscaling was premature and resplendent with problems.

Recently, it has become apparent that we still do not have the mineral nutrient formulation optimized for woody perennial production in culture. In a project where five genotypes of *Populus* had to be established as shoot cultures, vast differences were observed as to the vigor and uniformity of the shoot growth of each genotype. In beginning to decipher the causes of such differences in performance, we discovered that for some genotypes, neither of two standard mineral formulations (WPM, MS) were adequate for shoot growth. Total nitrogen levels, potassium levels, and total soluble salts were all involved. Other researchers have also come to similar conclusions with other species (2, 3).

One problem that has been elucidated is the cause of "shoot-tip necrosis" or the browning and subsequent death of the actively-growing tips of shoot cultures (10). This has been found to be a calcium deficiency problem and is amazingly closely related to similar calcium deficiency problems observed outside of culture. Again, further refining of the basic mineral salt formulation is in order and should be an emphasis of new research efforts.

2.2. Embryogenic systems

A recent encouraging aspect of the microculture of woody perennial plants has been a number of successes in the development of embryogenic systems based on mass cell culture and utilizing mature stock tissues as the original source (9; unpublished results from our lab on *Populus*). Interestingly, the direct source of tissue for the suspension cultures was shoot cultures. What appears to be possible is the restoration of some degree of juvenility (usually a requirement for most successful embryogenic systems using woody perennials) via the establishment of shoot cultures and the subsequent use of the resultant tissue in embryogenic culture systems. This approach has been proven useful in other research systems (11).

In our studies using *Populus*, competent cell suspension cultures can be generated. The source of these suspensions are roots derived from stabilized shoot cultures. The use of root tissue as a source tissue for cell suspensions appears to have several advantages. Roots placed into liquid culture where the medium is stimulatory to callusing readily initiate cell division along the length of the root axis and generate a mass of sloughing cells within a single subculture. The need for maintaining a cell suspension culture stock over many subcultures and risking the problems of genetic instability and loss of totipotency is averted. In addition, root tissue is relatively easily generated in shoot cultures and is of an inherent uniform nature.

Another observation that may have general application in relation to embryogenic/organogenic cell cultures of woody perennials is the intermediate developmental stage of nodule formation. Loose cell suspensions of *Populus* placed in a medium of lowered auxin/cytokinin ratio form dense clusters of small, highly cytoplasmic cells that are usually termed nodules. Such nodules appear to be highly competent as they can be induced to become organogenic. In a hormonally-balanced medium, embryoids form internally within the nodules and eventually grow-out of the confines of the original nodule (Fig. 1). In a medium favoring rhizogenesis, roots can be generated but these appear to form superficially on the surface of the nodule. In a medium favoring caulogenesis, shoots are formed and appear to originate deep within the nodule. Whether this developmental sequence is generalizable for woody perennial plants is as yet unknown, however nodule-type growth and organogenesis are closely associated in many culture systems (4, 6, 7, 8, 12).

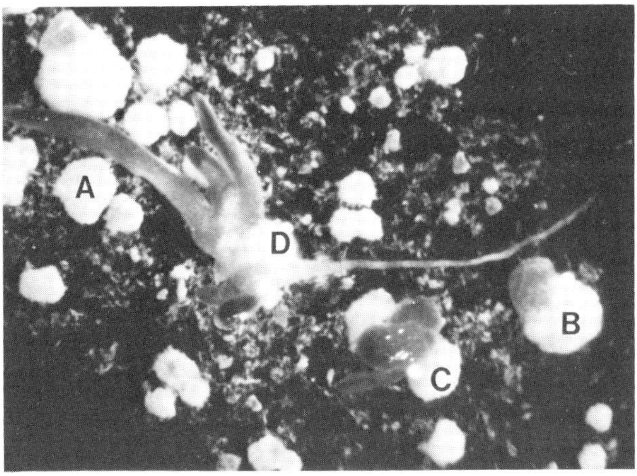

FIGURE 1. A photomicrograph of a nodule culture of Populus showing four stages of organogenesis: (a) nodule. (b) bud emerging from interior of nodule. (c) bud with first leaf development. (d) fully-formed plantlet with root.

A proposed working system for the development and maintenance of organogenic cell suspension cultures of woody perennials is given below: The ease and reliability of recovery of normal plants from such a system has yet to be fully evaluated.

MATURE WOODY PLANT SELECTION

↓

STABILIZED SHOOT CULTURE

↓

ROOT PIECES

↓

LIQUID CULTURE WITH HIGH AUXIN/CYTOKININ

↓

CELL SUSPENSION (MAINTAINED FOR ONLY ONE SUBCULTURE)

↓

LIQUID CULTURE WITH LOWERED AUXIN/CYTOKININ

↓

NODULE FORMATION

↓

EMBRYOGENESIS/ORGANOGENESIS

3. POST CULTURE ASPECTS

The production of rooted and acclimated micropropagules of woody plants that can be readily incorporated into crop production programs is highly feasible. The overwhelming variable is the quality of the original microcutting, which is dependent on the factors discussed above and elsewhere in this symposium.

With many woody plants, a much greater attention must be given to the maintenance of the regenerated root systems than with most herbaceous crops. Most woody perennials do not have highly fibrous root systems so that the problems of root system deformation and injury become important. Although many of the herbaceous plants may also not have fibrous root systems, the long-term consequences of poorly structured root systems are not as apparent as with the long-lived woody perennials where root girdling and predisposition to disease are major concerns.

Two important techniques that tend to minimize root problems are the rooting in non-*in vitro* environments and the utilization of plug systems where root pruning is employed. In addition to the substantially increased costs, rooting *in vitro* has not been satisfactory in our tests. Although roots may be more easily generated *in vitro*, moving such rooted and tender propagules into the next stage of production is difficult. Not only is root and shoot injury more likely, but it is very difficult to plant such rooted micropropagules without deformation of the root system. Luckily, most of the original roots formed in culture appear to function poorly in many systems and do not continue growth to form the main root system of the plants. Instead, rerooting often occurs and the *in vitro* rooting step may often function as a "pre-rooting" step. In fact, some investigators have taken advantage of this phenomenon and designed a much less costly semi-*in vitro* prerooting step into the production scheme (13). With conifers, the use of activated charcoal in the *in vitro* rooting environment may be particularly important (3).

The utilization of plugs for receiving micropropagules and for rooting and acclimation has been discussed earlier in this symposium.

Another factor that we have found important is the maintenance of continued growth of the propagule throughout the rooting/acclimation period. Microcuttings of woody perennials that are not in active growth tend to root poorly and less uniformly. In addition, with some plants such as tree peony and *Amelanchier*, stoppage of growth can lead to an onset of dormancy of the propagule and thus significant problems in the utilization of that propagule in subsequent growing operations.

Another caution that must be observed when working with woody perennials is the effect of the handling of the propagules on subsequent growth of the crop. This became apparent to us in some experiments with birch in which plugged, rooted and acclimated plants that were stored under refrigerated conditions produced trees in the field with a considerably different growth form than non-stored micropropagated plants.

4. CONCLUSIONS

The widespread application of microculture for the production of woody perennial plants awaits a better understanding of the controls of the growth dynamics of these plants. Undoubtedly, more woody plants will be utilized in micropropagation, but these will be the exceptions and will be outnumbered by the failures until our basic knowledge is improved enough to enable us to intelligently design culture systems.

REFERENCES

1. Amos RA: 1982. Micropropagation of members of the Pinales and Taxales; an overview of their adaptability of shoot culture. MS Thesis, Univ. of Wisconsin, Madison.

2. Driver JA, AH Kuniyuki: 1984. In-vitro propagation of Paradox walnut rootstocks. HortScience 18:506-509.

3. Gupta PK, DJ Durzan: 1985. Shoot multiplication from mature trees of Douglas fir (Pseudotsuga menziesii) and sugar pine (Pinus lambertiana). Plant Cell Reports (in press).

4. Hu CY, JD Ochs, FM Mancini: 1978. Further observations on Ilex embryoid production. Phytomorphology 21:103 107.

5. McCown DD, BH McCown: 1986. North American hardwoods. In: J.M. Bonga and D.J. Durzan (eds). Tissue culture in forestry, second edition. Martinus Nijhoff/W. Junk Publishers, the Netherlands, in press.

6. Radojevic L: 1979. Somatic embryos and plantlets from callus cultures of Paulownia tomentosa Steud. Z. Pflanzenphysiol. 91:57-62.

7. Radojevic L, R Vujicic, M Neskovic: 1975. Embryogenesis in tissue cultures of Corylus avellane L. Z. Pflanzenphysiol. 77:33-41.

8. Rao PS: 1965. In vitro induction of embryonal proliferation in Santalum album L. Phytomorphology 15:175-179.

9. Rao PS, P Ozias-Akins: 1985. Plant regeneration through somatic embryogenesis in protoplast cultures of sandalwood (Santalum album L.) Protoplasma 124:80-86.

10. Sha L, BH McCown, LA Peterson: 1985. Occurrence and cause of shoot-tip necrosis in shoot cultures. J. Amer. Soc. Hort. Sci. 110:631-634.

11. Smith MAL, BH McCown: 1983. A comparison of source tissue for protoplast isolation from three woody plant species. Plant Sci. Lett. 28:149-156.

12. Staritsky G: 1970. Embryoid formation in callus tissues of coffee. Acta Bot. Neerl. 19:509-514.

13. Zimmerman RH, I Fordham: 1985. Simplified method for rooting apple cultivars in vitro. J. Amer. Soc. Hort. Sci. 110:34-48.

ORCHID TISSUE CULTURE

R. J. Griesbach

U. S. Department of Agriculture, Agricultural Research Service,
Florist and Nursery Crops Laboratory, Beltsville, MD 20705, U.S.A.

1. HISTORICAL PERSPECTIVE

The first attempt at plant tissue culture began in 1902 when Haberlandt (3, 4) placed the epidermal, pith parenchyma and palisade cells of various monocots in culture. The culture medium contained various sugars, inorganic salts and amino acids. None of the cells, however, grew. The first successful attempt came in 1904 when Hanning (3, 4), using similar media, was able to grow whole plants from isolated embryos. During the next two decades work centered around improving embryo culture techniques. It was found that several complex additives such as yeast extract, coconut milk and peptone greatly stimulated growth (3).

Orchids were the first plants for which embryo culture techniques were widely applied commercially. Orchid seeds are quite small weighing between 0.3 and 14 μg. The seed generally contains no endosperm or cotyledon and consists only of a small, undifferentiated, spherical, embryo-like mass of cells. In most instances, this embryo-like structure contains fewer than 50 cells (1).

In nature, a fungal mycorrhizal association is needed for germination of the orchid seed. Upon germination, the embryo-like seed structure swells and bursts out of the seed coat. An egg-shaped, green seedling is soon formed. This stage has been named the protocorm stage by Bernard (1). On the upper surface, the first leaf primordium forms as a small bulge. The seedling now looks like a child's top. Soon after, absorbing hairs form on the lower surfaces. The first leaf then appears, which is followed by the first root. The entire process requires several months (Figure 1) (1).

Zimmerman, R.H. et al. (eds), Tissue Culture as a Plant Production System for Horticultural Crops.
 ISBN 90-247-3378-2.

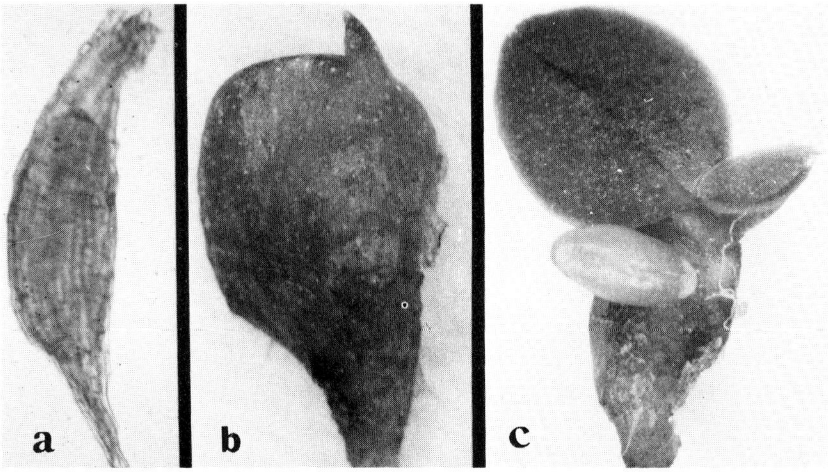

FIGURE 1. Stages in the germination of a typical orchid (Phalaenopsis). The orchid seed (a) contains an undifferentiated, 30-celled mass of cells. When the seed is placed on the appropriate culture medium, the 30 cells will swell and start to divide. After 1 month of growth, the protocorm stage is reached (b). On the top of the protocorm an apical meristem has formed. The protocorm can be vegetatively propagated through division. In general, a protocorm can be cut into 4 parts, each of which will regenerate a new, whole protocorm when placed on the appropriate culture medium. If the protocorm is placed on a basal medium, it will continue development into a small plantlet (c). This process takes an additional 3 to 6 weeks. In many instances, the very young, differentiated leaves and roots can be sectioned and placed on an appropriate culture medium to induce multiple protocorm formation.

Orchid growers first germinated orchid seeds in the middle 1800's, when Harris (1) discovered that the seeds could germinate if sown at the base of the mother plant. From one seed capsule that could contain up to 1,000,000 seeds, a handful of seedlings could be raised. In the late 1800's, Bernard (1) discovered the mycorrhizal requirement for seed germination and devised methods for germinating orchid seeds in the presence of the mycorrhizal fungi. This procedure was more successful than sowing seed at the base of the mother plant. Hundreds of seedlings could be raised using this method of symbiotic seed germination. In 15,000 trials, however, successful germination occurred in only a few hundred cases. In the 1920's, Knudson (1) applied embryo culture techniques to orchid seeds. Using this asymbiotic procedure for seed germination thousands of seedlings could now be raised from a single seed capsule. Embryo culture techniques revolutionized the orchid industry since it made large scale hybridization and seedling production possible. Orchids are very slowly propagated vegetatively. Until this time almost all of the commercially available orchid plants were species obtained through jungle collection. Many orchid companies very quickly built laboratories to germinate orchid seed aseptically via embryo culture techniques (1).

During the 1930's, emphasis in the research laboratories shifted from embryo culture to tissue culture. In 1934, White (3, 4) placed excised tomato root tips in aseptic culture. When yeast extract was added to the medium, the roots were able to divide and grow indefinitely. In 1936, Brown and Gardener (3, 4) found that tumors, similar to the tumors induced by crown gall bacteria, could be produced by exogenous auxin applications. This information was quickly applied to tissue culture. Gautheret, Nobecourt and White (3, 4), each independently demonstrated in 1939 that the auxin indoleacetic acid, when incorporated into a culture medium, stimulated the production of undifferentiated callus. This was the first time hormones were added to tissue culture media. In 1956, Miller and others (3, 4) discovered that the active ingredient in coconut milk and yeast extract, which promoted growth in culture, was another class of hormones, the cytokinins. Skoog and Miller (3, 4) further reported in 1957 that the auxin/cytokinin balance was very important for the growth of cells in culture. If the auxin level were relatively high and the cytokinin level relatively low, undifferentiated callus growth occurred. High levels of cytokinin relative to auxin induced the appearance of differentiated shoots. Thus, it was now possible to regenerate a whole plant from a single undifferentiated cell (3).

At this time, orchids entered the scene again. In 1955, Morel and Martin (6) were able to successfully propagate potatoes, carnations and dahlias through the isolation and tissue culture of meristematic domes. In the late 1950's, Morel turned his attention to orchids. It was known for some time that the seedlings from virus-infected orchids were apparently virus-free. Morel (6) had hoped to use tissue culture to isolate small meristematic domes that might not be infected with virus and regenerate virus-free, whole plants from the meristems. Excision and culture of very small meristem domes less than 0.2 mm in length resulted in virus-free plants, but when larger meristem-tips up to 1 mm in length were cultured, the plants that were produced remained virus-infected. Although virus-free plants were not produced from the large meristem tips, these experimental results had important practical applications. As the size of the meristem tip increased, the number of propagules also increased. Meristems that are placed in culture on the proper medium will proliferate into a mass of protocorms. These protocorms can then be further proliferated or treated like seed-derived protocorms. Morel states "If each protocorm is sliced into four and each piece regenerates only two protocorms (and this is a minimum), the clone can be increased every month by a factor of eight. This means that it is possible to obtain more than one billion plants from a single bud in only nine months" (6). The practical significance of tissue culture changed the entire orchid industry for most orchids could only be doubled in number every 1 to 2 years using classical vegetative propagation techniques. In addition, cultivated orchids are highly heterozygous and do not breed true from seed. Several large orchid firms were already performing aseptic embryo culture and very easily started meristem propagation of selected cultivars (6, 7).

2. RECENT PROGRESS

There have been a number of excellent reviews that describe specific methods and media formulations for the tissue culture of various orchid species and hybrids (1, 2, 6). Orchid meristems, like the embryos, are highly adaptable and can grow on a great variety of media. In general, meristem culture requires a relatively high potassium concentration (6.7mM, 500 mg/l KCl). Many orchid embryos and protocorms lack nitrate reductase. Therefore, a medium which has a low nitrate ion concentration (2.1mM, 500 mg/l $Ca(NO_3)_2.4H_2O$) and a high ammonium ion concentration (3.8mM, 500 mg/l $(NH_4)_2SO_4$) results in better growth. Auxins are generally not required for meristem culture, but an increased degree of proliferation is obtained when indoleacetic acid (IAA) (5.7-11.4μM, 1-2 mg 1), naphthaleneacetic acid (NAA)(2.7-10.7μM, 0.5-2 mg/l) or indolebutyric acid (IBA) (2.5-9.8μM, 0.5-2 mg/l) are added. Cytokinins are also not generally required for meristem tip proliferation. Marked increases in proliferation are, however, seen when isopentenyladenine (2iP) (4.9μM, 1 mg/l), kinetin (4.6μM, 1 mg/l) or benzyladenine (BA) (4.4μM, 1 mg/l), is used in conjuction with an auxin. Several complex additives are routinely used: banana (40 g/l), coconut milk (100-150 ml/l), peptone (1-2 g/l). The only general trend in vitamin requirements is a growth-enhancing effect for nicotinamide (4.1μM, 0.5 mg/l).

The orchid family is extremely large with over 25,000 different species. This taxonomic diversity is reflected in the specific requirements for the meristem culture of given orchid species and cultivars. One should refer to the excellent reviews (1, 2, 6) on the subject for specific details. In one review (2), specific protocols were described for the culture of 22 different genera.

3. SOCIOECONOMIC CONSIDERATIONS

New and radical ideas are not easily accepted. That was the case of tissue culture in the orchid industry. In the 1920's Knudson (2) suggested that a mycrorrhizal association was not essential for seed germination and subsequent growth and development of the seedlings. This idea was not accepted until 10 years later when Knudson proved that mature, flowering orchid plants could be obtained via his asymbiotic procedures.

Meristem culture was not initially accepted as a suitable method for rapid propagation by many commercial orchid growers. The growers at first did not believe that the plants produced from meristem culture would be identical to the parent plant. The first company to try meristem propagation was the French concern of Vacherot & Lecoufle. After seeing the first flowering of Morel's meristem derived plants, they decided in 1963 to propagate a plant of *Vuylstekeara* Rutilant via meristem culture. These plants derived from meristems flowered in 1965 (5). All were identical to the parent plant. Vacherot and Lecoufle then decided to meristem propagate plants on a commercial basis (5).

Still, many commercial firms were hesitant to produce or sell meristem-derived plants (7). They wanted to see a large block of such propagated plants in flower. Vacherot and Lecoufle in 1967 flowered several hundred meristem derived plants of *Odontonia* Boussole. All were identical to the parent plant (5).

Selected orchid clones were very rare and exceedingly expensive until the technical breakthrough of tissue culture propagation. A prime example is Sophiolaeliocattleya Falcon 'Alexanderi'. This plant flowered for the first time in 1921. Vegetative divisions from the original plant sold for $500.00 in 1965 (7). In 1966, small meristem propagated plants could be bought for only $5.00 (7). Quality clones could now be mass-produced allowing the average person, not only the wealthy, to purchase the best clones of a given cross. Many commercial firms accepted meristem culture as a means of propagation following the proven success of the French firm. In the late 1960's, a flood of commercial tissue culture laboratories were established to meristem progagate selected clones (7).

Tissue culture caused another problem in the orchid industry. In the late 1960's, a popular belief was that all meristem-propagated orchids from virus-infected plants were virus-free. Because of this belief, commercial firms inadvertently began distributing virus-infected plants. Very quickly this problem was identified and commercial screening procedures were developed to identify virus-free clones for use in meristem propagation.

Tissue culture propagation of orchids is now a widely accepted practice. This technique has changed the orchid industry in a number of ways. In some instances, classical orchid breeding in several genera has greatly decreased. In other cases, tissue culture has indirectly favored the breeding of other genera. For example, the number of commercial and amateur growers that are currently breeding standard cut-flower cattleyas has decreased due to tissue culture. In general, it takes 5 to 8 years to bloom a standard 6" pot, cut flower type cattleya from seed. Of the seedlings, about 10% are of sufficiently high quality to be economically valuable. The expense to raise a meristem-propagated plant to maturity is the same as a sexually propagated seedling. Which would you purchase if the price was equal? Which plant would you grow? Many commercial orchid companies now cannot economically breed genera that require a long time to reach maturity and are easy to mass propagate. These companies have shifted breeding emphasis to those genera, such as Phalaenopsis, that reach maturity in a short period of time and are very difficult to tissue culture propagate.

REFERENCES

1. Arditti J: 1967. Factors affecting the germination of orchid seeds. Bot. Rev. 33:1-97.

2. Arditti J: 1977. Clonal propagation of orchids by means of tissue culture - a manual, pp. 203-293. In: J Arditti (ed). Orchid biology: reviews + perspectives, Vol. I. Cornell Univ. Press, Ithaca, NY.

3. Griesbach RJ: 1985. Methods in plant tissue culture, pp. 100-115. In: D Knorr (ed). Impact of biotechnology on food production and processing. Marcel Dekker, NY.

4. Krikorian AD: 1982. Cloning higher plants from aseptically cultured tissues and cells. Biol. Rev. 57:151-218.

5. Lecoufle M: 1967. A mass blooming of mericlones. Amer. Orchid Soc. Bull. 36:405-406.

6. Morel G: 1974. Clonal multiplication of orchids, pp. 169-222. In: C Withner (ed). The orchids: scientific studies. John Wiley and Sons, NY.

7. Scully, R.: 1964. Meristem propagation - 1964's topic. Amer. Orchid Soc. Bull. 33:105-107.

LABORATORY DESIGN

O. C. BROOME

U. S. Department of Agriculture, Agricultural Research Service, Fruit Laboratory, Beltsville, MD 20705, U.S.A.

1. INTRODUCTION

The success of a tissue culture laboratory is affected by the design of the facility and its furnishings. Specific elements of both the facility and its furnishings can be selected to promote sterility, ease of operation, plant growth and work comfort. The intended use of the laboratory is a major factor influencing its design. Tissue culture laboratories are used for teaching, research or commercial production and possibly for hobby. Each category of user has different perspectives, different goals that would affect design. The commercial laboratory will be used as a model. It should be efficiency-oriented because of profit motives. The same principles used by the commercial laboratory could be applied to research and teaching laboratories at the appropriate level of sophistication. This discussion is intended to aid the reader in planning a tissue culture facility. In recent years there have been other publications on the subject (1,2,3,4,5) that can be consulted for further details.

2. FACILITY

2.1. Location

If one has the option of starting with property selection and constructing of a laboratory on a specific site, some items must be considered that would already be taken care of if just adapting an existing structure. These would include land cost and availability of electricity, water and sewer. Where water and sewer are not available,

Zimmerman, R.H. et al. (eds), Tissue Culture as a Plant Production System for Horticultural Crops.
 ISBN 90-247-3378-2.

wells must be drilled and soil percolation tests done to determine suitability for septic tank drainage fields, with special emphasis on managing a large amount of waste water. Factors relative to both new construction and an existing structure would be proper zoning where it applies, proximity to an adequate work force and transportation for the personnel, and a responsible shipping service. Another consideration is whether there is a heavy particle load in the air from industry, saw mill, unpaved roads, etc., because of the need for stringent air filtration to safeguard cleanliness.

2.2. Structure

Future problems could be averted if in the early stages of planning a construction engineer and an experienced tissue culture consultant coordinate construction details with special design needs relative to tissue culture. Points of significance would include an air tight structure with insulation against heat and cold plus adequate air conditioning to handle the heavy heat load from the media preparation, transfer and growth room areas. Good air filtration is necessary throughout the entire laboratory to aid cleanliness, and efficient air flow would be especially necessary where chlorine solutions and alcohol are used. Some heating and cooling would also be required in the shipping area to promote personnel comfort and plant survival during plantlet handling. Careful ducting and air directing would help maintain even temperatures throughout the growth rooms. Wall and floor surfaces should be chosen for ease in keeping clean as well as for durability and economy. Sliding, see-through doors save space and promote safety. Automated doors would further assist in moving loaded carts. All laboratory functions and related storage must be on the same floor since there is both a safety hazard and an efficiency factor in carrying cultures and supplies up and down stairs.

2.3. Floor plan

The designation of work areas and the layout within each one should promote cleanliness and efficient work-flow to prevent back-tracking within and between areas. Figure 1 illustrates one possible floor plan allowing for a "clean zone". Positive air pressure in this "clean" zone, and a tacky floor mat at the entrance to it, helps keep out dust.

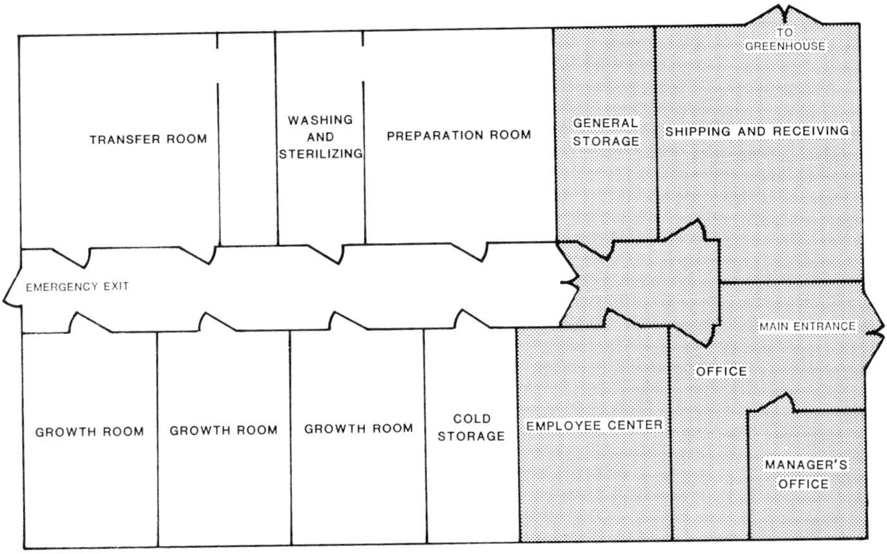

FIGURE 1. Floorplan of laboratory designating work areas, entrances/exits, and "clean" ▭ vs "dirty" ▭ zones.

Although hinged doors are indicated in the figure, see-through sliding doors are preferred for all interior doorways except the double doors in the hallway. The exclusion of windows except in the office area may aid in temperature control. Any effect on internal temperature would depend on the climate where the laboratory is located. Also, the external light cycle would not be desired in the growth rooms.

As noted in Figure 1, the work areas are the office, shipping and receiving, the employee center, general storage, cold storage, washing and sterilizing, preparation, transfer and growth.

Main access though the office provides controlled entry into the laboratory.

Shipping and receiving is actually a multipurpose area. It and the office are the only parts of the laboratory with direct outdoor contact and thus, are the dirtiest. If the laboratory is to include a greenhouse, it should be accessible only through shipping and receiving. In that case, S & R would also serve as a greenhouse-headhouse for potting operations.

Actual measurements are not given for the work areas. These must be individually determined based on crop selection and annual output desired. Crop transfer cycle time and increase rate used with annual output and culture vessel size could be used to estimate the amount of growth room space needed. The amount of time needed to divide and transfer a container of propagules would help determine the number of operators and transfer hoods required. It should be noted here that some laboratories run multiple shifts for transfers. Calculating backwards to glassware cleaning and medium preparation would provide some basis for determining space and equipment. A generous rather than conservative approach is recommended since it is less expensive and easier to build in extra space initially than to add on for future expansion.

3. FURNISHINGS

3.1. Office

Partitioning of the office would allow separate spaces for the laboratory manager, a secretary and a time clock or sign-in area. Furnishings needed for a specific laboratory office could best be determined by consulting an office manager but should definitely include a word processor or personal computer. A personal computer with word processing capability would be useful for keeping culture and shipment records, supply inventory, and ordering. Depending on the labeling system used for the cultures, labels could be computer generated. Carpeting should be antistatic treated to prevent inteference with the electronics. The choice of a time-keeping system, whether timeclock cards, sign-in sheet or honor system, would be determined to some extent by the size of the staff. A larger staff would be more manageable with a more formal system.

3.2. Shipping and receiving

As specified earlier, this area should be furnished for several functions---shipping and receiving, planting and rooting, water heating, water purification and storage, janitorial functions and some storage.

3.2.1. Shipping and receiving activities would require a variety of shipping cartons, a dispenser for packing material, strapping tape and scissors and a table to work on. If glass vessels are to be shipped, air bubble cushioning or corrugated dividers should be used to protect them from breakage. Labels, packing slips and invoices would be provided by the office.

3.2.2. Packing, planting and direct rooting would require tables, stools, adequate light, and a sink. Plantlets could be taken from the culture containers and washed in the preparation area and then taken to shipping and receiving for further handling. Processing the jars in the preparation area would preclude dirty jars coming back into a clean area from shipping and receiving. The plantlet handling would include separation, grading, counting and packing for shipping bare-root plantlets as well as in-house planting and direct rooting.

3.2.3. Adequate water heating for the lab would best be determined by a reliable plumbing contractor supplied with information on the glassware washer used and the volume of glassware to be cleaned in addition to usual water needs. There would be space for the heaters in shipping and receiving.

3.2.4. Water purification and storage would logically be placed in the shipping and receiving area because of the space needed for storage tanks. Water purification options include distillation, ion exchange and reverse osmosis or purchase of distilled water. The choice of system is partly influenced by the quality of water available; the more particles, minerals and organics present, the more stringent the processing required. Excessive water cleaning is detrimental, however, since ultrapure water can pit a stainless steel tank. Water in the range of 200,000 ohms-cm resistivity or greater is adequate for propagation needs. Whichever system is chosen, all monitors and controls should be easily visible and readily accessible.

3.2.5. Janitorial functions, such as major floor stripping and waxing, may be partly contracted out, but there would still be a need for some equipment and supplies: mops, brooms, vacuum, cleaning solutions and buckets. The bulk of these would probably be used only weekly and, thus, should be out of the way. Daily cleaning would be facilitated by having the entire building prefitted with a central vacuum system. The pumps could also be located in shipping and receiving.

3.2.6. Storage of planting and potting materials can be located in shipping and receiving.

3.3. The employee center

The center should be large enough to accommodate the maximum number of employees present during a shift. Since employee needs are a matter of health, safety, security and comfort, the center should include not only adequate lockers, toilets and showers, but also first aid supplies, fire extinguisher, phone and folding tables and chairs for breaks, lunch and staff conferences. If a greenhouse is included with the laboratory, the shower would become a matter of necessity in the event of a chemical spill on an employee. Since food must not be stored in refrigerators used for laboratory chemicals, employees should be provided with a refrigerator and would enjoy having a microwave oven. A pleasantly decorated, employee center would provide a retreat from the pressure and noise of a busy transfer room.

3.4. General storage

Since the room would contain a variety of supplies from pencils and bathroom items to bulk chemicals, extra culture vessels, transfer room prefilters, growth room fluorescent bulbs, etc., some thought should be given to the space requirements of the different supplies so that shelving could be set up to allow adequate but efficient spacing. The size of the room should be adjusted to the overall size of the laboratory and to frequency of deliveries. At some locations more supplies need to be stored then at others. Acids and bases have special storage requirements and ethyl alcohol should be kept in a locked cabinet. Lighting should be good and the room well organized so that time would not be lost in searching for supplies.

3.5. Cold storage

Cold storage should be cooled to 4°C and controlled by a sensitive, accurate thermostat. Adequate air movement will be necessary for evenness of temperature. Constant low level light is required by some plants for good cold storage and should be provided for. Cultures and harvested plantlets could be stored here on mobile carts. Shelves along the walls could be used for longterm storage (3 to 6 months) of plantlets and meristems.

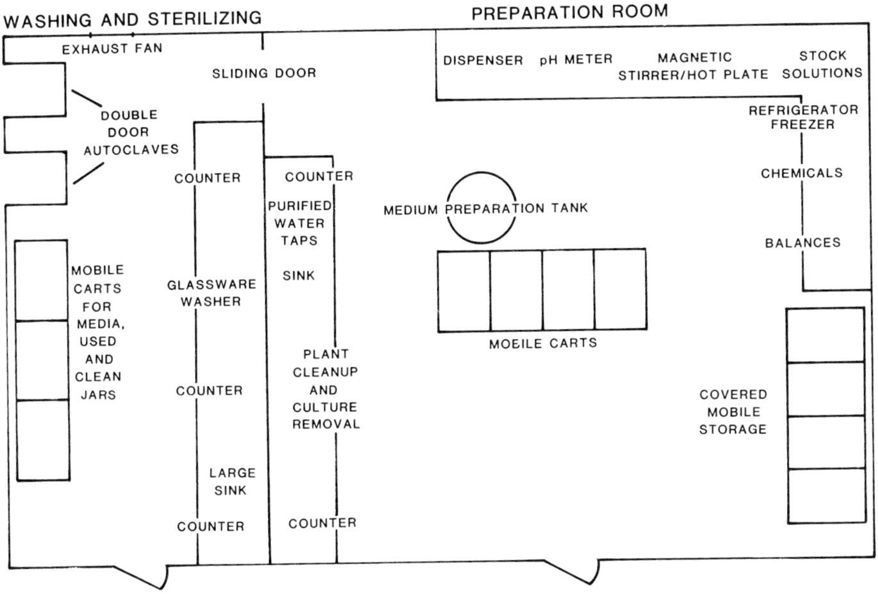

FIGURE 2. Layout within Washing and Sterilizing and Preparation Rooms. Equipment is situated to promote a positive work flow.

3.6. The preparation room

Figure 2 shows a possible floor plan for separating glassware cleanup and sterilizing from medium preparation and tissue "entry-exit". Counter surfaces in the rooms should be heat resistant and stand up to a lot of cleaning. Under-counter and wall cabinets would be needed in part of the laboratory for storage. See-through doors for the wall cabinets facilitate locating items inside them.

The room should contain purified water taps, a sink, counter top work space, balances, chemicals, stock solutions, a pH meter, medium mixer/heater, a dispenser, a refrigerator/freezer and storage space.

A large reservoir of purified water, with 2 or 3 taps would allow concurrent uses and cut down on the time required to collect any volume. There should be a water quality monitor at the storage tank (and someone assigned to check it daily); there should also be an indicator in the laboratory so that changes in water quality could be detected easily. It would be convenient for the taps to be part of a moderately large sink in the room.

It is possible to purchase a balance with a weighing range from milligrams to hundreds of grams. The greatest amount of weighing would be in the hundred gram range; therefore a top loading balance would be most efficient. The milligram weights usually pertain to critical substances such as growth regulators and micronutrients and merit an analytical balance. Since there may be a lot of vibration-producing activity on the bench top near the analytical balance, it may be necessary to use a vibration damping block or to place the balance on a free-standing table in a space cut out from the counter.

Certain chemicals must be frozen, others and some stock solutions need to be refrigerated. Therefore, a standard refrigerator/freezer is needed. Other stock solutions can be stored at room temperature and can be kept in bottles with variable or fixed volume dispenser tops to facilitate medium preparation.

Medium preparation requires mixing the solution while adding ingredients and, sometimes, heating at the same time. For smaller batches, this can be accomplished using a magnetic stirrer/hot plate. Large batches can be prepared in a commercial food processor tank with the medium agitated by a propeller-type mixer inserted from above. The processors are designed to be bolted to the floor but could be secured to a dolly for ease in moving around the preparation room. If mixing without heat is preferred, the medium can be prepared in plastic barrels. In this case, a good mixer is important to keep the unmelted gelling agent in suspension during dispensing.

A crude adjustment of medium pH could be made using nitrazine paper. However, using a pH meter is more accurate and therefore preferable. A meter accurate to one decimal place is adequate and to conserve counter space, wall-mounted models are available. If the food processing tank is used, a longer than normal pH electrode lead may be necessary.

There are several options for medium dispensing. For both small and large aliquots one can pour to measure and then dispense, but it is more efficient to use one of the dispensing pumps. If an automatic syringe dispenser is used, one should weigh the cost of replacing broken glass syringe barrels against an all stainless steel apparatus. For volumes that do not have to be exact, a peristaltic dispenser pump with hand grip control provides for rapid dispensing, but a very simple pump is good enough.

Mobile, multishelved, wire-construction carts such as those used for laundry and cold room facilities make excellent storage "cabinets" when covered by heavy clear plastic sheeting. Tailored covers can be made by securing the sheeting at the corners with velcro from top to bottom. Then the contents are fully visible and the "cabinets" can be moved around in the room to adjust space as needed. Carts without covers would be used to transport racks of dirty glassware from the transfer room for cleanup and to hold vessels of medium while they await sterilizing.

A work area near the sink could accommodate plant entry-exit, i.e., initial cleanup before tissue culturing or removing from culture containers for rooting, planting or shipping. The lab bench here should include some sit down space.

3.7. Washing and sterilizing room

A major furnishing for this room is the autoclave. While pressure cookers used on a stove top have served small laboratories, sterilizing can easily become a bottleneck even with a moderate-sized autoclave. Thus purchasing several large ones is a reasonable investment for the future. Both steam and electric heated autoclaves are available. Unless steam heat is used in the building and is thus accessible to heat the autoclave, an electric one is the choice. An automatic model is more efficient than a manual one because it requires less attention. While table top versions are available, the larger autoclaves are floor models. For a compromise on price, used autoclaves are available both from sterilizer companies and hospital supply companies. A double door autoclave allows loading from the preparation room and unloading into the transfer room where the sterile medium will be cooled, stored and used. Whatever autoclave is chosen, it should be placed with adequate access for servicing and it should be one that can be serviced by a nearby company on short notice. Good insulation of the autoclave chamber is needed to minimize heat escape into the room.

Another large item in the preparation room is the glassware washer, unless the washing is to be done by hand. A home or scientific variety washer would do the job, but a commercial blast-type washer with a scrap trap would allow for rapid processing of culture containers without first removing plant debris and agar. New and used washers would be available through a restaurant supplier. A stainless steel cafeteria-style counter

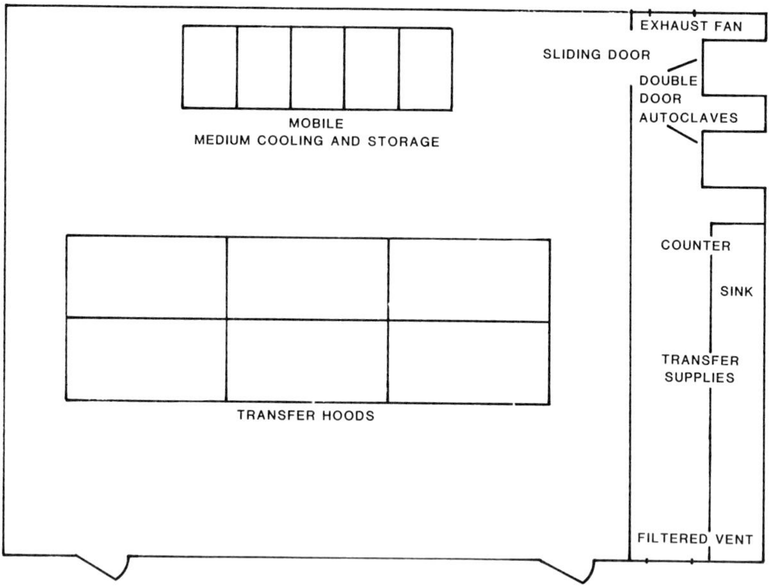

FIGURE 3. Layout within the Transfer Room.

top with raised sides has been used with the sink and glassware washer. It has proven durable, contains spills and is easily rinsed down.

A single large sink has advantages over several smaller ones. With multiple faucets and drains it could be subdivided by wash pans when needed, while smaller sinks make washing of large pots, beakers, graduated cylinders, etc., difficult and thus contribute to breakage.

Use of the sink, washer and autoclave will create excessive heat and humidity in the room. Therefore, a steam exhausting system is desirable.

3.8. Transfer room

Figure 3 shows a possible floorplan for the transfer room. The major fixed elements, the autoclaves, sink, and a few cabinets are partitioned off in a separate area to allow control of the steam. An exhaust fan must be provided to vent the steam when the autoclave is opened. A filtered vent at the hallway end of the room would allow for minimal disturbance of the transfer room air during the exhausting. Although the transfer hoods could be moved, their size and weight make it desirable to locate them permanently. As mentioned in the previous section there are advantages to having an autoclave that opens into the transfer room as well as the preparation room since medium can be moved readily to an appropriate area for cooling. Adequate space should be set aside in the transfer room for medium cooling and short-term storage. Covered carts such as those described for preparation room storage would suffice. A section of undercounter and wall cabinets next to the sink would be useful for storing transfer supplies, such as surgical blades, and labeling materials. Large rapid turn-over items like dispensers of sterile paper towels and dissection pans can be kept on covered carts as already described. Much of the area in this room is open to allow space for carts of medium beside the hoods during transfer.

Several variations of transfer hoods are available. They include individual or multiple operator stations and floor or raised air intake. Multiple operator stations provide more versatility. A floor air intake tends to pull in more dust than a raised one, necessitating more frequent cleaning of the prefilter. If the hoods are fitted with supports to suspend sterilizers and instrument racks below the lights, more bench space is available for culture jars, etc.

A variety of instrument sterilizers are currently in use. These include an electrically heated ceramic tube, an ultrasonic cleaner with cleaning solution and a heated container of glass beads. Also used in some laboratories is a series of dilute hypochlorite solutions. Open flame sterilization is not recommended because of the fire hazard, particularly in combination with alcohol.

Air conditioning and ventilation in the transfer room is important because of the heat load from the transfer hoods and instrument sterilizers.

The comfort of the transfer technician can influence his work. For that reason a good chair is important. Characteristics of the chair would be a well-cushioned seat, an adjustable back, a foot rail and wheels. Another aid to efficient transferring is proper size and height of waste can to make it easy to discard the plant refuse.

3.9. Growth rooms

Multiple growth rooms would allow others to function in the event that one breaks down; they would also provide for the use of different light periods and temperatures for different crops. Each room should be outfitted with open shelves permitting air convection. An alternative is using shelves that are simply frames for containerized trays or racks of cultures. Shelves along the walls should be no deeper than the average reach; those in the middle of the room can be twice that wide since they will be accessible from both sides. Mobile warehouse-type shelves are used by some laboratories. They conserve space but are expensive. The shelves can go up as high as desired, but the higher ones will be more awkward to use even with a good safety ladder.

Fluorescent lights can be mounted under the shelves but should be hung far enough below so that the cultures above are not heated by them. Enough lights should be used to provide about 35-40 μmol $s^{-1}m^{-2}$ at the culture jar level. The lamp ballasts need to be remote wired in a bank outside the room. Doing so reduces the heat load in the growth room. If the ballasts are remote, they should be cooled by a thermostatically controlled fan. If they are inside the building, there is no advantage in the overall building heat load, but it should be easier to maintain an even temperature throughout the growth room. The total building heat load could be reduced by placing the ballasts in a panel outside the building. If the ballasts are left in place, then extra precautions must be taken to prevent heating cultures on the shelves above. All the lights in a given room may as well be on a single time cycle control since there would be light leaks between areas of a room on different cycles. Using more than one electrical circuit to light a room provides some light in case one circuit goes out. Cool and warm white deluxe fluorescent bulbs are adequate for culture lighting needs. Incandescent bulbs have been added by some laboratories, whereas others use all or part Gro-lux lamps. No definitive studies have been published comparing results from the different lights. There are times when a reduced light intensity is desired; this change can be readily effected by inserting a diffuser panel between the lights and the cultures.

The cooling need for the growth rooms can be estimated from the numbers of fixtures and their heat output. Room cooling should hold the temperature within plus or minus 1°C and carefully routed continuous air movement should be provided to prevent hot spots. The use of a max-min thermometer or thermograph will show daily extremes.

A peel-off type tacky mat just inside the door of each growth room will reduce dust that would normally be tracked in. This is advisable since it would not be good to have to clean the growth rooms frequently; however, they should be fitted with the central vacuum system for occasions when they must be cleaned.

4. OTHER CONSIDERATIONS

Reduction of handling is a time and money saver. A tray or rack that holds a uniform quantity of culture vessels, and fits the autoclave, the glassware washer, the mobile carts and the growth room shelves would achieve this purpose through containerized movement. Thus, the culture vessels need not be handled individually but by the trayful. This economy of movement is amplified by transporting the trays through all phases of handling on the large carts already described.

Survival of plant cultures in the growth rooms and in cold storage is dependent on temperature control. While light is needed for growth, the proper temperature is critical for survival. In areas where power outages are expected, it becomes very important to have one or more backup systems in the event of electrical failure.

If acclimated plants are to be produced by the laboratory, then a greenhouse must be part of the facility. This structure would need heat, air cooling (with a drip pad system where summers are hot), running water, mist area, lights and shading. Screened vents and a double-door entry would help protect against insect borne diseases. Soil potting and a pesticide storage cabinet could be located in shipping and receiving.

5. CONCLUSION

There are many ways and a variety of equipment types for performing a given task in the tissue culture process. No one system is without its faults and workers must be trained in any system used. It has been the goal of this article to provide some guidelines in designing a tissue culture laboratory and to stimulate the reader into careful consideration of the many needs of such a facility.

REFERENCES

1. Biondi S, TA Thorpe: 1981. Requirements for a tissue culture facility, pp. 1-20. In: TA Thorpe (ed). Plant tissue culture. Methods and applications in agriculture. Academic Press, New York.

2. Brown DCW, TA Thorpe: 1984. Organization of a plant tissue culture laboratory, pp. 1-12. In: IK Vasil (ed). Cell culture and somatic cell genetics of plants, Volume 1. Academic Press, Orlando.

3. de Fossard RA: 1976. Tissue culture for plant propagators. The University of New England Printery, Armidale, N.S.W., Australia.

4. Kyte L: 1983. Plants from test tubes. An introduction to micropropagation. Timber Press, Portland.

5. Wetherell DF: 1982. Introduction to in vitro propagation. Avery Publishing Group, Inc., Wayne, NJ.

SUMMARY OF PANEL DISCUSSION ON LABORATORY DESIGN

O. C. BROOME (moderator), B. BRIGGS, R. EVANS, AND G. HENNEN

U. S. Department of Agriculture, Beltsville, MD 20705, U.S.A; Briggs Nursery, Olympia, WA 98501, U.S.A.; Twyford Plant Laboratories, Inc., Santa Paula, CA 93060, U.S.A.; and Oglesby Plant Laboratories, Altha, FL 32421, U.S.A., respectively.

Design of a laboratory is a function of current and projected production. Factors to be considered in designing a laboratory include methods (a) to improve materials flow through the laboratory, (b) to strictly isolate clean areas from other areas of the facility, (c) to improve movement of personnel in and out of clean areas, and (d) to plan so that future expansion can be accomodated readily. Reasons for building a new laboratory include the need to increase production capacity, desire to improved air quality within the laboratory, and the desire to scale up production without increasing supervision. Constraints on laboratory design are cost and preexisting facilities with which one must work, e.g. an existing building.

A concept being used on some laboratories is to construct the laboratory as a separate building within another building. An air space of 0.3 to 1.5 m (1 to 5 feet) or more exists between the walls and ceilings of the two structures and helps insulate the laboratory from external temperature extremes.

The crops produced and procedures used will have an important influence on laboratory design. When the crops being produced are rooted outside of the laboratory, the amount of growth room space will be reduced in relation to space required in greenhouses or other facilities for rooting cuttings. Heat pumps can be used for heating and cooling in appropriate climates and the design can include separate heat pumps for separate rooms in the laboratory to provide greater temperature control.

Inclusion of a cold storage room (1 to 4 C) in a laboratory is highly recommended to hold cultures temporarily or as a repository for long term storage of stock cultures. Windows in the transfer and media preparation rooms can be valuable for improving employee morale. It is highly recommended to maintain all clean areas of the laboratory under positive pressure to prevent non-filtered air from entering the clean area when doors are opened.

Instrument sterilization techniques and systems vary but the consensus is not to use alcohol dips and open flames in the hoods because of the fire hazard created. Systems in use are either heated or non-heated. In the former category are small ceramic heating tubes designed for bacteriological instruments and metal tub-like heaters containing glass beads. Non-heated systems include soaking instruments in hypochlorite solutions of varying strength and use of bactericidal solutions in ultrasonic cleaners. Each system has advantages and disadvantages. The ceramic heating tubes and hypochlorite solutions both cause deterioration of forceps and scalpel blades and handles fairly rapidly.

Fogging systems seem to be preferred to mist for direct rooting but location of lines for the system is important. Control devices are still not completely satisfactory and time clock controls may work better than humidistats in some cases.

LARGE SCALE TISSUE CULTURE PRODUCTION FOR HORTICULTURAL CROPS

RANDALL A. STRODE AND GERRI ABNER

Agri-Starts, Inc., Apopka, FL 32703, U.S.A.

Formerly, there was mentioned a great potential for the use of plantlets from tissue culture for horticultural crop production, but not until recently has this potential for rapid expansion become a reality. Although there are many reasons for this, they can be readily grouped into two main areas. First, there is a long list of sound advantages of tissue culture for the growers. The rapid propagation of new and highly demanded crops, year round shipments, uniformity, specific pathogen free, enhancement of juvenile vigor and suckering are just a few. Second, the cost of producing plants conventionally continues to go up while the cost to produce tissue cultured plants has not gone up nearly as much, and in some cases has come down. This is mainly a result of improved efficiency in many areas of the process.

Large scale production of a number of horticultural crops using conventional propagation has been going on for a long time. The truth of this last statement might depend, however, on how one defines large scale in the industry, but let us assume that the statement is true. The ability to produce has pretty much kept pace with the demand. For the most part, mass overproduction of any item has not occurred. The industry has primarily been made up of small family-owned operations using these conventional propagation techniques; therefore, there has been no real ability to mass overproduce, or at least with any relative ease or low cost. The scenario today has changed now that there are large corporations, large sums of venture capital, and new technologies for rapidly producing plants in large numbers. What this can mean for the industry as a whole can be very positive or negative depending on how this new power, money, and technology are managed. For the first time in the history of the industry, it will have the ability to produce well beyond the demand. This is the case with most industries in the United States today and more emphasis needs to be placed on sales and less on the

Zimmerman, R.H. et al. (eds), Tissue Culture as a Plant Production System for Horticultural Crops.
© 1986. Martinus Nijhoff Publishers, Dordrecht. Printed in The Netherlands. ISBN 90-247-3378-2.

production. Ours for the most part has been a production-emphasized industry.

A large production facility has several economic advantages by scaling production upward. Initially equipment is needed throughout the production process whether a tissue culture facility is designed to produce small or large numbers. A media preparation area needs a pH Meter, a balance or scale, a stirrer, and an autoclave, etc. Cash outlay could be the same initially in either case, but if one is planning on large production at the start, one can go ahead and purchase large or more heavy-duty equipment that can grow into and save money later on. For example, a stirrer with a 1/3 HP motor lists for $197.35, a 1 HP for $296.45, and a 1 3/4 HP for $359.70. The second motor has 3X the horsepower for half again as much money while the third has over 5X for 75% more.

By producing large numbers, one is also able to take advantage of cash discounts or obtain bulk rate prices for production supplies. By ordering agar in bulk instead of small quantities, one can cut the cost almost in half. Cell pack trays are another example where the cost is dramatically reduced by ordering large quantities. A small company may have to pay as much as 60% more for a small order.

Because of necessity, a large operation is forced to buy some equipment that a small production facility can do without. An example is an automatic dishwasher. The company has to put out a large sum of money initially but it benefits in two ways. One way is that over time the cost of dishwashing goes down, because the dishwasher can wash the containers so much faster than by hand. Secondly, the dishwasher uses much hotter water and can dispense sterilizing agents that hands could not stand therefore doing a better job than hand washing.

So, what is large scale tissue culture production and what will it do for us? Large scale tissue culture production might be looked at as the "state of the art" of production technologies, methodologies and systems combined with skilled people to produce plants with great efficiency. The large scale of yesterday may be small scale today and today's scale may be small by tomorrow's standards. We are talking about a relative term that will change as the needs of the industry change and as new technologies are introduced. What is large scale in terms of total number for one crop item may be small for another. Large scale production of a new exciting orchid might be a few thousand plants while large scale production of a new

Spathiphyllum could be many, many times that. The size of the "large scale" will be determined by market demand. To overproduce or produce in mass some crops would defeat the very purpose and advantages of tissue culture. The foliage industry alone produces more than 1,000 kinds of foliage plants. The different kinds of plants in other plant industries in which tissue culture could become a part are even larger in terms of cultivars and total numbers. Large scale tissue culture production will serve all of us well by strengthening our entire horticultural industry, if all of the advantages are utilized. The main advantage is getting more and newer cultivars into the marketplace quicker. To mass overproduce a few cultivars will surely hurt everyone in the marketplace.

What might a large scale or "state of the art" tissue culture facility or business look like today? The facility would be located best within a new tightly sealed building. Positive air pressure and an efficient air handling system throughout will solve many problems with which older facilities struggle. The facility would have an open floor plan concept. The actual work areas and arrangement would be somewhat varied depending on the comfort levels of those running the facility. The work areas (rooms) that are somewhat universal include 10 areas; office, breakroom, restrooms, storage rooms, shipping area, greenhouse prep area, media prep area, culture rooms, transfer area, and a safety valve short-term free space. The interaction between the different work areas is great. What one would see though is a straight flow from receiving to wash area to media preparation area, through a double door sterilizer, to storage to transfer room to the culture room(s). From the culture room(s) a good accessible flow to the greenhouse prep to the greenhouse or shipping and also back to the transfer room. There are two points to be made here. One, the entire facility needs to be in balance, each of its areas sized to the other. To make an analogy, it will operate much like a sewing machine, with each part in balance with another so that when one part moves faster the others do also. There is a maximum that a tissue culture facility can produce just as there is a maximum that a sewing machine can produce. To get more production one needs to buy another sewing machine or build a second facility. One cannot just add culture rooms to get more production just as you cannot add more needles to a sewing machine to make it sew more. We can make both more efficient, which may have a net increase in production, but the balanced "sewing machine principle" will hold if the facility was built in balance to start with. The second point falls right in line here in that the facility must be flexible. Equipment, crops, and people will change and technologies will improve. The facility must have those things built into it so that it can adapt. The point is, in this rapidly growing and changing new technology, one must be able to adapt to the more efficient ways of doing things to stay competitive. So how does one build flexibility and adaptability into a facility? Simply try not to get locked into any one system. To accommodate change, a certain amount of free space is planned along with plenty of storage.

To be more specific, in this "state of the art" facility one would see great simplicity, starting with the most important decision, the vessel in which the plants are to be produced. There would be one kind of vessel only. That is the important part. We prefer the medium-sized baby food jar, which is a "one-handed vessel" built for mechanical manipulation. It is referred to as a one-handed vessel because with one hand the tops can be put on and taken off during the plant manipulation process. The vessel would have one tray which it would never leave. The vessel would be stored, filled, washed, and shelved in it. There would be one kind of well ventilated shelving with uniform light and temperature. The entire facility, including the culture rooms, would have wide aisles and hallways for ease of moving efficiently large carts to carry the trays of vessels. The need to keep the handling of the vessels and trays to a minimum is of paramount importance.

In large scale production, for smooth continuous production, the number of variables must be reduced to an absolute minimum. This statement is not to be taken lightly. Plant crops are alive and dynamic, and keen observation and attention to detail cannot be replaced by a machine. The energy, time, and attention the crops require necessitate excluding the luxury of dealing with many other variables.

There are problems in any kind of production process, and large-scale plant tissue culture is no exception. Time-motion studies must be done on many areas to improve efficiency. Quality and continuous flow in the production requires well-trained, motivated personnel. Training and motivation are an ongoing process that each business must constantly address and is an entire subject in itself. The point to be made is simple: a motivated group of people can make a poor facility work and a great facility work better.

There are many problems with the crops themselves, which also is another large area for discussion, but most of the problems are eliminated by proper explant source selection and regular explanting procedures.

In summation, a large plant tissue culture facility with its sight on large scale production has to make considerations and decisions on the vessel, equipment, building, floor plan, flow tools, technology, and people. When they are considered, the question must be asked: "Does it add a variable or does it eliminate one? What will be the limiting factor?" Mass production is defined as the production of machinery and other articles in standard sizes and large numbers. Mass production makes it possible to manufacture more things faster because mass production also means whenever any part breaks down, a new part that will fit perfectly can be obtained to replace the broken part. With time, these practices of mass production will become more a part of large scale tissue culture production. Eli Whitney would be pleased to see his "assembly line" applied more and more in different parts of the tissue culture process. The net effect is to improve efficiency, but in the Ornamental Horticulture Industry, like other industries, it is possible to produce too much of any one item. We now have that potential. The challenge before us is to improve efficiency <u>and</u> to know our marketplace.